# Livestock Handling and Transport
## 2nd Edition

# Livestock Handling and Transport
# 2nd Edition

*Edited by*

**T. Grandin**
*Department of Animal Sciences*
*Colorado State University*
*Fort Collins*
*USA*

CABI *Publishing*

**CABI *Publishing* is a division of CAB *International***

CABI Publishing
CAB International
Wallingford
Oxon OX10 8DE
UK

CABI Publishing
10 E 40th Street
Suite 3203
New York, NY 10016
USA

Tel: +44 (0)1491 832111
Fax: +44 (0)1491 833508
Email: cabi@cabi.org
Web site: http://www.cabi.org

Tel: +1 212 481 7018
Fax: +1 212 686 7993
Email: cabi-nao@cabi.org

A catalogue record for this book is available from the British Library, London, UK.

**Library of Congress Cataloging-in-Publication Data**
Livestock handling and transport / edited by T. Grandin.-- 2nd ed.
    p.  cm.
   Includes bibliographical references.
   ISBN 0-85199-409-1 (alk. paper)
   1. Livestock--Handling.   2. Livestock--Transportation.   3. Livestock--Effect of
stress on.
  I. Grandin, Temple.
  SF88 .L58 2000
  636.08'3--dc21                              99-058044

ISBN 0 85199 409 1

Typeset by AMA DataSet Ltd, UK.
Printed and bound in the UK by Biddles Ltd, Guildford and King's Lynn, UK.

# Contents

# Contributors

**J.L. Albright,** *Purdue University, West Lafayette, IN 47907, USA*

**A. Barber,** *Department of Agriculture, PO Box 81, Keith, South Australia*

**D.M. Broom,** *Department of Clinical Veterinary Medicine, University of Cambridge, Madingley Road, Cambridge CB3 OES, UK*

**L. Coppinger,** *School of Cognitive Science, Hampshire College, Amherst, MA 01002, USA*

**R. Coppinger,** *School of Cognitive Science, Hampshire College, Amherst, MA 01002, USA*

**R. Ewbank,** *19 Woodfield Road, Ealing, London, UK*

**R.B. Freeman,** *Agricultural Engineering Section, University of Melbourne, Melbourne, Victoria, Australia*

**H.W. Gonyou,** *Prairie Swine Centre Inc., PO Box 21057, 2105 – 8th Street East, Saskatoon, Saskatchewan, Canada, S7H 5N9*

**T. Grandin,** *Department of Animal Sciences, Colorado State University, Fort Collins, CO 80523, USA*

**W.B. Gross,** *Department of Large Animal Clinical Sciences, Virginia Polytechnic Institute and State University, Blacksburg, VA 24061, USA*

**P.H. Hemsworth,** *Animal Welfare Centre, Agriculture Victoria and University of Melbourne, Victorian Institute of Animal Science, Private Bag 7, Snedyes Road, Werribee, Victoria 3030, Australia*

**K.A. Houpt,** *Department of Physiology, College of Veterinary Medicine, Cornell University, Ithaca, NY 14853-6401, USA*

**G.D. Hutson,** *Department of Animal Production, Institute of Land and Food Resources, University of Melbourne, Parkville, Victoria 3052, Australia*

**T.G. Knowles,** *School of Veterinary Science, University of Bristol, Langford, Bristol BS40 5DU, UK*

**E. Lambooij,** *Institute for Animal Science and Health, PO Box 65, 8200 AB Lelystad, The Netherlands*

**S. Lieb,** *Department of Animal Science, University of Florida, Gainesville, FL 32611, USA*

**L.R. Matthews,** *Animal Behaviour and Welfare Research Centre, Ruakura Agricultural Centre, Hamilton, New Zealand*

**C. Nicol,** *University of Bristol, Department of Clinical Veterinary Science, Langford, Bristol BS40 5DU, UK*

**P.B. Siegel,** *Department of Animal and Poultry Sciences, Virginia Polytechnic Institute and State University, Blacksburg, VA 24061, USA*

**V. Tarrant,** *The National Food Centre, Teagasc, Dunsinea, Castleknock, Dublin 15, Eire*

**P.D. Warriss,** *School of Veterinary Science, University of Bristol, Langford, Bristol BS40 5DU, UK*

**C. Weeks,** *University of Bristol, Department of Clinical Veterinary Science, Langford, Bristol BS40 5DU, UK*

**H.Ll. Williams,** *Formerly of Department of Farm Animal and Equine Medicine and Surgery, Royal Veterinary College, University of London, Potters Bar, Hertfordshire EN6 1NB, UK*

# Preface

The purpose of this book is to serve as a source of the latest scientific research information and as an archive of practical information. In this second edition the goal is to bring together in one book the latest research data and practical information on animal handling and the design of facilities. Some of the most valuable contributions to the knowledge of animal handling and transport are often located in producer publications that are difficult to obtain.

During the last 7 years since the first edition was published, there have been many new studies to determine the stressfulness of handling and transport. Data from these new studies have been incorporated into the revised chapters and a new chapter has been added on the stress physiology of transport. Scientific information on animal stress will help people make sensible decisions to improve animal welfare during handling and transport. Many people who read the first edition stated that they thought the diagrams of handling facilities were useful and they liked being able to have access to both scientific and practical information in one book. Each chapter contains an extensive reference list of both practical and scientific publications.

As agriculture and raising animals become more and more concentrated, valuable knowledge is sometimes lost. This book is intended to help preserve some of the best practical knowledge and provide access to the latest scientific studies. All aspects of animal handling and transport are covered, including handling for veterinary and husbandry procedures, restraint methods, transport systems, corral and stockyard design, handling at slaughter plants and welfare. Principles of animal behaviour and how behavioural patterns related to handling are covered for cattle, sheep, pigs, horses, deer and poultry.

# Introduction: Management and Economic Factors of Handling and Transport

<br>
## Temple Grandin

*Department of Animal Sciences, Colorado State University, Fort Collins, CO 80523, USA*

## Management Attitude

Observations on hundreds of ranches, farms and feedlots and at auctions and slaughter plants in the USA, Canada, Mexico, Australia, New Zealand and Europe indicate that the single most important factor which determines how animals are handled is the attitude of the manager (Grandin, 1988). Even though this reference is over 10 years old, subsequent observations have strengthened the author's original observation. Operations with efficient humane handling and transport practices have a manager who is committed to animal care. Operations where abuse occurs almost always have lax management or management that does not care. A good manager can train most employees to handle animals properly. Observations by the author at numerous livestock operations indicate that handling sometimes greatly improves with a change in management and sometimes deteriorates when management changes. In order for a new manager to improve handling in a poor operation, he/she must have the power to hire and fire employees.

Good handling equipment and well-designed transport vehicles provide the tools which make efficient and humane handling and transport easier. However, abuses will occur in good facilities that are poorly managed. Some engineers mistakenly believe that engineering can solve all the problems. Engineering and equipment is only one-third of the animal-handling equation. Employee training and good management are the other two-thirds. Progressive managers will find this book useful for educating both themselves and their employees.

The author has observed that improving handling methods must start with training managers. A manager who recognizes the value of good handling practices will be motivated to supervise his/her employees. If

©CAB *International* 2000. *Livestock Handling and Transport,* 2nd edn
(ed. T. Grandin)

employees are trained while the manager stays in the office, handling is likely to deteriorate back to pretraining levels after the training sessions are over. On large operations where hundreds or thousands of animals are handled, employees need a strong manager to act as their 'conscience'. People who handle hundreds of animals can become numb and callous. They need the supervision of a management person who is involved enough to care but not so involved that he/she becomes numb and desensitized. A person at the general manager or superintendent level is usually the most effective.

## Quality of Stockmanship

Between 1970 and 1990, there has been a gradual improvement in handling and stockmanship at auctions, feedlots and slaughter plants (Grandin, 1990). In both Europe and the USA large supermarket chains and restaurant companies, such as the McDonalds Corporation, are now auditing stockmanship on farms and in slaughter plants. Their tremendous purchasing power can motivate great change. In 1996 the percentage of US beef plants that were able to render 95% of the cattle insensible with a single shot from a captive bolt gun was only 30% (Grandin, 1997a). Audits conducted by the McDonalds Corporation and the author in 1999 indicated that the percentage of plants that stunned 95% of cattle correctly rose to 90%. Since 1995 the USA has had a tremendous expansion in the numbers of large dairies and pig farms. The quality of handling and stockmanship has declined on a few large dairies and pig farms because they are either unable or unwilling to hire enough qualified stockpersons. About 10% are chronic abusers who allow overt cruelty to occur, such as throwing calves, abuse of cripples or the use of brutal restraint methods, where live cattle are hung upside down prior to religious slaughter. A recent survey of 24 US slaughter plants indicated that 59% had excellent handling and low usage of electric prods and two plants (9%) had very abusive employees (Grandin, 1997a). In all sectors of the industry there is a need to treat stockpeople as professionals (Hemsworth and Coleman, 1998). This will help motivate them to have higher levels of stockmanship. Good stockmanship will significantly improve productivity (Rushen *et al.*, 1999).

## Importance of Continuous Measurement of Handling

Some managers allow rough handling to occur because their operations are understaffed. They can easily measure the number of animals handled per person, but they are not measuring the consequences of rough handling, such as bruises, injuries, lower weight gains, reduced milk yield or lowered pregnancy rates. To monitor these losses, there must be continuous measurement of both production losses and the methods used to handle animals. Grandin (1997b, 1998) has developed an objective scoring system for assessing handling of animals at slaughter plants. An auditor determines the percentage of cattle or pigs that are stunned correctly, are prodded with an electric prod, vocalize during handling (squeal, moo or bellow) and slip or fall during handling. Vocalization is correlated with physiological stress measurements (Dunn, 1990; White *et al.*, 1995). It is one of the most important measurements for pinpointing severe stress (see Chapter 20). Each item is scored on a yes/no basis for each animal. Scoring systems for measuring each animal's reactions, such as kicking or restlessness during milking or breeding, could also be developed. The scores could then be correlated with productivity measures. These scoring systems are modelled after hazard analysis critical control points (HACCP) programmes used for food safety. Reducing electric prod use, vocalization and slipping and falling during handling are all critical control points for reducing stress during handling. Continuous measurement is required to prevent a return of rough handling. It works the same way as microbiological testing for food safety. A food-processing facility would gradually get dirty unless microbiological counts were done on a regular basis. It is much easier to manage things that are measured. For further information on scoring systems, see Chapter 20.

## Regional Differences

There are regional differences in the quality of animal handling. These differences are related to basic attitudes towards animals. For example, handling in Scandinavian countries, Canada and the northern USA is often better than handling in the southern USA and Mexico (Grandin, 1988). Curtis and Guither (1983) reported that in Europe there was less interest in animal welfare in the southern countries. Kellert (1978, 1980) has also observed regional differences in attitudes. People with the 'macho' attitude are more likely to try to overpower an animal with force instead of trying to 'out-think' it. Progressive feedlots, slaughter plants and auctions are hiring more women to handle animals. The managers report that they are often gentler and more careful with the animals. Hertzog *et al.* (1991) reported that women were also more concerned about animal welfare issues. People who like animals and have a positive attitude towards them will have more productive pigs (Hemsworth *et al.*, 1994; Hemsworth and Coleman, 1998). Braithwaite and Braithwaite (1982) give further information on attitudes towards animals.

## Economic Losses

Losses which occur during handling and transport cause large monetary losses to the livestock industry. In the USA the National Beef Quality Audit calculated that $4.03 is lost due to bruises on every fed animal marketed. Boleman *et al.* (1998) and Smith *et al.* (1995) report that 48% of US-fed steers and heifers have bruised carcasses. Bruises on cull cows and bulls would add up to additional losses of $12.00 per animal (Smith *et al.*, 1994). In Australia, bruises on cattle cost $36 million dollars annually (Blackshaw *et al.*, 1987) and in New Zealand bruises on cattle represent an economic loss of 1% of the country's annual export earnings (Marshall, 1977). Cockram and Lee (1991) found that sheep sold through markets had more bruises than sheep sold direct to the slaughter plant. More recent studies with cattle confirm the finding that animals which go through auctions have more bruises (McNally and Warriss, 1996; Hoffman *et al.*, 1998). A recent Canadian study showed that 15% of the cattle had severe bruising and 78% of the carcasses were bruised (Van Donkersgoed *et al.*, 1999). Smith *et al.* (1999) found that 22% of cull cows in the USA had severe bruising and 2.2% of these animals had extreme bruises that destroyed major portions of the carcass. Selling cull cows when they are still in good body condition will provide the greatest economic benefit (Apple *et al.*, 1999a,b). A survey of cull sows in Minnesota indicated that 67% had foot lesions and 4.6% had shoulder lesions (Ritter *et al.*, 1999). The shoulder lesions cause extensive meat damage.

Stress-induced meat-quality problems, such as dark cutters, cause even greater losses. The National Beef Quality Audit estimates that dark cutters cost the beef industry $6.08 for every fed animal slaughtered (Boleman *et al.*, 1998). Dark-cutting beef is darker and drier than normal and has a shorter shelf-life. Good reviews on dark-cutting beef can be found in Fabiansson *et al.* (1988), Hood and Tarrant (1981) and Scanga *et al.* (1998).

Research at Oklahoma State University (1999) indicated that withdrawing feed from fed feedlot cattle for 24 h prior to slaughter resulted in a loss of $5.00 per animal due to carcass shrinkage and increased dark cutters. Feedlot managers sometimes do this so that an extra steer can be transported without violating truck weight limitations, but it is a false economy.

Carcass shrinkage (loss of weight) due to rough handling or long hours in transport causes additional losses. Shorthose and Wythes (1988) review numerous studies which quantify shrinkage in cattle and sheep. Large economic losses also occur due to death losses and morbidity in calves which are transported long distances (Hails, 1978). Death losses in US cattle are approximately 1% of the fed cattle (Jensen *et al.*, 1976; Irwin *et al.*, 1979; Bartlett *et al.*, 1987). Even though these studies are 15–25 years old, the death loss percentages are still the same today. A high percentage of the death loss is caused by shipping fever, a respiratory disease caused by a combination of shipping stress and viral and bacterial agents. Shipping fever (bovine respiratory disease) costs the US cattle industry $624 million annually (National Agriculture Statistics Service, 1992–1998). Sickness occurs in about 5% of yearlings (Jensen *et al.*, 1976) and 15% of calves (Bartlett *et al.*, 1987). About 70% of all losses occur in calves weighing less than 225 kg (Noon *et al.*, 1980). Vaccinating animals at the ranch of origin can reduce death losses from 1.6% to 1% and reduce sickness from 19% to 15% (Bartlett *et al.*, 1987). A more recent study shows that preweaning calves and vaccinating 35–45 days prior to shipment to a feedlot resulted in reducing death losses due to respiratory disease (shipping fever) at the feedlot from 0.98% to 0.16% (National Cattlemen's Association, 1994). A combination of preshipment vaccination and good trucking practices can keep death losses on 35-h non-stop trips at under 0.1% (Mills, 1987; Grandin, 1997c). Savings in medical costs and losses from reduced weight gains on sick cattle would be even higher. Progressive cattle feedlot operators have found that quiet handling during vaccinating enables cattle to go back on feed more quickly. Cattle which become agitated while being handled in a squeeze chute will have lower weight gains and tougher meat (Voisinet *et al.*, 1997a,b). Improving vaccinating and handling practices to reduce sickness could improve profitability. Researchers at Texas A&M University (1998–1992) found that healthy feedlot cattle provide $49.55–$117.42 more profit per animal.

In pigs, deaths during transport and pale, soft, exudative (PSE) meat cause large financial loss. PSE is a pork quality defect which is caused by a combination of factors, such as pigs with stress-susceptible genes, rough handling shortly before slaughter and poor carcass chilling. Good reviews on PSE and stress-susceptible pigs can be found in Smulders (1983), Grandin (1985), Sather *et al.* (1991) and Tarrant (1993). A recent Canadian study showed that, even when the stress gene had been bred out of 90% of the pigs, there was still 14.8% PSE (Murray and Johnson, 1998). The author visited the plant where this study was conducted and observed very excessive use of electric prods. In another Canadian plant with good handling, similar pigs had only 4% PSE. Pork from heavyweight pigs with the stress gene was judged by a taste panel to be tougher and drier than pork from pigs free of the stress gene (Monin *et al.*, 1999).

The US pork industry loses an average of 34 cents on every pig marketed due to PSE (Morgan *et al.*, 1993). Shrink losses due to PSE during transportation of chilled pork cause a loss of $10^6$ kg of meat annually in the USA (Kaufman, 1978). Much of their loss is caused by genetic factors. PSE causes further losses because consumers dislike shrinkage during cooking (Topel, 1976).

A British study showed that, at the retail level, each PSE carcass causes an additional $2.00 (£1.13) loss (Smith and Lesser, 1982). The USA has high levels of PSE pork (Cassens *et al.*, 1992). Morgan *et al.* (1993), reports that 9.1% of all hams and loins processed in the USA have PSE. In Denmark, pig breeding and handling are closely monitored: PSE levels in Denmark are about 2% (Barton-Gade, 1989). In Switzerland, retailers stated that they would be willing to pay up to $15.00 (23 SwFr) more per pig for pork that had good colour and low drip loss (Von Rohr *et al.*, 1999).

## Economic Incentives Reduce Losses

The structure of the marketing system can provide either an incentive or a disincentive to reduce losses. Denmark has used the latest information on genetics, handling and transport to greatly reduce PSE. Their vertically integrated industry consists of producer-owned slaughter-plant cooperatives (Fensvig, 1989). Each individual pig is identified and losses caused by the producer can be traced back. Since the producers own the slaughter plants and processing factories, they are motivated to reduce losses.

In the USA, farms and slaughter plants are often owned by different people. A major reason why meat quality defects are high in the USA is their marketing system. It fails to provide economic incentives to reduce losses (Grandin, 1989). A high percentage of pigs and cattle in the USA are still sold on a live-weight basis, where the animals are paid for prior to slaughter. Losses due to bruises, dark cutters, deaths and PSE are absorbed by the slaughter plant. A survey (Grandin, 1981) indicated that bruises on cattle were greatly reduced when producers switched to a carcass-based selling system where bruise damage was deducted from their payments. Marketing systems which allow losses to be passed on to the next buyer provide little incentive to reduce losses. Segmented marketing systems where cattle pass through one or more middlemen, brokers or order buyers prior to reaching their final destination contribute to substantial losses in both the USA and Mexico.

Handling of pigs greatly improved when the USA entered the Japanese export trade. When slaughter-plant managers watched a Japanese grader reject up to 40% of their pork loins due to PSE, a strong economic incentive was created to improve handling. Observations in three different plants showed that simple changes in handling procedures, such as showering, reducing electric prod usage and resting pigs, enabled 10% more pork to be exported to Japan.

A cattle producer is not motivated to vaccinate his/her calves unless he/she gets a premium price. A high percentage of calves arriving at large feedlots in the USA have never been vaccinated. They cannot be traced back to the original owner because there is no nationwide identification system in the USA. Implementation of a nationwide identification system would provide a major economic incentive to provide better care and handling. Producers would be motivated to vaccinate calves if they knew they could get claims for sickness.

Fortunately, producer groups are working together to produce truck-load lots of calves that have been preweaned and vaccinated 5 weeks prior to selling. These calves are being sold at premium prices because buyers know that they will be less likely to get sick.

Insurance payments for livestock transport must be structured to motivate good practices. If an insurance policy pays for all bruises and deaths, a truck driver has little incentive to reduce losses. Insurance policies should protect a trucking company from a catastrophic loss, such as tipping a truck over, but the policies should not cover one or two dead pigs.

Several vertically integrated pork companies have implemented incentive payment programmes to reduce transport death losses in pigs. A well-managed incentive programme can reduce losses by more than half. Incentive programmes for cattle truck drivers also help reduce losses in cattle shipped long distances (Mills, 1987). People handling livestock or poultry should never be paid based on the number of animals they can run through a race or the number of trucks they can load. This will result in careless work and increased injuries because it provides the wrong incentive. Payment should always be based on the quality of work. Incentive programmes have greatly reduced losses in the poultry industry. Poultry-catching crews and employees at some poultry companies receive a bonus for minimizing broken legs and wings.

Contracts for buying and selling livestock should have built-in incentives to reduce losses. The Australian sheep-shipping industry (Fig. 1.1) provides three examples of a lack of financial incentives to reduce losses. In Australia, death losses on sheep ships sailing to the Middle East average 1–2.5% and can rise to 6% (Higgs, 1991; Higgs *et al.*, 1991). Grandin (1983) reported that ships' officers stated that very low death losses of 0.47% are possible if sheep are carefully acclimatized in assembly feedlots and prepared prior to loading. A contributing factor to high death losses is contracts based on the number of live sheep loaded instead of the number of live sheep delivered at the destination (Grandin, 1983). There was little economic incentive to prepare sheep properly and train them to eat pelleted feed prior to transport or to identify the groups of sheep that are likely to have high death losses. Some lines of sheep have very high death losses and overall death losses could be greatly reduced if susceptible sheep could be identified (Norris *et al.*, 1989a). One of the main contributors to sheep deaths is refusal to eat prior to loading (Norris and Richards, 1989; Norris *et al.*, 1989b; Higgs *et al.*, 1991). Death losses during

**Fig. 1.1.** Huge ships with multiple decks of sheep pens transport millions of sheep from Australia to the Middle East. Large ships hold 60,000–100,000 sheep.

loading are very low, but death losses during unloading may reach 20% (Norris *et al.*, 1990). High discharge death rates occur at ports which have poor facilities and slow unloading. If the people receiving the sheep had to pay for the shipboard losses, they would be motivated to install better unloading facilities. Sheep deaths have also increased when oil prices are high, because the ships sail at a slower speed to save fuel (Gregory, 1992). This is an unfortunate example of an economic incentive that has increased death losses.

## Change in Marketing Systems

As marketing systems around the world change to provide greater accountability and trace-back, all segments of the livestock industry will become increasingly interested in improving handling and transport procedures. Increasing food-safety concerns will be a major impetus to implement trace-back programmes. Individuals will be motivated to improve practices if they are held accountable for losses.

However, to determine where the losses are occurring, they must be measured. One vertically integrated chicken company had rough chicken-catching practices because they had never measured the percentage of broken wings. The first time they measured broken wings they were shocked at the high percentage. Measuring damage on a regular basis will provide a further incentive to reduce losses.

Livestock industry leaders, legislators and animal welfare advocates should work to modify marketing systems so as to provide economic incentives to reduce losses during transport and handling. Economic incentives are powerful tools for improving treatment of animals during handling and transport, but they are less effective for improving animal welfare in housing systems. Unfortunately, crowding too many pigs in a pen or too many chickens in a battery cage sometimes provides an economic advantage.

## Genetic Problems

Overselection of animals for traits such as rapid weight gain or increased milk production can cause serious welfare problems. Increased selection for rapid growth and a high percentage of lean meat has resulted in weaker pigs, where more are susceptible to death during transport (Grandin and Deesing, 1998). Very lean pigs which have the halothane stress gene will have higher death losses. Murray and Johnson (1998) report that death losses are 9.2% in homozygous-positive pigs, 0.27% in carriers and 0.05% in homozygous-negative pigs. This is especially a problem if market pigs are grown to very heavy slaughter weights of 120 kg or greater. The author has observed that lean hybrids selected for rapid growth and heavy muscling often have double and triple death losses when grown to heavy weights. British pigs which are slaughtered at lighter weights of 100 kg and taken directly from the farm to the slaughter plant have an average death loss of only 0.072% during transport and lairage (Warriss and Brown, 1994). The best average death loss percentage in a British slaughter plant was 0.045% (Warriss and Brown, 1994). One large, vertically integrated pork company in the USA pays an incentive payment of $15 per week to truck loading crews who achieve a 0.02% death loss. The best crews are often able to collect their bonuses. This company raises moderately muscled 110–120 kg lean hybrids.

There is also evidence that selection for greater and greater growth and yield in pigs has resulted in decreased disease resistance (Meeker *et al.*, 1987; Rothschild, 1998). Continuous selection for greater and greater yields of meat and milk provides economic benefits in the short run, but it may ultimately cause a disaster when an epidemic occurs in high producing animals with weakened immune systems. In the USA porcine respiratory and reproductive syndrome (PRRS) has increased in lean hybrid, high-producing pigs. Halibur *et al.* (1998) reports that there are genetic effects on the incidence of infection with PRRS. The author is concerned that some of the worst animal welfare problems in the future may be caused by overselection for a narrow range of production traits.

## Conclusions

The most powerful method for reducing losses during transport and handling is financial incentives. Marketing systems should be reorganized in countries where losses are passed from one seller to the next buyer. Direct marketing from the producer to the slaughter plant reduces losses if people are held accountable for them. The importance of good management must always be emphasized. Management commitment to good practices is the single most important factor which determines the quality of handling and transport. Measurement of losses caused by handling and transport should be done on a regular basis and used as a basis for a system of financial incentives for people who are handling and transporting animals.

## References

Apple, J.K., Davis, J.C., Stephenson, J., Hankins, J.F., Davis, J.R. and Beaty, S.L. (1999a) Influence of body condition scores on the carcass characteristics of subprimal yield in cull beef cows. *Journal of Animal Science* 77, 2660–2669.

Apple, J.K., Davis, J.C. and Stephenson, J. (1999b) Influence of body condition score on by product yield and value from cull beef cows. *Journal of Animal Science* 77, 2670–2679.

Bartlett, B.B., Rust, S.R. and Ritchie, H.D. (1987) Survey results of the effects of a pre shipment vaccination program on sale price, morbidity and morbidity of feeder cattle in northern Michigan. *Journal of Animal Science* 65 (Suppl. 1), 106 (abstract).

Barton-Gade, P. (1989) Pre-slaughter treatment and transportation research in Denmark. In: *35th International Congress of Meat Science and Technology.* Danish Meat Research Institute, Roskilde, Denmark, pp. 140–145.

Blackshaw, J.K., Blackshaw, A.W. and Kusano, T. (1987) Cattle behavior in a sale yard and its potential to cause bruising. *Australian Journal of Experimental Agriculture* 27, 753–757.

Boleman, S.L., Boleman, S.J., Morgan, W.W., Hale, D.S., Griffin, D.B., Savell, J.W., Ames, R.P., Smith, M.T., Tatum, J.D., Field, T.G., Smith, G.C., Gardner, B.A., Morgan, J.B., Northcutt, S.L., Dolezal, H.G., Gill, D.R. and Ray, F.K. (1998) National Beef Quality Audit – 1995: survey of producer-related defects and carcass quality and quantity attributes. *Journal of Animal Science* 76, 96–103.

Braithwaite, J. and Braithwaite, V. (1982) Attitudes toward animal suffering: an exploratory study. *International Journal Study of Animal Problems* 3(1), 42–49.

Cassens, R.G., Kauffman, R.G., Sherer, A. and Meeker, D.L. (1992) Variations on pork quality: a 1991 survey. In: *38th International Congress of Meat Science and Technology.* Département de Technologie de la Viande, INRA, St Genes Champanelle, France.

Cockram, M.S. and Lee, R.A. (1991) Some preslaughter factors affecting the occurrence of bruising in sheep. *British Veterinary Journal* 147, 120–125.

Curtis, S.E. and Guither, H.D. (1983) Animal welfare an international perspective. In: Baker, F.D. (ed.) *Beef Cattle Science Handbook.* Westview Press, Boulder, Colorado, pp. 1187–1191.

Dunn, C.S. (1990) Stress reactions of cattle undergoing ritual slaughter using two methods of restraint. *Veterinary Record* 126, 522–525.

Fabiansson, S.U., Shorthose, W.R. and Warner, R.D. (1988) (eds) Dark cutting in cattle and sheep. In: *Proceedings of an Australian Workshop.* Australian Meat and Livestock Development Corporation, Sydney, Australia.

Fensvig, A.T. (1989) Strategies of the Danish meat industry and their relation to future research and development. In: *35th International Congress of Meat Science and Technology. Danish Meat Research Institute,* Roskilde, Denmark, pp. 3–8.

Grandin, T. (1981) Bruises on southwestern feedlot cattle. *Journal of Animal Science* 53(Suppl. 1), 213 (abstract).

Grandin, T. (1983) *A Survey of Handling Practices and Facilities Used in the Export of Australian Livestock.* Department of Primary Industry, Australian Bureau of Animal Health, Australian Government Publishing Service, Canberra, Australia. Also reprinted as Paper No. 84–6533, American Society of Agricultural Engineers, St Joseph, Michigan.

Grandin, T. (1985) Improving pork quality through handling systems. In: *Animal Health and Nutrition.* Watt Publishing, Mt Morris, Illinois, July/August, 14–26.

Grandin, T. (1988) Behavior of slaughter plant and auction employees towards animals. *Anthrozoos* 1, 205–213.

Grandin, T. (1989) Pre-slaughter treatment and transportation research in the United States. In: *International Congress of Meat Science and Technology.* Danish Meat Research Institute, Roskilde, Denmark, pp. 133–139.

Grandin, T. (1990) Handling practices in US feedlots and packing plants. In: *Proceedings Livestock Conservation Institute.* Livestock Conservation Institute, Madison, Wisconsin, pp. 115–120.

Grandin, T. (1997a) *Survey of Stunning and Handling in Federally Inspected Beef, Veal, Pork and Sheep Slaughter Plants.* USDA/Agricultural Research Service Project 3602–32000–002–08G, Beltsville, Maryland.

Grandin, T. (1997b) *Good Management Practices for Animal Handling at Stunning.* American Meat Institute, Washington, DC.

Grandin, T. (1997c) Assessment of stress during transport and handling. *Journal of American Science* 75, 249–257.

Grandin, T. (1998) Objective scoring of animal handling and stunning practices at slaughter plants. *Journal of American Veterinary Medical Association* 212, 36–39.

Grandin, T. and Deesing, M. (1998) *Genetics and the Behavior of Domestic Animals.* Academic Press, San Diego, California.

Gregory, N.G. (1992) Pre-slaughter handling and stunning. In: *38th International Congress of Meat Science and Technology.* Danish Meat Research Institute, Roskilde, Denmark, pp. 133–139.

Hails, M.R. (1978) Transportation stress in animals: a review. *Animal Regulation Studies* 282–343.

Halibur, P.G., Rothschild, M.F., Paul, P.S., Thacker, B.J. and Meng, X.J. (1998) Differences in susceptibility of Duroc, Hampshire and Meishan pigs to infection with a high virulence strain (VR2385) of porcine respiratory and reproductive syndrome virus (PRRSV). *Journal of Animal Breeding and Genetics* 115, 181–189.

Hemsworth, P.H. and Coleman, G.J. (1998) *Human–Livestock Interactions.* CAB International, Wallingford, UK.

Hemsworth, P.H., Coleman, G.J. and Barnett, J.L. (1994) Improving the attitude and behavior of stockpeople towards pigs and the consequences on the behavior and reproductive performance of commercial pigs. *Applied Animal Behaviour Science* 39, 349–362.

Hertzog, H.A., Bethcard, N.S. and Pittman, R.B. (1991) Gender sex role orientation and attitudes towards animals. *Anthrozoos* 4, 184–191.

Higgs, A.R. (1991) *National Recording System for the Livestock Sheep Export Industry* Miscellaneous Publication 12/91, NDRS, Canberra, Australia.

Higgs, A.R., Norris, R.T. and Richards, R.B. (1991) Season, age and adiposity influence death rates in sheep exported by sea. *Australian Journal of Agricultural Research* 42, 205–214.

Hoffman, D.E., Spire, M.F., Schwenke, J.R. and Unrah, J.A. (1998) Effect of source of cattle and distance transported to a commercial slaughter facility on carcass bruises in mature beef cows. *Journal of Animal Science* 212, 668–672.

Hood, D.E. and Tarrant, P.V. (eds) (1981) *The Problem of Dark Cutting Beef.* Martinus Nijhoff, Boston.

Irwin, M.R., McConnell, S., Coleman, J.D. and Wilcox, G.E. (1979) Bovine respiratory disease complex: a comparison of potential predisposing and etiological factors in Australia and in the United States. *Journal of the American Veterinary Medical Association* 175, 1095–1099.

Jensen, R., Peirson, R.E., Braddy, P.M., Saari, D.A., Lauerman, L.H., England, J.J., Horton, D.P. and McChesney, A.E. (1976) Diseases of yearling cattle in Colorado. *Journal of the American Veterinary Medical Association* 169, 497–499.

Kaufman, R.G. (1978) Shrinkage of PSE, normal and DFD hams during transit and processing. *Journal of Animal Science* 46, 1236–1240.

Kellert, S. (1978) *Policy Implications of a National Study of American Attitudes and Behavioral Relationships with Animals*. Stock No. 024–101–00482–7, US Fish and Wildlife Service, US Department of the Interior, US Government Printing Office, Washington, DC.

Kellert, S. (1980) American attitudes towards, and knowledge of animals: an update. *International Journal for the Study of Animal Problems* (Humane Society of the US, Washington, DC) 1(2), 87–119.

McNally, P.W. and Warriss P.D. (1996). Recent bruising in cattle at abattoirs. *Veterinary Record* 138(6), 126–128.

Marshall, B.L. (1977) Bruising in cattle presented for slaughter. *New Zealand Veterinary Journal* 25, 83–86.

Meeker, D.L., Rothschild, M.F., Christian, L.L., Warner, C.M. and Hill, H.T. (1987). Genetic control of immune response to pseudorabies and atrophic rhinitis vaccines. *Journal Animal Science* 64, 407–413.

Mills, B. (1987) Arrive alive. *Beef Today*, October, 19–20.

Monin, G., Larzul, C., LeRoy, P., Culioli, J., Mourot, J., Rousset-Akrim, S., Talmant, A., Touraille, C. and Sellier, P. (1999) Effects of halothane genotype and the slaughter weight on texture of pork. *Journal of Animal Science* 77, 408–415.

Morgan, J.B., Cannon, F.K., McKeith, D., Meeker, D. and Smith, G.C. (1993). *National Pork Chain Audit (Packer, Processor, Distributor)*. Final Report to the National Pork Producers Council, Colorado State University and University of Illinois, Champaign-Urbana.

Murray, A.C. and Johnson, C.P. (1998) Importance of halothane gene on muscle quality and preslaughter death in western Canadian pigs. *Canadian Journal of Animal Science* 78, 543–548.

National Agricultural Statistics Service (1992–1998) *Cattle and Calves Death Loss*. United States Department of Agriculture, Washington, DC.

National Cattlemen's Association (1994) *Strategic Alliances Field Study in Coordination with Colorado State University and Texas A&M University*. National Cattlemen's Beef Association, Englewood, Colorado.

Noon, T.N., Mare, C.J., Mazure, M., Prouth, F., Trautman, R.J. and Reed, R.J. (1980) *Selected Aspects of Respiratory Disease in New Feeder Calves*. Arizona Cattle Feeders Day, University of Arizona, Tucson, Arizona.

Norris, R.T. and Richards, R.B. (1989) Deaths in sheep exported from Western Australia – analysis of ship master's reports. *Australian Veterinary Journal* 66, 97–102.

Norris, R.T., Richards, R.B. and Dunlop, R.H. (1989) An epidemiological study of sheep deaths before and during export by sea from Western Australia. *Australian Veterinary Journal* 66, 276–279.

Norris, R.T., Richards, R.B. and Dunlop, R.H. (1989) Pre embarkation risk factors for sheep deaths during export by sea from Western Australia. *Australian Veterinary Journal* 66, 309–314.

Norris, R.T., Richards, B. and Higgs, T. (1990) The live sheep export industry. *Journal of Agriculture Western Australia* 31(4), 131–148.

Oklahoma State University (1999) *Department of Animal Science Research Report p-965*. Stillwater, Oklahoma.

Ritter, L.A., Xue, J., Dial, G.D., Morrison, R.B. and Marsh, W.E. (1999) Prevalence of lesions and body condition scores among female swine at slaughter. *Journal of the American Veterinary Medical Association* 214, 525–528.

Rothschild, M.F. (1998) Selection for disease resistance in the pig. In: *Proceeding National Swine Improvement Federation.* North Carolina State University, Raleigh, North Carolina.

Rushen, J., Taylor, A.A. and de Passillé, A.M. (1999) Domestic animals' fear of humans and its effect on their welfare. *Applied Animal Behaviour* 65, 285–303.

Sather, A.P., Murray, A.C., Zawadski, S.M. and Johnson, P. (1991) The effect of the halothane gene on pork product quality of pigs reared under commercial conditions. *Canadian Journal of Animal Science* 71, 959–967.

Scanga, J.A., Belk, K.E., Tatum, J.D., Grandin, T. and Smith, G.C. (1998) Factors contributing to the incidence of dark cutting beef. *Journal of Animal Science* 76, 2040–2047.

Shorthose, W.R. and Wythes, J.R. (1988) Transport of sheep and cattle. In: *34th International Congress of Meat Science and Technology.* CSIRO Meat Research Laboratory, Cannon Hill, Queensland, Australia.

Smith, G.C., Morgan, J.B., Tatum, J.D., Kukay, C.C., Smith, M.T., Schnell, T.D., Hilton, G.G., Lambert, C., Cowman, G. and Lloyd, B. (1994) *Improving the Consistency and Competitiveness of Non-fed Beef, and Improving the Salvage Value of Cull Cows and Bulls.* The final report of the National Cattlemen's Beef Association, Colorado State University, Fort Collins.

Smith, G.C., Savell, J.W., Dolezal, H.G., Field, T.G., Gill, D.G., Griffin, D.B., Hale, D.S., Morgan, J.B., Northcutt, S.L., Tatum, J.D., Ames, R., Boleman, S., Gardner, B., Morgan, W., Smith, M., Lamber, C. and Cowman, G. (1995) *Improving the Quality, Consistency, Competitiveness and Market Share of Beef – a Blueprint for Total Quality Management in the Beef Industry.* The final report of the National Beef Quality Audit, 1995, final report to the National Cattlemen's Beef Association, Colorado State University, Fort Collins, Oklahoma State University, Stillwater, and Texas A&M University, College Station.

Smith, G.C., Belk, K.E., Tatum, J.D., Field, T.C., Scanga, J.A., Roeber, D.L. and Smith, C.D. (1999) *National Market Cow and Bull Beef Audit.* Colorado State University for the National Cattlemen's Beef Association, Englewood, Colorado.

Smith, W.C. and Lesser, D. (1982) An economic assessment of pale soft exudative musculature in the fresh and cured pig carcass. *Animal Production* 34, 291–299.

Smulders, F.J.M. (1983) Pre stunning treatment during lairage and meat quality. In: Eikelenboom, G. (ed.) *Stunning of Animals for Slaughter.* Martinus Nijhoff, The Hague, pp. 90–95.

Tarrant, P.V. (1993) An overview of production, slaughter and processing factors that affect pork quality: general review. In: Puolanne, E., Demeyer, D.I., Ruusunen, M. and Ellis, S. (eds) *Pork Quality: Genetic and Metabolic Factors.* CAB International, Wallingford, UK.

Texas A&M University (1998–1992) *Ranch-to-rail Statistics.* Department of Animal Science, College Station, Texas.

Topel, D.G. (1976) Palatability and visual acceptance of dark, normal and pale colored porcine m. longissimus. *Journal of Food Science* 41, 628–630.

Van Donkergoed, J., Jewison, G., Mann, M., Cherry, B., Altwasser, B., Lower, R., Wiggins, K., Dejonge, R., Thorlakson, B., Moss, E., Mills, C. and Grogen, H. (1997) Canadian Beef Quality Audit. *Canadian Journal of Animal Science* 38, 217–225

Voisinet, B.D., Grandin, T., Tatum, J.D., O'Connor, S.F. and Struthers, J.J. (1997a) *Bos indicus*-cross feedlot cattle with excitable temperaments have tougher meat and a higher incidence of borderline dark cutters. *Meat Science* 46, 367–377.

Voisinet, B.D., Grandin, T., Tatum, J.D., O'Connor, S.F. and Struthers, J.J. (1997b) Feedlot cattle with calm temperaments have higher average daily gains than cattle with excitable temperaments. *Journal of Animal Science* 75, 892–896.

Von Rohr, P., Hofer, A. and Kunzi, N. (1999) Economic values for meat quality traits in pigs. *Journal of Animal Science* 77, 2633–2640.

Warriss, P.D. and Brown, S.N. (1994) A survey of mortality in slaughter pigs during transport and lairage. *Veterinary Record* 134, 513–515.

White, R.G., deShazer, J.A. and Trassler, C.J. (1995) Vocalization and physiological responses of pigs during castration with and without a local anesthetic. *Journal of Animal Science* 73, 381–386.

# Behavioural Principles of Animal Handling and Transport

## Harold W. Gonyou

*Prairie Swine Centre Inc., PO Box 21057, 2105 – 8th Street East, Saskatoon, Saskatchewan, Canada S7H 5N9*

## Introduction

Animal behaviour has been defined as 'the overt and composite functioning of animals individually and collectively ...[and] the means whereby the animal mediates dynamically with its environment, both animate and inanimate' (Fraser, 1980). This definition is very appropriate when considering animal handling and transport, as it identifies many of the key components of the process. When designing our management procedures, we must consider whether we are going to handle animals as individuals or as groups, and what type of animate (humans) and inanimate (facilities) environments are involved. Handling and transport may involve two distinct types of actions: directed movement and restraint. As neither of these actions is part of the normal maintenance behaviours of the animals, handling and transport can be among the most stressful events in an animal's life. Even our first astronauts were told to expect weightlessness during their flights, but our animals must move through chutes and ride in vehicles without prior explanation. Several chapters in this book refer to the relevant behavioural features of each species of livestock. This chapter will address the types of behavioural responses desired during handling and transport and how the characteristics of the animals and their interaction with humans and facilities may affect these behaviours.

## Movement and Restraint

Handling and transport involve two distinct types of actions: movement to a new location and remaining stationary. The first may involve movement out of a pen, down an alley or through a narrow race or chute. In some cases, a single animal is moved; in others, an entire herd may be involved. Holding an animal or animals stationary, which I will refer to as restraint, may involve a severe restriction to movement, as in a squeeze chute (crush), or relative freedom of movement within a confined area, as in a transport vehicle. Some procedures involve several steps including both movement and restraint. Kenny and Tarrant (1987) divided the transportation process into the component of repenning in a new environment (movement and restraint), loading/unloading (movement), confinement on a stationary vehicle (restraint) and confinement on a moving vehicle (restraint).

Movement is accomplished by making the target location, or the route to it, more attractive than the starting location. This involves balancing the forces of attraction and repulsion. The starting position can be made aversive in some way by instilling fear or threatening pain or discomfort. Extreme forms of this process may involve shouting, prodding or slapping the animal in a pen or alley. A subtler, and often just as effective, method is to move into the flight zone of the animal. Alternatively, the target location can be made more attractive by providing more space, better lighting, social contact or other features. Provided animals are not excited, they display a degree of natural curiosity that can be used to direct movement. Handlers may move in such a way as to attract the animals' attention toward an open gate or new pen. Calm cattle can be moved out of a pen by standing near the open gate as opposed to moving around the animals.

It is generally advisable to use the minimal amount of attractive or repulsive force as possible in moving animals. This is particularly true if the animals are to be returned to their original location. Using excessive force to move an animal out of its pen will make it more difficult to return it to the same pen after it has been handled and treated. For this reason, many animal facilities limit the use of aversive handling, such as the use of stock prods, to the load-out area and then only if no other means of encouraging movement are successful. The use of fear-producing handling methods may be counter-productive when moving certain animals. Some animals react to fear by becoming immobile. This reaction is evident in the tonic immobility response of frightened chickens (Gallup *et al.*, 1970; Jones, 1986). As fear levels increase, some animals will react violently, perhaps even aggressively, toward the handler. Grandin (1987) points out that penetrating an animal's flight zone too deeply may cause it to attempt to escape over the sides of the race. Penetration of the flight zone when no other means of escape is possible may cause animals to turn back on handlers (Hutson, 1982). Another reason to maintain nearly balanced levels of attractive and repulsive forces is that most procedures involve one or more shifts from movement to restraint to movement again.

## Animal Restraint

The second aspect of handling and transport is restraining the animal in one location or position. This is often necessary to perform various management or health-related procedures. These procedures vary in aversiveness, but few, if any, would be perceived as pleasant. As most animals will be restrained several times during their lifetime, it is important to recognize that an aversive procedure will make subsequent handling more difficult (Rushen, 1986). It is important to make each restraint as comfortable as possible and reduce the risk of fleeing during the procedure. Ewbank (1968) suggests four ways by which animals may be restrained: (i) direct force or physical barrier; (ii) knowledge and anticipation of animal behaviour; (iii) training; and (iv) drugs. Squeeze chutes and tilt tables are examples of physical means of restraint. These must be sized to accommodate the animal being restrained and strong enough to thwart any attempts to escape.

Many of our means of restraint rely on behavioural features of the animals. For many procedures, restraint of the head may be as effective as restraining the entire animal. Holding the head of a sheep is an effective means of restraint for many procedures. The use of a nose snare on pigs is effective because of the animal's natural tendency to pull back on the snare. However, this method of restraint can be very stressful. We can also effectively restrain animals by placing them in specific positions. Sheep will remain relatively still, and even enter a state of lowered awareness (Ruckebusch, 1964), if up-ended and held at an angle of approximately 60° from vertical. A more or less erect angle results in the animal struggling to escape (Kilgour and Dalton, 1984). Pigs will lie still in a sling with their feet off of the ground (Panepinto *et al.*, 1983) or on their back in a 'V'-trough. A similar system of restraint is used in conveyor restrainers used in some slaughter plants (Grandin, 1980).

Drugs or other physiological means may be used to restrain animals. The chemical xylazine is commonly used, under veterinary supervision, to sedate cattle during restraint (Ewbank, 1993). A mild electric current passed through the body of an animal can be an effective means of restraint. However, this procedure has been shown to be aversive to the animal and to have no analgesic effect (Grandin *et al.*, 1986; Rushen and Congdon, 1987). When using any means of restraint, it is necessary to weigh the benefits of a controlled animal against the distress it causes the animal.

## Behavioural Features of Animals

Several behavioural features are related to handling and transport and should be considered when working with animals or designing systems. Among the most important of these features are the flocking instinct, visual field and flight distance. Genetics, sex and previous experience also influence the response of animals.

The flocking instinct is particularly strong in sheep and less so in cattle and pigs. Hutson (1993) considers flocking a key aspect of sheep handling. It is very difficult to separate one sheep from a flock, and even cattle and pigs will attempt to rejoin the group when frightened. It is generally easier to move a group of animals than individuals. The flocking nature of sheep has been utilized in the use of leader sheep trained to lead other sheep through yards and alleys (Bremner et al., 1980). However, the formation of groups of unfamiliar individuals for handling and transport may lead to excessive fighting and be deleterious to the animals' welfare (Price and Tennessen, 1981; McGlone, 1985).

The response of animals to handling and transport depends on their sensory capacities. Most domestic animals are able to hear higher-frequency sounds than humans can (Kilgour and Dalton, 1984) and it is possible for them to become fearful of sounds of which their handlers are unaware. The visual field of the animal is also important. Grandin (1979a) recommends that handlers position themselves 45–60° from directly behind animals in a race, on the edge of their visual field. The size of the visual field is affected by the size and location of wool and horns (Hutson, 1980a). The concepts of visual field and flocking behaviour combine to facilitate movement if the lead animal has another animal within its field of vision (Hutson, 1980a). Restricting the ability of birds to see their surroundings is effective in reducing their struggling during restraint in slaughter plants (Jones et al., 1998).

Another important behavioural feature of an animal is its flight zone. Animals react to an approaching human by observing the person's movements and then turning away to escape. If no escape is possible, such as in a blind alley or the corner of a pen, the animal will run back past the handler, sometimes creating a dangerous situation. The distance at which animals react by fleeing is the flight distance (Hutson, 1982), and the area inside that distance is the flight zone. Entering the fight zone of an animal from behind causes the animal to move forward, whereas entering from the front causes it to back away or turn before fleeing. Handlers are advised to work on the edge of the flight zone, so that animal movement can be controlled by approaching and withdrawing from the animal (Grandin, 1987).

Within a species, there will be differences among breeds in ease of handling. Breeds of sheep have traditionally been raised in different environments, and it is likely that the resulting selection pressures have resulted in differences in behaviour among mountain, hill and lowland breeds (Arnold and Dudzinski, 1978). Breeds of sheep differ in their ease of handling during routine farming procedures (Whateley et al., 1974). Some apparent breed differences may in fact be due to different rearing environments or social inheritance of traits from their real or adoptive mother (Key and McIver, 1980). Fear of humans in sows is somewhat heritable (Hemsworth et al., 1990) and such genetic differences probably contribute to the variation in other handling traits in pigs (Lawrence *et al.*, 1991). Breed differences in 'temperament' or the reaction of cattle to being held in a squeeze chute are quite noticeable and moderately heritable (Heisler, 1979). However, the method of rearing of cattle may have a greater influence on their response to handling than does their genotype (Boivin *et al.*, 1992a).

Age, sex and physiological condition will also affect the behaviour of animals during handling and transport. Age is often confounded with size, physiological condition and experience. Young animals are often more flighty than mature animals, but their handling is also made more difficult by the lack of appropriately sized facilities in most operations. Many operations have only one set of handling facilities, which must accommodate the largest animals present. Smaller animals, just entering a feedlot or finishing barn, must be processed through these same facilities. The increased possibility of these animals turning or escaping from such ill-fitting facilities adds to the problems of handling young animals.

It is generally assumed that intact males are more difficult to handle than castrates. However, this difference may be age-dependent, as there is little difference in the ease of handling between castrate and intact cattle that are less than 2 years of age (Hinch, 1980; Vanderwert *et al.*, 1985; Tilbrook *et al.*, 1989). Kilgour and Dalton (1984) reported that bulls only become dangerous at older ages. Differences in the behaviour of intact and castrate males may be reduced by the use of anabolic implants (Vanderwert *et al.*, 1985).

Previous experience has a marked effect on the handling ability of animals. Previous experience with relatively non-aversive procedures, such as weighing (Hutson, 1980b) or regular moving (Geverink *et al.*, 1998), will facilitate subsequent movement through similar facilities. However, more aversive procedures, such as restraint in a tilt table, electroimmobilization (Grandin *et al.*, 1986) or hoof trimming (Lewis and Hurnik, 1998), will make subsequent handling more difficult. Provision of a reward following an aversive procedure may negate some of this negative reaction (Hutson, 1985).

## Interactions with Humans

There has recently been a great deal of interest in stockmanship, that is, the importance of human interactions with agricultural animals (Hemsworth and Coleman, 1998). A great deal of stockmanship involves the handling and transport of animals. It is important to know how animals react to our behaviour in order to effectively move and restrain them. Our intentions may not always be understood, and animals may become fearful and react strongly in a negative manner. In keeping with the principle that animals should not be pushed to an extreme emotion, our control must be subtle. Animals perceive specific features in their environment and will react to them by appropriate behavioural and emotional responses (Kendrick, 1991). Pigs perceive a human standing erect to be more threatening than one that is squatting

(Hemsworth *et al.*, 1986a). Therefore, when inspecting pigs in a pen, it is appropriate to squat. When driving pigs and a low level of fear is desired, it is appropriate to stand. Animals should not be rushed during movement. Dairy workers may try to speed the movement of cows, even though they are going in the right direction, and in so doing cause them to misplace their feet, injure themselves and become lame (Chesterton *et al.*, 1989). Rough and abusive handling at saleyards increases injuries to the animals (Blackshaw *et al.*, 1987). Impatience in pig herders may result in them slapping the pigs, which has negative consequences in terms of their reproductive performance (Hemsworth *et al.*, 1989). Effective handling is best achieved when stockpersons realize that their attitude and consequent behaviour affects the animals' productivity and welfare (Hemsworth and Coleman, 1998).

A second area of interest is the possibility of improving the human/animal relationship, and thus making handling easier, by exposing animals to human contact before it is required for management routines. Hargreaves and Hutson (1990) and Mateo *et al.* (1991) reported that several exposures of sheep to gentle handling, including speaking to and touching the animal, reduced the fearfulness of the animals and improved their approachability during subsequent routine handling. Repeated restraint, as opposed to gentle handling, had no effect on the response to future handling. Other research has exposed animals to handling during what are believed to be sensitive periods during which human/animal bonds should develop quickly. It is known that such a period exists for companion animals during the first few weeks of life (Freedman *et al.*, 1961). Hemsworth *et al.* (1986b) handled pigs up to 8 weeks of age and found that the pigs were less fearful of humans later in life than pigs that had not been handled. Another potentially sensitive period examined by Hemsworth *et al.* (1987) is the time of maternal responsiveness that follows parturition in cattle. Heifers exposed to humans at this time reacted less negatively to the herdsperson in the milking parlour for several weeks thereafter. Foals handled shortly after birth showed improvement in their response to handling later in life compared with non-handled animals (Waring, 1983). Although these studies demonstrate that exposure to humans can improve their subsequent response during handling, none confirm that the periods of treatment are more sensitive than others in the animals' lives. Boivin *et al.* (1992b) handled cattle during two potentially sensitive periods, after birth and after weaning, and reported that both affected subsequent response to handling similarly. Sensitive periods may exist for the establishment of good human/animal relationships with agricultural animals, but existing results do not confirm such periods.

Another area of interest in human/animal interactions is the ability of animals to recognize individuals who have treated them positively or negatively in the past. Calves will demonstrate a preference for individuals who have treated them well, compared with those who have treated them poorly (de Passille *et al.*, 1996). Similar studies with pigs have yielded conflicting results (Hemsworth *et al.*, 1994; Tanida and Nagano, 1998). Nevertheless, whenever possible, it is recommended that a few individuals be responsible for the most aversive procedures and that others be responsible for the day-to-day management of the animals.

## Interaction with the Physical Environment

The interaction of the animal with the physical environment is also a critical part of handling and transport. Movement is enhanced if the equipment and penning are attractive to the animal and do not provoke fear. The facilities must fit the animal's sensory capacity and avoid natural fears. Solid panels in handling facilities will block out visual and auditory disturbances in species as diverse as cattle (Grandin, 1979a) and chickens (Jones *et al.*, 1998). The result is a calmer animal that can be more easily controlled. The flocking instinct of animals allows for rapid movement of groups in wide alleys (Hutson and Hitchcock, 1978). However, when environmental features cause a hesitation in animal movement, animals in a wide alley bunch up and slow down more than animals in single-file races. Thus, if sheep are moved around a sharp corner, they should be in a single-file race (Hutson and Hitchcock, 1978). Similarly, if animals must enter a building before being caught in a squeeze chute, the single-file race should begin well before the entrance to the building (Grandin, 1979b). Group movement can be accomplished, even in difficult situations, by having multiple single-file races, divided by a wire panel so that visual, olfactory and auditory contact can be maintained (Grandin, 1979b).

Curved races can be very effective in handling facilities, as they require animals to move forward to maintain contact with animals ahead of them (Grandin, 1979a). In a straight race, the following animal can maintain visual contact even if a large gap exists between itself and the lead animal. The greatest advantage of curved races occurs if animals are being worked slowly or allowed to pause in the race while preceding animals are restrained or handled. If animals are being moved quickly and never stop walking, a straight race can be equally effective (Grandin, 1980).

With proper design, animal-handling facilities can reduce or eliminate the need for human involvement in some procedures. Cattle and sheep can be 'herded' into corrals by the use of one-way gates at water sources (Cheffins, 1987). The addition of hock boards to sheep races encourages one-way movement and results in a self-feeding system (Hutson and Butler, 1978). Morris and Hurnik (1990) designed a system of group housing for sows in which gates open and shut on schedule to move the animals to and from the feeding area. Electronic sow feeders often include the provision of sorting animals as they pass through the feeder. Robotic milking of cows requires not only that animals move through the system, but also that they stand in such a way that the udder and teats are exposed. A small step in the milking stall is effective in achieving the desired posture (Mottram *et al.*, 1994).

## Conclusions

A combination of the animal's previous experience, the personnel and facilities involved and the normal behavioural characteristics of the species determines the ease of animal handling. All of these factors should be considered in the management of animals that will be handled and transported. The remaining chapters in this book address these factors and others in more detail as they examine specific situations for each species of agricultural animal.

## References

Arnold, G.W. and Dudzinski, M.L. (1978) Social organization and animal dispersion. In: *Ethology of Free Ranging Domestic Animals.* Elsevier Scientific Publishing Company, Amsterdam, the Netherlands, pp. 51–96.

Blackshaw, J.K., Blackshaw, A.W. and Kusano, T. (1987) Cattle behaviour in saleyard and its potential to cause bruising. *Australian Journal of Experimental Agriculture* 27, 753–757.

Boivin, X., Le Neindre, P., Chupin, J.M., Garel, J.P. and Trillat, G. (1992a) Influence of breed and early management on ease of handling and open-field behaviour of cattle. *Applied Animal Behaviour Science* 32, 313–323.

Boivin, X., Le Neindre, P. and Chupin, J.M. (1992b) Establishment of cattle–human relationships. *Applied Animal Behaviour Science* 32, 325–335.

Bremner, K.J., Braggins, J.B. and Kilgour, R. (1980) Training sheep as 'leaders' in abattoirs and farm sheep years. *Proceedings of the New Zealand Society of Animal Production* 40, 111–116.

Cheffins, R.I. (1987) Self mustering cattle. *Queensland Agricultural Journal* 113, 329–336.

Chesterton, R.N., Pfeiffer, D.U., Morris, R.S. and Tanner, C.M. (1989) Environmental and behavioural factors affecting the prevalence of foot lameness in New Zealand dairy herds – a case–control study. *New Zealand Veterinary Journal* 37, 135–142.

de Passille, A. M., Rushen, J., Ladewig, J. and Petherick, C. (1996) Dairy calves' discrimination of people based on previous handling. *Journal of Animal Science* 74, 969–974.

Ewbank, R. (1968) The behavior of animals in restraint. In: Fox, M.W. (ed.) *Abnormal Behavior in Animals.* W.B. Saunders, Philadelphia, Pennsylvania, pp. 159–178.

Ewbank, R. (1993) Handling cattle in intensive systems. In: Grandin, T. (ed.) *Livestock Handling and Transport,* 1st edn. CAB International, Wallingford, UK, pp. 59–73.

Fraser, A.F. (1980) *Farm Animal Behaviour,* 2nd edn. Baillière Tindall, London, pp. 1–7.

Freedman, D.G., King, J.A. and Elliot, O. (1961) Critical period in the social development of dogs. *Science* 133, 1016–1017.

Gallup, G.G. Jr, Nash, R.F., Potter, R.J. and Donegan, N.G. (1970) Effect of varying conditions of fear on immobility reactions in domestic chickens (*Gallus gallus*). *Journal of Comparative Physiology and Psychology* 73, 442–445.

Geverink, N.A., Kappers, A., van de Burgwal, J.A., Lambooij, E., Blokhuis, H.J. and Wiegant, V.M. (1998) Effects of regular moving and handling on the behavioral and physiological responses of pigs to preslaughter treatment and consequences for subsequent meat quality. *Journal of Animal Science* 76, 2080–2085.

Grandin, T. (1979a) Handling livestock. *Agripractice* 20, 697–706.

Grandin, T. (1979b) Designing meat packing plant handling facilities for cattle and hogs. *Transactions of the America Society of Agricultural Engineers* 22, 912–917.

Grandin, T. (1980) Designs and specifications for livestock handling equipment in slaughter plants. *International Journal for the Study of Animal Problems* 1, 178–200.

Grandin, T. (1987) Animal handling. In: Price, E.O. (ed.) Farm animal behavior. *Veterinary Clinics of North America: Food Animal Practice* 3, 323–338.

Grandin, T., Curtis, S.E., Widowski, T.M. and Thurmon, J.C. (1986) Electro-immobilization versus mechanical restraint in an avoid–avoid choice test for ewes. *Journal of Animal Science* 62, 1469–1480.

Hargreaves, A.L. and Hutson, G.D. (1990) The effect of gentling on heart rate, flight distance and aversion of sheep to a handling procedure. *Applied Animal Behaviour Science* 26, 243–252.

Heisler, C.E. (1979) Breed differences and genetic parameters of temperament of beef cattle. MS thesis, University of Saskatchewan, Saskatoon, Canada.

Hemsworth, P.H. and Coleman, G.J. (1998) *Human–Livestock Interactions: The Stockperson and the Productivity and Welfare of Intensively Farmed Animals*, 1st edn. CAB International, Wallingford, UK, 156 pp.

Hemsworth, P.H., Gonyou, H.W. and Dziuk, P.J. (1986a) Human communication with pigs: the behavioural response of pigs to specific human signals. *Applied Animal Behaviour Science* 15, 45–54.

Hemsworth, P.H., Barnett, J.L., Hansen, C. and Gonyou, H.W. (1986b) The influence to early contact with humans on subsequent behavioural response of pigs to humans. *Applied Animal Behaviour Science* 15, 55–63.

Hemsworth, P.H., Hansen, C. and Barnett, J.L. (1987) The effect of human presence at the time of calving of primiparous cows on their subsequent behavioural response to milking. *Applied Animal Behaviour Science* 18, 247–255.

Hemsworth, P.H., Barnett, J.L., Coleman, G.J. and Hansen, C. (1989) A study of the relationships between the attitudinal and behavioural profiles of stockpersons and the level of fear of humans and reproductive performance of commercial pigs. *Applied Animal Behaviour Science* 23, 301–314.

Hemsworth, P.H., Barnett, J.L., Treacy, D. and Madgwick, P. (1990) The heritability of the trait fear of humans and the association between this trait and subsequent reproductive performance of gilts. *Applied Animal Behaviour Science* 25, 85–95.

Hemsworth, P.H., Coleman, G.J., Cox, M. and Barnett, J.L. (1994) Stimulus general-isation: the inability of pigs to discriminate between humans on the basis of their previous handling experience. *Applied Animal Behaviour Science* 40, 129–142.

Hinch, G.N. (1980) Handling strategies for entire and castrated male cattle. In: Wodzicka-Tomaszewska, M., Edey, T.N. and Lynch, J.J. (eds) *Behaviour in Relation to Reproduction, Management and Welfare of Farm Animals*. Reviews in Rural Science, University of New England, Armidale, Australia, pp. 157–160.

Hutson, G.D. (1980a) Visual field, restricted vision and sheep movement in laneways. *Applied Animal Ethology* 6, 175–187.

Hutson, G.D. (1980b) The effect of previous experience on sheep movement through yards. *Applied Animal Ethology* 6, 233–240.

Hutson, G.D. (1982) 'Flight distance' in Merino sheep. *Animal Production* 35, 231–235.

Hutson, G.D. (1985) The influence of barley food rewards on sheep movement through a handling system. *Applied Animal Behaviour Science* 14, 263–273.

Hutson, G.D. (1993) Behavioural principles of sheep handling. In: Grandin, T. (ed.) *Livestock Handling and Transport*, 1st edn. CAB International, Wallingford, UK, pp. 127–146.

Hutson, G.D. and Butler, M.L. (1978) A self-feeding sheep race that works. *Journal of Agriculture* 76, 335–336.

Hutson, G.D. and Hitchcock. D.K. (1978) The movement of sheep around corners. *Applied Animal Ethology* 4, 349–355.

Jones, R.B. (1986) The tonic immobility reaction of the domestic fowl: a review. *World's Poultry Science Journal* 42, 82–96.

Jones, R.B., Satterlee, D.G. and Cadd, G.G. (1998) Struggling responses of broiler chickens shackled in groups on a moving line: effects of light intensity, hoods, and 'curtains'. *Applied Animal Behaviour Science* 58, 341–352.

Kendrick, K.M. (1991) How the sheep's brain controls and visual recognition of animals and humans. *Journal of Animal Science* 69, 5008–5016.

Kenny, F.J. and Tarrant, P.V. (1987) The physiological and behavioral responses of crossbred Friesian steers to short-haul transport by road. *Livestock Production Science* 17, 63–75.

Key, C. and MacIver, R.M. (1980) The effects of maternal influences on sheep: breed differences in grazing, resting and courtship behaviour. *Applied Animal Ethology* 6, 33–48.

Kilgour, R. and Dalton, D.C. (1984) *Livestock Behaviour: a Practical Guide*. Collins Technical Books, Glasgow, UK.

Lawrence, A.B., Terlouw, E.M.C. and Illuis, A.W. (1991) Individual differences in behavioural responses of pigs to non-social and social challenges. *Applied Animal Behaviour Science* 30, 73–86.

Lewis, N.J. and Hurnik, J.F. (1998) The effect of some common management practices on the ease of handling of dairy cows. *Applied Animal Behaviour Science* 58, 213–220.

McGlone, J.J. (1985) A quantitative ethogram of aggressive and submissive behaviors in recently regrouped pigs. *Journal of Animal Science* 61, 559–565.

Mateo, J.M., Estep, D.Q. and McCann, J.S. (1991) Effects of differential handling on the behaviour of domestic ewes (*Ovis aries*). *Applied Animal Behaviour Science* 32, 45–54.

Morris, J.R. and Hurnik, J.F. (1990) An alternative housing system for sows. *Canadian Journal of Animal Science* 70, 957–961.

Mottram, T.T., Caroff, H. and Gilbert, C. (1994) Modifying the posture of cows for automatic milking. *Applied Animal Behaviour Science* 41, 191–198.

Panepinto, L.M., Phillips. R.W., Norton, S.E., Pryor, T.C. and Cox, R. (1983) A comfortable, minimum stress method of restraint for Yucatan miniature swine. *Laboratory Animal Science* 33, 95–97.

Price M.A. and Tennessen, T. (1981) Preslaughter management and dark-cutting in the carcasses of young bulls. *Canadian Journal of Animal Science* 61, 205–208.

Ruckebusch, Y. (1964) Etude électrophysiologique et comportementale de l'immobilisation réflexe, chez les petits ruminants. *Revue Médecine Vétérinaire* 115, 793.

Rushen, J. (1986) Aversion of sheep to electro-immobilization and physical restraint. *Applied Animal Behaviour Science* 15, 315–324.

Rushen, J. and Congdon, P. (1987) Electro-immobilisation of sheep may not reduce the aversiveness of a painful treatment. *Veterinary Record* 120, 37–38.

Tanida, H. and Nagano, Y. (1998) The ability of miniature pigs to discriminate between a stranger and their familiar handler. *Applied Animal Behaviour Science* 56, 149–159.

Tilbrook, A.J., Hemsworth, P.H., Barnett, J.L. and Skinner, A. (1989) An investigation of the social behaviour and response to humans of young cattle. *Applied Animal Behaviour Science* 23, 107–116.

Vanderwert, W., Berger, L.L., McKeith F.K., Baker, A.M., Gonyou, H.W. and Bechtel, P.J. (1985) Influence of zeranol implants on growth, behavior and carcass traits in Angus and Limousin bulls and steers. *Journal of Animal Science* 61, 310–319.

Warning, G.H. (1983) *Horse Behavior: The Behavioral Traits and Adaptations of Domestic and Wild Horses, Including Ponies.* Noyes Publications, Park Ridge, New Jersey, pp. 236–240.

Whateley, J., Kilgour, R. and Dalton, D.C. (1974) Behaviour of hill country sheep breeds during farming routines. *Proceedings of the New Zealand Society of Animal Production* 34, 28–36.

# General Principles of Stress and Well-being

3

## P.B. Siegel[1] and W.B. Gross[2]

[1]Department of Animal and Poultry Sciences, [2]Department of Large Animal Clinical Sciences, Virginia Polytechnic Institute and State University, Blacksburg, VA 24061, USA

## Introduction

Responses of animals to their environments are easier to evaluate when viewed as aims and strategies for survival. Nature is dynamic and maintenance of variation within and between populations enhances their adaptation to environmental changes and associated stressors. Thus, conservation of genetic variation allows for ecological niches to be filled, both short- and long-term.

In contrast to populations, the genome of an individual remains constant (barring mutation) throughout life. Factors influencing genetic variation of populations include selection, mutation, migration and chance. Although some individuals do not survive or reproduce and there is unequal reproduction among those that do, an individual's aim is to pass its genes on to subsequent generations. Those individuals best adapted to the current environment seem to have the greatest opportunity of accomplishing this objective. Owners of livestock and poultry may prefer uniformity and high productivity with no morbidity or mortality. These preferences, however, may be in conflict with maintaining genetic variation in populations and with allocation of resources by the animal as it passes through various environments during its life.

Genetic variation in animal populations allows some individuals to survive and exploit environmental changes, which results in differential reproduction. Therefore, within a population there is a range of structural, biochemical, behavioural and disease resistance factors that are under varying degrees of genetic influence and whose phenotypic (outward) expression may be modified by its environment. Over a period of generations, genetic changes within the population can result in an increasing frequency of individuals that adapt to environmental changes. These genetic changes may occur due to differences in husbandry practices influencing selection (e.g. Muir and Craig, 1998) and in variation among individuals in resistance to diseases (e.g. Sarkar, 1992). Genetic factors also influence the allocation of an animal's resources to its various components. In response to environmental changes, allocation of resources can also be altered by the stress system. Although this chapter explores relationships between stress and well-being of animals with particular emphasis on chickens, the results are also valid for other farm animals (e.g. Broom, 1988).

## How Animals Respond to Stress

Stress is a norm in social animals and our understanding of the stress system is based on the work of Selye (1950, 1976) where the major components include the cerebrum, hypothalamus, pituitary, adrenals, glucocorticoids and a cascading of responses as animals attempt to respond to stressors. Gluco-corticoids (cortisol and corticosterone) are carried by the blood to all cells of the body where they enter the nucleus. They then regulate the translation of active genes of the cell into messenger RNA (mRNA), which migrates to the mitochondria, where encoded proteins are produced. Understanding the stress system, however, is elusive and suffers from confusion and controversy. The stress response may be viewed as a physiological mechanism that links the stressor to a target organ, and target organ effects may be positive or negative. Thus, stress *per se* may or may not be damaging, and positive aspects of stress were reviewed by Zulkifli and Siegel (1995). In their review they cite the Yerkes–Dodson law, which relates degree of stress and performance efficiency. The law states that performance will be enhanced as arousal increases, but only up to a certain point or optimum level. Exceeding the optimum leads to inefficiency. This concept of 'optimum stress' is discussed throughout this chapter. The nebulous concept of stress and ramifications of stress in regard to well-being have been lucidly described in the delightful book, *Why Zebras Don't Get Ulcers* (Sapolsky, 1994).

Responses to a stressor can include anatomical, physiological and/or behavioural changes. Although normal values of various criteria differ not only among livestock and poultry species and among stocks within classes of livestock and poultry, patterns of response are similar. Long-term responses to a stressor result in an increase in size of the adrenal glands and a reduction of lymphoid mass. Both short- and long-term responses to stressors or the feeding of corticosterone reduce sizes of thymus, bursa of Fabricius and spleen, which rapidly return to their prior size after removal of a stressor or curtailment of feeding corticosterone (Gross *et al.*, 1980). Chickens differ in their threshold of response to dietary corticosterone and in degree of response once thresholds are reached (Siegel *et al.*, 1989). These differences were observed in immunoresponsiveness, efficiency of food utilization, growth, feathering and relative weights of liver, spleen, testes, breast muscle and abdominal fat.

Blood plasma levels of cortisol and/or corticosterone are frequently used as criteria for measuring response to stressors. Because the utilization of a given blood level of corticosterone differs among individuals, it is sometimes difficult to correlate blood levels of glucocorticoids with other manifestations of a stressor (Panretto and Vickery, 1972). Blood levels of the thyroid hormone tri-iodothyronine($T_3$) have also been used as a measurement of the stress response (e.g. Friend *et al.*, 1985; Yahav, 1998). When plasma corticosterone and thyroid hormones were used to measure effects of long-term stress of chickens under various housing conditions, Gibson *et al.* (1986) concluded that results were equivocal and that these hormones were not especially useful measures for long-term stress. However, cortisol and/or corticosterone are useful measures of the response to acute short-term stress induced by handling or restraint.

In response to stressors, plasma and tissue levels of ascorbic acid may be reduced (Kechik and Sykes, 1979) and it is not surprising that effects of ascorbic acid on stress responses have been studied (e.g. Gross, 1992b). Morphological changes of avian thrombocytes reflect changes in stress levels which are not affected by corticosterone (Gross, 1989). Alternatives to plasma $T_3$ and glucocorticoid levels as measures of stressors include various physiological or immunological responses. The number of blood lymphocytes decreases and the number of polymorphs increases after a stressful event (Maxwell and Robertson, 1998). The difficulty of large normal variation in the numbers of leucocytes can be largely circumvented by use of polymorph:lymphocyte (P:L) ratios, which provide a more sensitive measure of the stress response than levels of plasma glucocorticoids (Gross and Siegel, 1983). Recent advances in technology have allowed for increased use of changes in endogenous opioids as measures of stress (e.g. Sapolsky, 1992; Zanella *et al.*, 1998).

The stress system allows animals to allocate resources based on their perception of the environment as well as direct physical insults from the environment. An animal can be stressed by any change in its internal and external environments. Examples include rate of growth, reproductive state, climate, unusual sound or light, social interactions, availability of food and water, handling and moving, injected materials such as killed bacteria and some vaccines, and a disease in progress (e.g. Gross, 1974, 1995; Siegel, 1980; Freeman, 1987; Pierson *et al.*, 1997). Criteria of response may vary with time and profiles of responses may be described (see review by Mitchell and Kettlewell, 1998). For example, exposure to a short-term stressor, such as a brief sound (Gross, 1990b), resulted in changes in blood profiles, such as heterophil:lymphocyte (H:L), ratios within 18 h. The response peaked at about 20 h and returned to normal in about 30 h. In contrast, peak H:L ratios were observed 4 h after corticosterone administration (Gross, 1992a). Following exposure to a short-term stressor, antibody responsiveness is reduced, while resistance to bacterial diseases is increased.

There may be considerable variation among individuals in their perception of the stressfulness of an event, absorption of glucocorticoids from the blood and response of tissues to glucocorticoids. However, on a group basis, H:L ratios, feed efficiency and antibody response to an antigen seem to be associated (Fig. 3.1). Improved environmental quality influences these associations (Hester *et al.*, 1996a,b) and increases the correlation between the antibody titre responses of individuals to two different red blood cell (RBC) antigens (Gross and Siegel, 1990), with sensitivity of responses being better when lower dosages of RBCs are employed (Gross, 1986). It appears that at higher dosages the system becomes overloaded (Ubosi *et al.*, 1985). Differences in sheep RBC antibody titres of lines selectively bred for either high or low antibody responsiveness to this antigen (Siegel and Gross, 1980) increased when environmental quality was improved (Gross, 1986).

Levels of stress can also be estimated by presence of diseases. When stress levels are too high, viral diseases and other diseases which stimulate a lymphoid response are more common (Gross and Siegel, 1997). Cell-mediated immunity is reduced, resulting in an increased incidence of tumours and coccidiosis (Gross, 1972, 1976). When levels of stress are too low, bacterial and parasitic diseases are more common and responses to some toxins are more severe (Brown *et al.*, 1986). At an 'optimum stress' level, incidence of essentially all diseases is reduced. 'Optimum stress', however, may vary among genetic stocks and among individuals within a population because of differences in their background, as well as prior experiences.

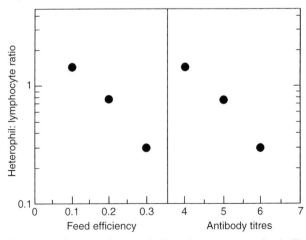

**Fig. 3.1.**   Relationship between heterophil:lymphocyte ratios, feed efficiency (body weight change/weight of feed consumed) and antibody responsiveness to sheep red blood cells in chickens.

## Resource Allocations

Resource allocations should convey the concept that, at any particular time, resources available to an individual are finite. Therefore, there will always be competition for resources between body functions, such as growth, maintenance, reproduction and health. Added to this mix are responses to stressors, which result in a redistribution of resources. A hypothetical example of redistribution of resources is seen in Fig. 3.2, where a comparison between a sick and a healthy animal is presented. In this example a healthy animal has a reserve of 10% for maintaining health and an equal division of resources for growth, reproduction and maintenance. When it becomes sick, however, resources allocated for growth and reproduction become nil and there is a reduction in those available for maintenance, because resources have been directed toward improvement of health status. For redistribution of resources to occur, it may be essential in some cases that changes occur quickly, while in other cases changes may be more gradual. In addition, the magnitude of

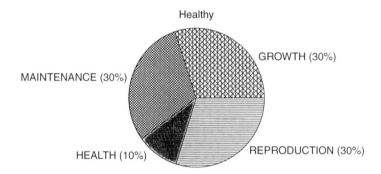

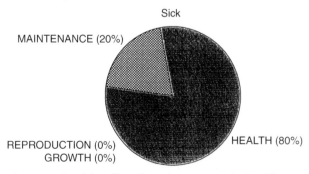

**Fig. 3.2.** An example of the allocation of resources of a healthy (top) and a sick (bottom) animal.

response to a stressor may be influenced by not only the animal's perception of its current environment, but also how the current environment differs from prior environments. Therefore, animals may also acquire, allocate and redistribute resources based on past history and their perceptions of the environment.

The following is an example of resource allocations (Gross and Siegel, 1997). Chickens fed on alternate days exhibit greater resistance to an *Escherichia coli* challenge than those fed *ad libitum*. When subsequently allowed *ad libitum* feeding, their rate of weight gain was greater and their resistance to *E. coli* challenge was less than that of those fed *ad libitum*. It is possible that the restricted feed supply yielded a stress response appropriate for conditions likely to include a bacterial or parasitic challenge. The chickens thus allocated finite resources to defence instead of growth. Adequate feeding then yielded a response appropriate for conditions where a bacterial or parasitic challenge was unlikely. At this point, the need to regain a genetically desired body-weight had priority over maintaining a high level of well-being defence.

Gross (1995) provided another example. When chicks were transferred from brooders to their experimental cages, there was no change in availability of feed. There was, however, variation among chickens in their allocation of resources between antibacterial defence and growth. This resulted in a negative correlation between their initial increase in body weight and resistance to an *E. coli* challenge, which was dependent on the individual's resource allocation.


## Long-term Higher, Lower and 'Optimum Stress'

'Optimum stress' is a relative term and when we write of higher, lower and optimum we do not wish to imply that if a little stress can be beneficial more is even better. The optimum will vary with genetic stock, prior experience of the animals and the environment. In one experiment with male chickens, Gross and Siegel (1981) characterized environments as providing high, medium and low social stress. The low-social-stress environment consisted of maintaining chickens in individual cages with solid sides and wire fronts, floors and backs, with water and feed available continuously. These chickens could hear but not see each other. The high-social-stress environment consisted of larger cages housing five males. Each day, however, one individual per cage was moved into another cage according to a plan that reduced the possibility of contact with previous cage mates. The medium-social-stress environment consisted of caging nine males in a series of cages throughout.

When chickens were continuously exposed for over 3 months to the high social stressors, feed consumption was not affected, whereas body-weight gain, feed efficiency and the correlation between antibody titres and resistance to *Mycoplasma gallisepticum* challenge infection were reduced. Even though there was a reduction in lymphoid mass, antibody responsiveness to RBCs was not changed. Under extended periods of higher levels of stress, H:L ratios ranged between 0.6 and 1.2. An H:L ratio above 1.3 usually indicates a disease in progress. When animals are exposed to stressful environments, growth potential is reduced, adaptability increases, even though senses are less acute, discrimination is improved and activity is increased. As an animal becomes better adapted to a harsher environment, resources are diverted from growth and reproduction to respond to the stressor.

Chickens exposed to the low-social-stress environment became lethargic and exhibited less preening and their vocalizations suggested contentment. Initially, weight gain and feed efficiency were increased, but after 3 months in this environment feed consumption, growth rate and feed efficiency were greatly reduced (Gross and Siegel, 1981). Cockerels maintained in the medium-social-stress environment maintained their body-weight throughout and had the best feed efficiency of the three groups. Also, they had the highest antibody titres to sheep RBC antigen. Ranking of the three environments according to the stress hormone corticosterone was high > medium > low. These findings are generalized by other research reports.

At low levels of environmental stress phenotypic variability was reduced and the chickens became unusually susceptible to bacterial infection (Larsen *et al.*, 1985). Unusually low levels of environmental stimuli are characterized by H:L ratios between 0.2 and 0.3, demonstrating the need for some stress (stimulation) in order to maintain more efficient biochemical activity. Because greater or lesser levels of stress seem to be detrimental, the aim of good husbandry should be to provide an 'optimum stress' level. It is probable that the optimum may vary according to genetic stock and prior experience. Reviews of this topic have been provided by Jones (1987), Zulkifli and Siegel (1995) and Gross and Siegel (1997).

## Environmental Stress and Disease Defence

Activation of an animal's defence against disease requires that resources be diverted from growth and reproduction. The stress system allows animals to maintain disease defence at a level commensurate to the risk. When at a low population density, the chances of an animal encountering infective levels of bacteria or parasites are reduced and the need for a phagocytic defence against them is reduced. As population densities increase, the probability of encountering pathogenic concentrations of bacteria and parasites in the environment also increases. Higher levels of stress enhance defence against bacteria and parasites; however, there is a cost in growth and reproduction. As stress (glucocorticoid) levels increase, levels of superoxide radical in polymorphs increase, which, in turn, enhances the ability to destroy bacteria (Som *et al.*, 1983). Although there is enhanced resistance to bacterial infection, further increases will reduce defences (Gross, 1984), implying an optimum. In contrast, low-stress environments increase susceptibility to

opportunistic bacteria such as coliforms, faecal streptococci and staphylococci
(Larsen *et al.*, 1985; Siegel *et al.*, 1987), and to internal and external parasites,
such as mites and coccidia (Hall *et al.*, 1979; Gross 1985). In higher-stress
environments numbers of chickens susceptible to viral infections and
tumours increased (Gross and Colmano, 1969; Mohamed and Hanson, 1980;
Thompson *et al.*, 1980). Such environments enhanced sensitization of cell-
mediated immunity, while effectiveness of cell-mediated immunity was
reduced (Gross, 1985). In an 'optimal stress' environment chickens are not
highly susceptible to bacteria, parasites, viruses or tumours. Once again, we
emphasize that optimum is a relative term that will vary from flock to flock. For
those interested in further detail, examples from experiments conducted in our
laboratory are provided in a summary review (Gross and Siegel, 1997).

## Modifying the Stress Response

Within a population, individuals may differ genetically in their perception of
stressors, resulting in considerable variation in the response to the same
stressor. Because responses of individuals to different stressors also vary
(Fig. 3.3), it is possible, through artificial selection, to develop populations

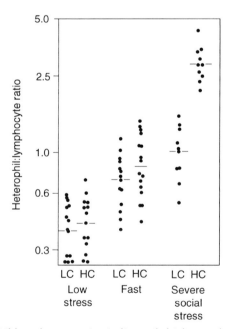

**Fig. 3.3.**    Heterophil:lymphocyte ratios in lines of chickens selected for high (HC)
or low (LC) corticosterone response to social strife when housed in low- or
high-social-stress environments or following the stress of fasting.

which have reduced or increased responses to specific stressors (Gross and Siegel, 1985; Jones and Satterlee, 1996; Faure and Mills, 1998).

When animals are repeatedly exposed to the same stressor and the magnitude of each succeeding response is reduced, it can be said that habituation has occurred. The habituation may involve memory and/or physiological manifestations (Siegel, 1989). Prior exposure of young chickens to acute thermal stressors appears to improve heat tolerance later in life. This preconditioning does not have to be the same stressor, but what is needed is the synthesis and liberation of glucocorticoids (Zulkifli and Siegel, 1995; Zulkifli *et al.*, 1995). These findings are consistent with *in vitro* and *in vivo* studies suggesting that macrophages respond to thermal and non-thermal stressors by producing similar kinds of 'stress proteins' (Miller and Qureshi, 1992).

Effects of stress may be alleviated by chemicals which inhibit the production of adrenal glucocorticoids. One such chemical is ascorbic acid (Gross, 1988, 1992b; Gross *et al.*, 1988b). Others are adrenal-blocking chemicals, such as metyrapone (Zulkifli *et al.*, 1995) and 1,1-dichloro-2,2*bis*-*p*-chlorophenyl ethane (Gross, 1990a). After the administration of an optimal dose of such compounds, the physiological manifestations of stress are reduced. Examples include weight loss associated with transportation, inhibiting viral infections and tumours, increasing feed efficiency, reducing effects of heat stress and reducing the stress inhibition of antibody responsiveness (Gross, 1989, 1990a).

## Human–Animal Relationships

Relationships between animals and their handlers can greatly affect responses of animals to a range of factors and it is not surprising that the roles of the stockperson and veterinarian have been much studied in numerous livestock and poultry species (e.g. Hemsworth and Barnett, 1987; Gross and Siegel, 1997; Hemsworth and Gonyou, 1997; Odendaal, 1998; Sambraus, 1998). Socialized animals welcome the presence of their handlers. The process of socialization may be accomplished by frequent exposure to kind care and handling, beginning at the youngest possible age.

Long-term effects of the human–animal relationship (Jones, 1987; Barnett *et al.*, 1992), coupled with background genotype and prior experiences, influence subsequent responses to various situations (e.g. Gross and Siegel, 1981, 1982; Nicol, 1992). Positive socialization with humans can result in animals approaching caretakers. Negative socialization can result in escape behaviours, and ignored individuals exhibit fear when exposed to humans. Although gentle handling may exert its strongest influence by facilitating habituation to humans (Jones and Waddington, 1992), feed efficiencies, body-weights and antibody responses to RBC antigens are higher for positively socialized chickens than for those held under similar environments and ignored. Responses to stressors and to the administration of corticosterone are greatly reduced (Gross and Siegel, 1982) and resistance to most diseases is enhanced in socialized chickens. When socialized and ignored chickens were challenged with *E. coli*, the incidence of severe lesions was reduced among socialized chickens (31% vs. 6%), whereas incidence (46%) of initial air sac lesions was similar (Gross and Siegel, 1982). Variability of many responses was reduced when chickens were socialized.

Both the stressfulness of the environment and socialization influence the responses of chickens (Fig. 3.4). Socialized chickens which are in an 'optimal stress' environment seem to have the most favourable responses.

| | STRESS LEVEL | | | |
| | HIGH | OPTIMAL | LOWER | SOCIALIZED |
|---|---|---|---|---|
| FEED EFFICIENCY | ↓ | ↑ | ↓ | ↑ |
| BODY-WEIGHT | ↓ | ↔ | ↓ | ↑ |
| ADAPTABILITY | ↑ | ↔ | ↓ | ↑ |
| GENETIC SELECTION | ↓ | ↑ | ↓ | ↑ |
| ANTIBODY RESPONSE | ↓ | ↔ | ↑ | ↑ |
| CMI SENSITIZATION | ↑ | ↔ | ↓ | |
| CMI MANIFESTATION | ↓ | ↔ | ↑ | |
| BACTERIAL DEFENCE | ↑ | ↔ | ↓ | ↑ |
| VIRAL DEFENCE | ↓ | ↔ | ↑ | ↑ |
| H:L RATIO | 0.8 | 0.5 | 0.2 | VARIABLE |

↓ INFERIOR RESPONSE

↑ SUPERIOR RESPONSE

↔ IMPLIES EXPRESSION OF GENETIC POTENTIAL MAY VARY

**Fig. 3.4.** A general summary of expected responses of chickens to levels of stress and to socialization.

## Genetic–Environment Interactions

The response of an animal to environmental factors is determined by its genetic background as modified by prior environmental experiences. The first week after hatch or birth can be very important in regard to the human–animal relationship. Although the expression of traits may differ among stocks, in an environment where stress is optimal, genetic differences may be more evident because exposure to environmental stressors may modify genetic influences on expression of traits (Gross, 1986; Gross *et al.*, 1988a).

Genotype–environment interactions occur when, relative to each other, a series of genotypes do not respond similarly in a series of environments. Numerous reports appeared in the 1970s demonstrating behavioural involvement with genotype–environment interactions in chickens (see review by Siegel, 1979). The implications of these interactions can be considerable when viewed in the context of well-being (e.g. Mathur and Horst, 1994). In experiments measuring production and disease resistance in lines of chickens selected for high- or low-corticosterone response to social strife, extreme divergence of responses was observed between the high line in a higher-social-strife environment and the low line in a lower-social-strife environment (Siegel, 1989). The low–low combination was more susceptible to infections from endemic bacteria and external parasites and the high–high combination to viral infections (Gross and Siegel, 1997). Whether or not an extreme response was advantageous depended on the measurement criteria. These data are consistent with the view that overall well-being may be at an optimal level when the bird is neither under- nor overstressed (Gross *et al.*, 1984; Zulkifli and Siegel, 1995).

## Relationship between Environment and Well-being

Many components are involved in the development of an animal's well-being. One is adequate food and water. Another is protection from environmental insults, such as adverse weather. Protection from pathogenic organisms and predators is also important. Stable social (between animals) and physical environments are valuable and contribute to an 'optimum stress' level. These needs are met by good husbandry, which is rewarded by increased productivity and more uniform responses to experimental procedures.

In our opinion, the most important factor affecting the well-being of livestock and poultry is their relationship with their human associates. Kind care (socialization) has many well-being and production benefits. Socialized animals are easier to work with, have improved feed efficiency (productivity), are more adaptable to adverse environments, are more resistant to diseases and produce better immunity. These factors make genetic selection easier. The responses of individuals within groups are more uniform, thus reducing the number of animals needed for research. Socialization can be easily applied to large groups of animals.

All people who work with animals should be aware of physical, nutritional and behavioural needs and should be able to relate positively to the feelings of the animals. The attitude of handlers is an essential important factor in determining that level of stress which enhances the animal's well-being.

# References

Barnett, J.L., Hemsworth, P.H. and Newman, E.A. (1992) Fear of humans and its relationship with productivity in laying hens at commercial farms. *British Poultry Science* 33, 699–710.

Broom, D.M. (1988) Relationship between welfare and disease susceptibility in farm animals. In: Gibson, T.E. (ed.) *Animal Disease – a Welfare Problem*. British Veterinary Association Animal Welfare Foundation, London, pp. 20–29.

Brown, C.W., Gross, W.B. and Ehrich, M. (1986) Effect of social stress on the toxicity of malathion in young chickens. *Avian Diseases* 30, 679–682.

Faure, J.M. and Mills, A.D. (1998) Improving adaptability of animals by selection. In: Grandin, T. (ed.) *Genetics and the Behavior of Domestic Animals*. Academic Press, San Diego, California pp. 235–264.

Freeman, B.M. (1987) The stress syndrome. *World's Poultry Science Journal* 43, 15–19.

Friend, T.H., Dellomeier, G.R. and Gbur, E.E. (1985) Comparison of four methods of calf confinement. 1. Physiology. *Journal of Animal Science* 60, 1095–1101.

Gibson, S.W., Hughes, B.O., Harvey, S. and Dun, P. (1986) Plasma concentrations of corticosterone and thyroid hormones in laying fowl from different housing systems. *British Poultry Science* 27, 621–628.

Gross, W.B. (1972) Effect of social stress on the occurrence of Marek's disease in chickens. *American Journal of Veterinary Research* 33, 2275–2279.

Gross, W.B. (1974) Stressor effects of initial bacterial exposure of chickens as determined by subsequent challenge exposure. *American Journal of Veterinary Research* 35, 1125–1128.

Gross, W.B. (1976) Plasma steroid tendency, social environment, and *Eimeria necatrix* infection. *Poultry Science* 55, 1508–1512.

Gross, W.B. (1984) Effect of range of social stress severity in *Escherichia coli* challenge infection. *American Journal of Veterinary Research* 45, 2074–2076.

Gross, W.B. (1985) Effect of social environment and oocyst dose on resistance and immunity to *Eimeria tenella* challenge. *Avian Diseases* 29, 1018–1029.

Gross, W.B. (1986) Effect of dose antigen and social environment on antibody responses of high and low antibody response chickens. *Poultry Science* 65, 687–692.

Gross, W.B. (1988) Effect of environmental stress on the responses of ascorbic acid-treated chickens to *Escherichia coli* challenge infection. *Avian Diseases* 32, 432–436.

Gross, W.B. (1989) Effect of adrenal blocking chemicals on viral and respiratory infections of chickens. *Canadian Journal of Veterinary Research* 53, 48–51.

Gross, W.B. (1990a) Effect of adrenal blocking chemicals on the responses of chickens to environmental stressors and ACTH. *Avian Pathology* 19, 295–304.

Gross, W.B. (1990b) Effect of exposure to a short duration sound on the stress response of chickens. *Avian Diseases* 34, 759–761.

Gross, W.B. (1992a) Effect of short term exposure to corticosterone on resistance to challenge exposure with *Escherichia coli* and antibody response to sheep erythrocytes. *American Journal of Veterinary Research* 53, 291–293.

Gross, W.B. (1992b) Effects of ascorbic acid on stress and disease resistance in chickens. *Avian Diseases* 36, 388–392.

Gross, W.B. (1995) Relationship between body-weight gain after movement of chickens to an unfamiliar cage and response to *Escherichia coli* challenge infection. *Avian Diseases* 39, 636–637.

Gross, W.B. and Colmano, G. (1969) Effect of social isolation on resistance to some infectious agents. *Poultry Science* 48, 514–520.

Gross, W.B. and Siegel, H.S. (1983) Evaluating the heterophil/lymphocyte ratio as a measure of stress in chickens. *Avian Diseases* 27, 972–979.

Gross, W.B. and Siegel, P.B. (1981) Long term exposure of chickens to three levels of social stress. *Avian Diseases* 25, 312–325.

Gross, W.B. and Siegel, P.B. (1982) Socialization as a factor in resistance to infection, feed efficiency, and response to antigens in chickens. *American Journal of Veterinary Research* 43, 2010–2012.

Gross, W.B. and Siegel, P.B. (1985) Selective breeding of chickens for corticosterone response to social stress. *Poultry Science* 64, 2230–2233.

Gross, W.B. and Siegel, P.B. (1990) Genetic-environmental interactions and antibody response to two antigens. *Avian Diseases* 34, 843–847.

Gross, W.B. and Siegel, P.B. (1997) Why some get sick. *Journal of Applied Poultry Science* 6, 453–460.

Gross, W.B., Siegel, P.B. and DuBose, R.T. (1980) Some effects of feeding corticosterone. *Poultry Science* 59, 516–522.

Gross, W.B., Dunnington, E.A. and Siegel, P.B. (1984) Environmental effects on the well-being of chickens from lines selected for response to social strife. *Archiv für Geflugelkunde* 48, 3–7.

Gross, W.B., Domermuth, C.H. and Siegel, P.B. (1988a) Genetic and environmental effects on the response of chickens to avian adenovirus group II infection. *Avian Pathology* 17, 767–774.

Gross, W.B., Jones, D. and Cherry, J. (1988b) Effect of ascorbic acid on the disease caused by *Escherichia coli* challenge infection. *Avian Diseases* 32, 407–408.

Hall, R.D., Gross, W.B. and Turner, E.C. (1979) Population development of *Ornithonyssus sylvarum* on leghorn roosters inoculated with steroids or subjected to extreme social interaction. *Veterinary Parasitology* 5, 287–297.

Hemsworth, P.H. and Barnett, J.L. (1987) Human–animal interactions. In: Price, E.V. (ed.) *Farm Animal Behaviour*. W.B. Saunders, Philadelphia, pp. 339–356.

Hemsworth, P.H. and Gonyou, H.W. (1997) Human contact. In: Appleby, M.C. and Hughes, B.O. (eds) *Animal Welfare*. CAB International, Wallingford, pp. 205–217.

Hester, P.Y., Muir, W.M., Craig, J.V. and Albright, J.L. (1996a) Group selection for adaptation to multiple-hen cages: hematology and adrenal function. *Poultry Science* 75, 1295–1307.

Hester, P.Y., Muir, W.M. and Craig, J.V. (1996b) Group selection for adaptation to multiple-hen cages: humoral immune response. *Poultry Science* 75, 1315–1320.

Jones, R.B. (1987) Social and environmental aspects of fear. In: Zayan, R. and Duncan, I.J.H. (eds) *Cognitive Aspects of Social Behaviour in Domestic Fowl.* Elsevier, Amsterdam, pp. 82–149.

Jones, R.B. and Satterlee, D.G. (1996) Threat-induced behavioral inhibition in Japanese quail genetically selected for contrasting adrenocortical response to mechanical restraint. *British Poultry Science* 37, 465–470.

Jones, R.B. and Waddington, D. (1992) Modification of fear in domestic chicks, *Gallus gallus domesticus*, via regular handling and early environmental enrichment. *Animal Behaviour* 43, 1021–1033.

Kechik, I.T. and Sykes, A.H. (1979) Effect of intestinal coccidiosis (*Eimeria acervulina*) on blood and tissue ascorbic acid concentration. *British Journal of Nutrition* 42, 97–103.

Larsen, C.T., Gross, W.B. and Davis, J.W. (1985) Social stress and resistance of chickens and swine to *Staphylococcus aureus*. *Canadian Journal of Comparative Medicine* 48, 208–210.

Mathur, P.K. and Horst, P. (1994) Methods for evaluating genotype–environment interactions illustrated by laying hens. *Journal Animal Breeding and Genetics* 111, 265–288.

Maxwell, M.H. and Robertson, G.W. (1998) The avian heterophil leucocyte system: a review. *World's Poultry Science Journal* 54, 155–178.

Miller, L. and Qureshi, M.A. (1992) Heat shock protein synthesis in chicken macrophages: influence of *in vivo* and *in vitro* heat shock, lead acetate and lipopolysaccharide. *Poultry Science* 71, 980–998.

Mitchell, M.A. and Kettlewell, P.J. (1998) Physiological stress and welfare of broiler chickens in transit: solutions not problems. *Poultry Science* 77, 1803–1814.

Mohamed, M.A. and Hanson, R.P. (1980) Effect of social stress on Newcastle disease virus (LaSota) infection. *Avian Diseases* 24, 908–915.

Muir, W.M. and Craig, J.V. (1998) Improving animal well-being through genetic selection. *Poultry Science* 77, 1781–1788.

Nicol, C.J. (1992) Effects of environmental enrichment and gentle handling on behaviour and fear responses of transported broilers. *Applied Animal Behaviour Science* 33, 367–380.

Odendaal, J.S.J. (1998) The practicing veterinarian and animal welfare as a human endeavour. *Applied Animal Behaviour Science* 59, 85–91.

Panretto, B.A. and Vickery, M.R. (1972) Distribution of cortisol and its rate of turnover in normal and cold stressed sheep. *Journal of Endocrinology* 55, 519–531.

Pierson, F.W., Larsen, C.T. and Gross, W.B. (1997) The effect of stress on the response of chickens to coccidiosis vaccination. *Veterinary Parasitology* 73, 177–180.

Sambraus, H.H. (1998) Applied ethology – its task and limits in veterinary practice. *Applied Animal Behaviour Science* 59, 39–48.

Sapolsky, R.M. (1992) Neuroendocrinology of the stress-response. In: Becker, J.B., Breedlove, S.M. and Crews, D. (eds) *Behavioural Endocrinology.* MIT Press, Cambridge, Massachusetts, pp. 287–324.

Sapolsky, R.M. (1994) *Why Zebras Don't Get Ulcers: a Guide to Stress, Stress-Related Diseases, and Coping.* W.H. Freeman, New York.

Sarkar, S. (1992) Sex, disease and evolution – variations on the theme from J.B.S. Haldane. *BioScience* 42, 448–454.

Selye, H. (1950) *Stress: The Physiology and Pathology of Exposure to Stress.* Acta, Montreal.

Selye, H. (1976) *The Stress of Life*. McGraw-Hill, New York.

Siegel, H.S. (1980) Physiological stress in birds. *BioScience* 30, 529–534.

Siegel, P.B. (1979) Behaviour genetics in chickens: a review. *World's Poultry Science Journal* 35, 9–19.

Siegel, P.B. (1989) The genetic–behavioural interface and well-being of poultry. *British Poultry Science* 30, 3–13.

Siegel, P.B. and Gross, W.B. (1980) Production and persistence of antibodies 1. Directional selection. *Poultry Science* 59, 1–5.

Siegel, P.B., Katanbaf, M.N., Anthony, N.B., Jones, D.E., Martin, A., Gross, W.B. and Dunnington, E.A. (1987) Response of chickens to *Streptococcus faecalis*: genotype–housing interactions. *Avian Diseases* 31, 804–808.

Siegel, P.B., Gross, W.B. and Dunnington, E.A. (1989) Effect of dietary corticosterone in young leghorn and meat-type cockerels. *British Poultry Science* 30, 185–192.

Som, S., Raha, C. and Chatterjee, I.B. (1983) Ascorbic acid: a scavenger of superoxide radical. *Acta Vitamineral Enzymology* 5, 243–250.

Thompson, D.L., Elgert, K.D., Gross, W.B. and Siegel, P.B. (1980) Cell-mediated immunity in Marek's disease virus-infected chickens genetically selected for high and low concentrations of plasma corticosterone. *American Journal of Veterinary Research* 41, 91–96.

Ubosi, C.O., Dunnington, E.A., Gross, W.B. and Siegel, P.B. (1985) Divergent selection of chickens for antibody response to sheep erythrocytes: kinetics of primary and secondary immunizations. *Avian Diseases* 29, 347–355.

Yahav, S. (1998) Physiological responses of chickens and turkeys to hot climate. In: *Proceedings 10th European Poultry Conference*, vol. 1. Jerusalem, 84–91.

Zanella, A.J., Brunner, P., Unshelm, J., Mendl, M.T. and Broom, D.M. (1998) The relationship between housing and social rank, β-endorphin and dynorphin (1–13) secretion in sows. *Applied Animal Behaviour Science* 59, 1–10.

Zulkifli, I. and Siegel, P.B. (1995) Is there a positive side to stress? *World's Poultry Science Journal* 51, 63–76.

Zulkifli, I., Siegel, H.S., Mashaly, M.M., Dunnington, E.A. and Siegel, P.B. (1995) Inhibition of adrenal steroidogenesis, neonatal feed restriction, and pituitary–adrenal axis response to subsequent fasting in chickens. *General and Comparative Endocrinology* 97, 49–56.

# Welfare Assessment and Welfare Problem Areas During Handling and Transport

**4**

## D.M. Broom

*Department of Clinical Veterinary Medicine, University of Cambridge, Madingley Road, Cambridge CB3 0ES, UK*

## Introduction

In order to be able to improve the welfare of livestock which are being handled or transported, it is necessary to make measurements which allow the assessment of welfare. It is also important that the concept of welfare is conducive to scientific measurement and sufficiently broad to encompass as many as possible of the aspects ascribed to it. The welfare of an animal is its state as regards its attempts to cope with its environment (Broom, 1986). For each coping system the environment is that which is external to the system. One important part of this state is that which involves attempts to cope with pathology, i.e. the health of the animal, so health is part of welfare. Feelings are a part of many mechanisms for attempting to cope with good and bad aspects of life and most feelings must have evolved because of their beneficial effects (Broom, 1998), so they are also an important part of welfare (Broom, 1991b, 1996). The extent to which coping attempts are succeeding and the amount which has to be done in order to cope must both be considered as a part of welfare. The welfare of an individual could be very good or very poor, so it can vary over a wide range.

The scientific assessment of welfare must be quite separate from any moral judgement. Welfare is a characteristic of an individual at the time of observation or measurement, so it can be assessed in an entirely objective way. Once the range of measurements of the different welfare indicators has been made, people can judge whether such welfare is acceptable. Moral decisions of this kind can be made by anyone, given the scientific results and guidance in how to interpret them. There is variation among people with respect to how poor the welfare of a farm animal has to be before they consider it to be intolerable. This variation is a consequence of the background, including the country of origin, of the person. However, there has been a substantial change in attitudes towards animals, as research on their behaviour and physiology has been publicized in the media and has revealed their complexity and similarity to humans. Greater knowledge of the extent to which the welfare of animals can be poor because of pain, fear or other adverse effects of humans and of the environment which people impose on them, has resulted in pronouncements by Protestant, Catholic, Muslim and Buddhist religious leaders about human obligations to animals which we use. New legislation is also being enacted in many countries.

The many mechanisms which exist within most animals for trying to cope with their environment and the various consequences of failure to cope mean that there are many possible measures of welfare. However, any one measure can show that welfare is poor. Measures which are relevant to animal welfare during handling and transport are outlined in the second section of this chapter. Welfare problems which can occur in the course of handling and transport are discussed in the third section.

## Welfare Measures

### Welfare assessment in general

A variety of welfare indicators which can be used to assess the welfare of animals which are being handled or transported are listed in Table 4.1. Some of these measures are of short-term effects, whilst others are more relevant to prolonged problems. Where animals are transported to slaughter, it is mainly the measures of short-term effects, such as behavioural aversion or increased

**Table 4.1.** Measures of welfare.

Physiological indicators of pleasure
Behavioural indicators of pleasure
Extent to which strongly preferred behaviours can be shown
Variety of normal behaviours shown or suppressed
Extent to which normal physiological processes and anatomical development are
    possible.
Extent of behavioural aversion shown
Physiological attempts to cope
Immunosuppression
Disease prevalence
Behavioural attempts to cope
Behaviour pathology
Brain changes, e.g. those indicating self narcotization
Body damage prevalence
Reduced ability to grow or breed
Reduced life expectancy

heart-rate, which are used, but some animals are kept for a long period after transport and measures such as increased disease incidence or suppression of normal development give information about the effects of the journey on welfare. Details of these and other measures may be found in Broom (1988), Fraser and Broom (1990) and Broom and Johnson (1993).

## Behaviour measures

The most obvious indicators that an animal is having difficulty coping with handling or transport are changes in behaviour, which show that some aspect of the situation is aversive. The animal may stop moving forward, freeze, back off, run away or vocalize. The occurrence of each of these can be quantified in comparisons of responses to different races, loading ramps, etc. Examples of behavioural responses, such as cattle stopping when they encounter dark areas or sharp shadows in a race and pigs freezing when hit or subjected to other disturbing situations, may be found in Grandin (1980, 1982, 1989).

The extent of behavioural responses to painful or otherwise unpleasant situations varies from one species to another according to the selection pressures which have acted during the evolution of the mechanisms control-ling behaviour. Humans often elicit antipredator behaviour in farm animals. Social species which can collaborate in defence against predators, such as pigs or humans, vocalize a lot when caught or hurt. Species which are unlikely to be able to defend themselves, such as sheep, vocalize far less when caught by a predator, probably because such an extreme response merely gives information to the predator that the animal attacked is severely injured and hence unlikely to be able to escape. Cattle can also be relatively undemonstrative when hurt or severely disturbed. Human observers sometimes wrongly assume that an animal which is not squealing is not hurt or disturbed by what is being done to it. In some cases, the animal is showing a freezing response, and in most cases, physiological measures must be used to find out the overall response of the animal.

Individual animals may vary in their responses to potential stressors. Some studies have shown that animals can be categorized according to their responses. For example, Hessing (1994) divided piglets into active and passive animals according to their response to a handling procedure which involved placing the animals on their backs. However, others found more of a continuum in responsiveness (Jensen *et al.*, 1995) and in studies of agonistic interactions of sows, they were found to vary from high-success animals, which won most of their interactions, low-success animals, which lost most interactions and had high cortisol levels, and no-success animals, which avoided interactions but had lower cortisol levels and greater reproductive success (Mendl *et al.*, 1992). The coping strategy adopted by the animal can have an effect on responses to the transport and lairage situation. For example, Geverink *et al.* (1998) showed that those pigs which were most aggressive in their home pen were also more likely to fight during pretransport or pre-slaughter handling, but pigs which were driven for some distance prior to transport were less likely to fight and hence cause skin damage during and after transport. This fact can be used to design a test which reveals whether or not the animals are likely to be severely affected by the transport situation (Lambooij *et al.*, 1995).

The procedures of loading and unloading animals into and out of transport vehicles can have very severe effects on the animals and these effects are revealed in part by behavioural responses. Species vary considerably in their responses to loading procedures. Any animal which is injured or frightened by people during the procedure can show extreme responses. However, in most efficient loading procedures, sheep are not greatly affected, cattle are sometimes affected, pigs are always affected and poultry which are handled by humans are always severely affected. Broom *et al.* (1996) and Parrott *et al.* (1998b) showed that sheep show largely physiological responses and these are associated with the novel situation encountered in the vehicle rather than the loading procedure. Pigs, on the other hand, are much affected by being driven up a ramp into a vehicle, the effect being greater if the ramp is steep.

When pigs are loaded on to a vehicle, it is very common for the driver of the vehicle or others who are moving the animals to use electric goads, sticks or brooms and to do so whilst shouting or hitting the animals. This is seldom done gently as the people are almost always in a hurry (Geverink and Lambooij, 1994). Pigs may stop during the loading attempts, turn sideways across the loading ramp, vocalize loudly and, in extremes, collapse. If electric goads are used, the response of the animals may be to scream, stop moving or move rapidly away from the person who uses the goad. Animals which can easily move usually move readily if humans approach closely and make contact with them. Where animals will not move when there is patient human approach and stay still until there is use of electric goads, this is usually because they are weak, injured or very frightened. The electric goad often has a very severe effect on the animal and its use is inhumane and unacceptable. Goads are often used just to accelerate movement and compensate for inadequately designed loading facilities or lairage design, when what the animals need is to be able to observe their environment and walk easily without being frightened. If, for example, pigs are urged to move from a well-illuminated loading ramp to a dark vehicle, they may stop because they are afraid of the change in light level. In this circumstance, it is never excusable to force them to move using sticks or goads. Such actions are cruel. If passage-ways are used which are too narrow, all animals and especially pigs may hesitate to move along them. If there are projections into passageways, animals may hesitate to move past them. Steep loading ramps can lead to complete failure to move, although sheep can climb much steeper ramps than pigs without baulking. The angle of loading ramp which leads to baulking by pigs is 15–20° (van Putten and Elshof, 1978; Fraser and Broom, 1990), so it is safest if ramps no steeper than 13° slope are used. Descending a loading ramp of steeper than 20° is difficult for pigs and should be avoided (Grandin, 1982).

When hens are to be removed from a cage they move away from an approaching human. Broiler chickens or turkeys, which are much less mobile than hens, may not always move away from a person who is trying to catch them but, given time, their behaviour indicates that they are disturbed by close human approach. After poultry are picked up by humans, they may struggle but often hang limply and, if put down, often show a freezing response. The behavioural response to being caught and carried is generally one of passive fear behaviour and is frequently not recognized by the people involved in handling them as indicating the severe disturbance which is revealed by physiological measures.

Once journeys start some species of farm animals explore the compartment in which they are placed and try to find a suitable place to sit or lie down. Poultry often lie where they are placed, whilst pigs normally try to lie down. Sheep and cattle try to lie down if the situation is not disturbing but stand if it is. After a period of acclimatization of sheep and cattle to the vehicle environment, during which time they may stand for 2–6 h looking around at intervals, most of the animals will lie down if the opportunity arises. Unfortunately for the animals, many journeys involve so many lateral movements or sudden brakings or accelerations that the animals cannot lie down. Pigs whose body weight is 110 kg usually lie down within a short time of the commencement of a smooth journey if the stocking density is 235 kg $m^{-2}$ and always lie down when the vehicle is stationary and during the night. However, at a stocking density of 278 kg $m^{-2}$ or higher, pigs continually change positions and are not able to rest (Lambooij, 1988; Lambooij and Engel, 1991). The number of pigs, or other animals, which remain standing during transport is also a relevant measure when the roughness of the journey, ideally measured in terms of accelerations in three possible planes, is measured. For example, Bradshaw *et al.* (1996a) found that more pigs remain standing during a rough journey then during a smooth journey. In journeys with certain vibration characteristics, pigs show behavioural evidence of motion sickness in that they retch and vomit (Bradshaw *et al.*, 1996c; Randall and Bradshaw, 1998).

An important behavioural measure of welfare when animals are transported is the amount of fighting which they show. When male adult cattle are mixed during transport or in lairage, they may fight and this behaviour can be recorded directly (Kenny and Tarrant, 1987). Calves of 6 months of age may also fight (Trunkfield and Broom, 1991) and fighting can be a serious problem in pigs (Guise and Penny, 1989; Bradshaw *et al.*, 1996b). The recording of such behaviour should include the occurrence of threats, as well as the contact behaviours which might cause injury, e.g. those described by Jensen (1994) for pigs.

A further, valuable method of using behaviour studies to assess the welfare of farm animals during handling and transport involves using the fact that the animals remember aversive situations in experimentally repeated exposures to such situations. Any stock-keeper will be familiar with the animal which refuses to go into a crush after having received painful treatment there in the past or hesitates about passing a place where a frightening event, such as a dog threat, occurred once before. These observations give us information about the welfare of the animal in the past, as well as at the present time. If the animal tries not to return to a place where it had an experience, then that experience was clearly aversive. The greater the reluctance of the animal to return, the greater the previous aversion must have been. This principle has been used by Rushen (1986a,b) in studies with sheep. Sheep which were driven down a race to a point where gentle handling occurred traversed the race as rapidly or more rapidly on a subsequent day. Sheep which were subjected to shearing at the end of the race on the first day were harder to drive down the race subsequently and those subjected to electroimmobilization at the end of the race were very difficult to drive down the race on later occasions. Hence the degree of difficulty in driving and the delay before the sheep could be driven down the race are measures of the current fearfulness of the sheep and this in turn reflects the aversiveness of the treatment when it was first experienced.

## Physiological measures

The physiological responses of animals to adverse conditions, such as those which they may encounter during handling and transport, will be affected by the anatomical and physiological constitution of the animal. For example, the major aspects of transport which affect the welfare of pigs are loading and unloading procedures, including the effects of close proximity to humans, vehicle conditions, the way that the vehicle is driven, what happens during stops and the duration of the journey. The response of the pig to these different aspects will depend on the genetically controlled adaptability of the pig, the physical condition of the pig and the previous experience. Our modern breeds of pig have been selected for large muscle blocks, fast growth and efficient feed conversion and nutrient partitioning. When the wild boar was compared with modern breeds of pig, the modern German Landrace was found to have muscles with a greater distance from the centre to the nearest blood-vessel and more anaerobic fibres and also a relatively smaller heart (Dämmrich, 1987). The least well adapted pigs for the stresses of transport are those which are extreme in these effects, for example those with the halothane-positive gene, but all pigs have serious problems during transport, which are generally reflected in some impairment of meat quality. It may well be that the meat quality of all pigs which are transported is worse than it would be if no transport occurred. Part of the solution to this welfare and economic problem is to take account of what the pig will have to put up with during transport when developing genetic strains of pig. Pigs with smaller muscle blocks and less risk of susceptibility to high levels of exercise and to stress could be selected.

As pointed out by Broom (1995), whenever physiological measurement is to be interpreted, it is important to ascertain the basal level for that measure and how it fluctuates over time. For example, plasma cortisol levels in most species vary during the day and tend to be higher during the morning than during the afternoon. A decision must be taken for each measure concerning whether the information required is the difference from baseline or the absolute value. For small effects, e.g. a 10% increase in heart rate, the difference from baseline is the key value to use. For large effects where the response reaches the maximal possible level, e.g. cortisol in plasma in very frightening circumstances, the absolute value should be used. In order to explain this, consider an animal severely frightened during the morning and showing an increase from a rather high baseline of 160 nmol $l^{-1}$ but in the afternoon showing the same maximal response which is 200 nmol $l^{-1}$ above the lower afternoon baseline. It is the actual value which is important here rather than a difference whose variation depends on baseline fluctuations.

Heart rate can decrease when animals are frightened but, in most farm animal studies, tachycardia (increase in heart rate) has been found to be associated with disturbing situations. Van Putten and Elshof (1978) found that the heart rate of pigs increased by a factor of 1.5 when an electric prodder was used on them and by 1.65 when they were made to climb a ramp. Steeper ramps caused greater increases up to a maximum level (van Putten, 1982). Heart-rate increase is not just a consequence of increased activity; heart rate can be increased in preparation for an expected future flight response. Baldock and Sibly (1990) obtained basal levels for heart rate during a variety of activities by sheep and then took account of these when calculating responses to various treatments. Social isolation caused a substantial response, but the greatest heart-rate increase occurred when the sheep were approached by a man with a dog. The responses to handling and transport are clearly much lower if the sheep have previously been accustomed to human handling. Heart rate is a useful measure of welfare but only for short-term problems, such as those encountered by animals during handling, loading on to vehicles and certain acute effects during the transport itself. However, some adverse conditions may lead to elevated heart rate for quite long periods. Parrott *et al.* (1998a) showed that heart rate increased from about 100 beats $min^{-1}$ to about 160 beats $min^{-1}$ when sheep were loaded on to a vehicle and the period of elevation of heart rate was at least 15 min. During transport of sheep, heart rate remained elevated for at least 9 h (Parrott *et al.*, 1998b).

Direct observation of animals without any attachment of recording instruments or sampling of body fluids can provide information about physiological processes. Breathing rate can be observed directly or from good quality video recordings. The metabolic rate and level of muscular activity are major determinants of breathing rate, but an individual animal which is disturbed by events in its environment may suddenly start to breathe fast. Muscle tremor can be directly observed and is sometimes associated with fear. Foaming at the mouth can have a variety of causes, so care is needed in interpreting the observations, but its occurrence may provide some information about welfare.

Changes in the adrenal medullary hormones adrenaline (= epinephrine) and noradrenaline (= norepinephrine) occur very rapidly and measurements of these hormones have not been used much in assessing welfare during transport. However, Parrott *et al.* (1998a) found that both hormones increased more during loading of sheep by means of a ramp than by loading with a lift.

Adrenal cortex changes occur in most of the situations which lead to aversion behaviour or heart-rate increase, but the effects take a few minutes to be evident and they last for 15 min to 2 h or a little longer. Plasma corticosterone levels in hens at depopulation were three times as high after normal rough handling than after gentle handling (Broom *et al.*, 1986; Broom and Knowles, 1989) and those of broilers were three and a half times the resting level after 2 h of transport and four and a quarter times higher after 4 h of transport (Freeman *et al.*, 1984; review by Knowles and Broom, 1990b). Another example comes from work on calves (Kent and Ewbank, 1986; Trunkfield *et al.*, 1991; review by Trunkfield and Broom, 1990). Plasma or saliva glucocorticoid levels gave information about treatments lasting up to 2 h but were less useful for journeys lasting longer than this. The previous environment of the animals, as well as the treatment when handled and transported, affected the animals' adrenal cortex responses, calves reared in small crates being affected much more by loading and an hour's journey than calves reared in groups (Trunkfield *et al.*, 1991).

The use of saliva cortisol measurement is now considered. In the plasma, most cortisol is bound to protein, but it is the free cortisol which acts in the body. Hormones such as testosterone and cortisol can enter the saliva by diffusion in salivary gland cells. The rate of diffusion is high enough to maintain an equilibrium between the free cortisol in plasma and that in saliva. The level is ten or more times lower in saliva but stimuli which cause plasma cortisol increases also cause comparable salivary cortisol increases in humans (Riad-Fahmy *et al.*, 1982), sheep (Fell *et al.*, 1985), pigs (Parrott *et al.*, 1989) and some other species. The injection of pilocarpine and sucking of citric acid crystals, which stimulate salivation, have no effect on the salivary cortisol concentration. However, any rise in salivary cortisol levels following some stimulus is delayed a few minutes as compared with the comparable rise in plasma cortisol concentration.

The cortisol response to handling and transport depends upon the species and on the breed of animal studied (Hall *et al.*, 1998a). When sheep were loaded on to a vehicle for the first time, all showed elevated plasma and saliva cortisol for at least the first hour (Broom *et al.*, 1996; Parrott *et al.*, 1998b). Cortisol levels during a journey were affected by being on a lower rather than an upper tier (Barton-Gade *et al.*, 1996), being on a rough rather than a smooth journey (Bradshaw *et al.*, 1996a), being mixed with strangers and being on a moving vehicle rather than being stationary (Bradshaw *et al.*, 1996b).

Animals which have substantial adrenal cortex responses during handling and transport show increased body temperature (Trunkfield *et al.*, 1991). The increase is usually of the order of 1°C, but the actual value at the end of a journey will depend upon the extent to which any adaptation of the initial response has occurred. Hence, if the temperature of pig blood at the time of slaughter is measured, it is essential that the details of the journey are considered when interpreting the result. The body temperature can be recorded during a journey with implanted or superficially attached temperature monitors linked directly or telemetrically to a data storage system. Parrott *et al.* (1999) described deep body temperature in eight sheep. When the animals were loaded into a vehicle and transported for 2.5 h, their body temperatures increased by about 1°C and in males the temperatures were elevated by 0.5°C for several hours. Exercise for 30 min resulted in a 2°C increase in core body temperature, which rapidly returned to baseline when the exercise finished. It would seem that prolonged increases in body temperature are an indicator of poor welfare.

In humans, vasopressin increases in the blood when the individual reports a feeling of nausea associated with motion sickness. Pigs also show motion sickness, retching and ejecting gut contents, especially when travelling along winding roads. These physical signs of motion sickness occur at the same time as increases in the levels of lysine vasopressin in the blood. Bradshaw *et al.* (1996c) showed that increased vomiting and retching in pigs coincided with higher levels of lysine vasopressin.

The measurement of oxytocin has not been of particular value in animal transport studies (e.g. Hall *et al.*, 1998b). However, plasma β-endorphin levels have been shown to increase during the loading of pigs (Bradshaw *et al.*, 1996c). The release of corticotrophin-releasing hormone (CRH) in the hypothalamus is followed by release of pro-opiomelanocortin (POMC) in the anterior pituitary, which quickly breaks down into components, including adrenocorticotrophic hormone (ACTH) which travels in the blood to the adrenal cortex, and β-endorphin. A rise in plasma β-endorphin often accompanies ACTH increases in plasma, but it is not yet clear what its function is. Although β-endorphin can have analgesic effects via mu receptors in the brain, this peptide hormone is also involved in the regulation of various reproductive hormones. Measurement of β-endorphin levels in blood is useful as a backup for ACTH or cortisol measurement.

Creatine kinase is released into the blood when there is muscle damage, e.g. bruising, and when there is vigorous exercise. It is clear that some kinds of damage which affect welfare result in creatine kinase release, so it can be used in conjunction with other indicators as a welfare measure. Lactate dehydrogenase (LDH) also increases in the blood after muscle-tissue damage, but increases can occur in animals whose muscles are not damaged. Deer which are very frightened by capture show large LDH increases (Jones and Price, 1992). The isomer of LDH which occurs in striated muscle (LDH5) leaks into the blood when animals are very disturbed, so the ratio of LDH5 to total LDH is of particular interest.

When animals are transported they will be deprived of water to some extent. On long journeys, they will have been unable to drink for many times longer than the normal interval between drinking bouts. This lack of control over interactions with the environment may be disturbing to the animals and there are also likely to be physiological consequences. The most obvious and straightforward way to assess this is to measure the osmolality of the blood (Broom et al., 1996). When food reserves are used up, there are various changes evident in the metabolites present in the blood. Several of these, for example β-hydroxybutyrate, can be measured and indicate the extent to which the food-reserve depletion is serious for the animal. Another measure which gives information about the significance for the animal of food deprivation is the delay since the last meal. Most farm animals are accustomed to feeding at regular times and, if feeding is prevented, especially when high rates of metabolism occur during journeys, the animals will be disturbed by this. Behavioural responses when allowed to eat or drink (e.g. Hall et al., 1997) also give important information about problems of deprivation.

The haematocrit, a count of red blood cells, is altered when animals are transported. If animals encounter a problem, such as those which may occur when they are handled or transported, there can be a release of blood cells from the spleen and a higher cell count (Parrott et al., 1998b). More prolonged problems, however, are likely to result in reduced cell counts (Broom et al., 1996).

A change which can be mediated by increased adrenal cortex activity and which may provide information about the welfare of animals during transport is immunosuppression. One or two studies in which animal transport affected T-cell function are reviewed by Kelley (1985), but such measurements are likely to be of most use in the assessment of more long-term welfare problems. The ability of the animal to react effectively to antigen challenge will depend upon the numbers of lymphocytes and the activity and efficiency of these lymphocytes. Measures of the ratios of white blood cells, for example the heterophil–lymphocyte ratio, are affected by a variety of factors, but some kinds of restraint seem to affect the ratio consistently, so they can give some information about welfare. Studies of T-cell activity, e.g. in vitro mitogen-stimulated cell proliferation, give information about the extent of immuno-suppression resulting from the particular treatment. If the immune system is working less well because of a treatment, the animal is coping less well with its environment and the welfare is poorer than in an animal which is not immunosuppressed.

## Mortality, injury and carcass characteristics

The term welfare is relevant only when an animal is alive, but death during handling and transport is usually preceded by a period of poor welfare. Mortality records during journeys are often the only record which give information about welfare during the journey, and the severity of the problems for the animals are often only too clear from such records. The number of pigs which were dead on arrival at the slaughterhouse was 0.07% in the UK and the Netherlands in the early 1990s, although the situation was worse in the past, especially with the Pietrain and Landrace breeds. The level in the Netherlands in 1970 was 0.7%. Recent estimates of the numbers of broilers and laying hens dead on arrival at UK slaughterhouses are 0.4% and 0.5%, respectively, but mortality of laying hens has been reported to be up to 50 times higher on occasion (Knowles and Broom, 1990b).

Amongst extreme injuries during transport are broken bones. These are rare in cattle, sheep, pigs and horses, but poor loading or unloading facilities and cruel or poorly trained staff who are attempting to move the animals may cause severe injuries. It is the laying hen, however, which is most likely to have bones broken during transit from housing conditions to point of slaughter. Following up earlier studies, Gregory and Wilkins (1989) found that 29% of a sample of 3115 end-of-lay hens from battery cages in the UK had at least one broken bone by the time they reached the stunner on the slaughter line. Hens from percheries or free range were less likely (10%) to have bones broken at this time (Gregory *et al.*, 1990) and it is clear from the work of Knowles and Broom (1990a) that lack of exercise results in wing bones being only half as strong in battery-cage hens as in hens from a perchery, which could flap their wings. A combination of brittle bones and rough handling by the catching team causes the bone breakage (Knowles *et al.*, 1993). There is no doubt that, for a hen, bone breakage must cause substantial pain and generally poor welfare.

Measurements made after slaughter can provide information about the welfare of the animals during handling, transport and lairage. Bruising, scratches and other superficial blemishes can be scored in a precise way and when carcasses are downgraded for these reasons, the people in charge of the animals can reasonably be criticized for not making sufficient efforts to prevent poor welfare. There is a cost of such blemishes to the industry, as well as to the animals, which was calculated to be 3.3% for pigs (Guise, 1991). The mixing of pigs resulted in a 7.2% increase in the incidence of blemishes (Guise and Penny, 1989). The cost, in both senses, of dark, firm, dry (DFD) and pale, soft, exudative (PSE) meat is even greater than this. DFD meat is associated with fighting in cattle and pigs, but cattle which are threatened but not directly involved in fights also show it (P.V. Tarrant, personal communication). PSE meat occurs more in some strains of pigs than in others, but its occurrence is related in most cases to other indicators of poor welfare.

## Welfare Problem Areas

Animals which are regularly and carefully handled and transported will show much less response to the procedures when subjected to them. The horse which is disturbed when first coaxed into a transport vehicle may show various signs of disturbance, but most of these signs will disappear by, for example, the tenth time of transporting, provided that the loading procedure and physical conditions, including space allowance, are appropriate for such an animal. However, most animals which are transported and those which are moved down races on a farm for the first time, or the first time for a long period, are substantially affected by the procedures. Hence it is fair to say that, for other than the animal which is a frequent traveller, transport and associated handling always result in poorer welfare than no treatment of this kind. Hence they should be avoided whenever possible.

The problems of animals which are transported are summarized and reviewed by Hall and Bradshaw (1998) under the following headings: duration of journey, novelty, vibration, shocks and impacts, noise, orientation, loading, ventilation, temperature, stocking density, location in vehicle, lack of food and water, breaks during the journey and mixing. These are discussed in general terms below.

Loading and unloading cause the major problems for most animals, so the procedures should be carried out with minimal frequency, especially for poultry or pigs, and with care, irrespective of the value of the animals. Those responsible for animals tend to treat animals like old ewes or end-of-lay hens which have little economic value, in a much rougher way than animals which will be used for breeding or animals with a high carcass value. It is quite wrong to treat animals in a cruel or inconsiderate way for such a reason. To act as if it does not matter what is done to animals destined for slaughter because they do not have long to live is also immoral. The welfare of each individual should be considered until the point at which it loses consciousness immediately prior to death.

Conditions during journeys are of great importance for the welfare of animals. Animals must be able to stand in their natural position with the roof well above their head and, with the exception of very short journeys for cattle and sheep, all must be able to lie down at the same time. The idea that animals are protected in some way by being packed close together is erroneous unless the vehicle is driven badly. Quadrupeds prefer to stand with their legs somewhat spread out so that they will not stumble or fall when the vehicle moves. Hence they do not touch one another if they are able to avoid doing so. Poultry crouch down throughout most journeys. A rough guide to the minimum space requirement for animals is based on the formula $A = 0.021\ W^{0.67}$, where $A$ is the area in $m^2$ required by the animal and $W$ is the animal's live weight in kg (modified after Esmay, 1978). This formula is based on the concept that the amount of space required by an animal is proportional to its surface area. The actual space allowance required in order to avoid poor welfare during a journey will depend upon the species, sex, age and presence or absence of horns or fleece. It will also depend on the temperature, humidity, ventilation system and number of tiers on the vehicle.

Adequate ventilation and protection from temperature extremes are also very important in the avoidance of very poor welfare. Exposure to low temperature can cause problems to any species, the actual lower critical temperature being higher for pigs than for sheep and cattle and higher still for poultry. Air flow over the animals can substantially increase chilling and can result in severe problems at temperatures which could be readily tolerated in still air. If animals get wet, they will be affected much more by a drop in temperature. In practice, those transporting animals normally take account of the risks of low temperature but many fail to appreciate the risks of high temperature, especially when accompanied by high humidity in the vehicle. Animals whose welfare is good in a field at 30°C may die when transported in an unventilated vehicle at the same temperature. Hence vehicles which will be used at temperatures above 20°C require a ventilation system. The system may be to have open sides, properly designed openings which provide adequate air flow during movement or a fan system. On hot days, a ventilation system which depends on movement is of no use when the vehicle is stationary, so, unless it is certain that the whole vehicle will be kept in shade when stopped, supplementary ventilation is needed. The major change which is required during hot conditions without a forced ventilation system is that stocking density must be reduced. Most deaths due to transport in hot conditions occur because the stocking density is too high.

In general, journeys by road result in poorer welfare than journeys by rail, sea or air and longer road journeys result in more welfare problems than shorter journeys (Hails, 1978). Hence any road journey should be as short as possible and the option of slaughtering near the point of rearing and of transporting meat should be taken wherever this is possible. The greatest problems arise when animals become dehydrated or energy-depleted and this can occur after as little as 4–6 h in domestic fowl and 6–8 h in horses.

All animals should be fit for the intended journey and proper preparation for the journey should be carried out, both in terms of preparing the animals and planning the sequence of human actions. Animals should be handled early in life, so that they are less disturbed by later handling. All staff who handle animals should have proper training before they are allowed to do so. A licensing scheme for animal handlers is needed so that, for new staff, adequate instruction is received and, for all staff, the possibility exists for the licence to be lost if cruelty or gross incompetence is demonstrated.

Driving vehicles carrying animals in such a way that the animals are thrown around will result in poor welfare and economic losses. Drivers have been found to be more careful in their driving if they were paid fuel economy bonuses (Guise, 1991).

Carcass characteristics are clearly important indicators of substantial problems for animals, and welfare could be improved by paying people concerned with the handling and transport of animals going to slaughter in inverse proportion to the incidence of damage. In particular, those transporting hens could be penalized if: (i) X-ray studies of a proportion of birds showed more than a very low level of broken bones; (ii) levels of recent bruising or scratches on the carcass were too high; or (iii) levels of DFD meat were too high. Some of these measurements can also be used in evidence in a prosecution under cruelty to animals laws.

Many of the above points apply to handling on farm and in lairage, as well as to handling associated with transport and slaughter. Properly designed animal-handling facilities and adequately trained staff with an incentive to make the welfare of animals as good as possible are needed. The design of lairage areas and of the whole slaughterhouse area, from vehicle stopping to stunning of animals, is of great importance in relation to animal welfare and to meat quality. A slaughterhouse may be very large and have a great through-put of animals, but the conditions for each individual as it leaves the vehicle, moves down races, waits in various places and finally goes to the point of stunning should be carefully considered. Provided that the facilities are well designed, if staff are paid according to the quality of the carcass after slaughter, the welfare of the animals will be much improved. The worst method of payment, in relation to animal welfare and carcass quality, is by speed of throughput.

As Hall and Bradshaw (1998) explain, information on the stress effects of transport is available from four kinds of study:

**1.** Studies where transport, not necessarily in conditions representative of commercial practice, was used explicitly as a stressor to evoke a physiological response of particular interest (Smart *et al.*, 1994; Horton *et al.*, 1996).

**2.** (a) Uncontrolled studies with physiological and behavioural measurements being made before and after long or short commercial or experimental journeys (Becker *et al.*, 1985, 1989; Dalin *et al.*, 1988, 1993; Knowles *et al.*, 1994).

(b) Uncontrolled studies during long or short commercial or experimental journeys (Lambooij, 1988; Hall, 1995).

**3.** Studies comparing animals that were transported with animals that were left behind to act as controls (Nyberg *et al.*, 1988; Knowles *et al.*, 1995).

**4.** Studies where the different stressors that impinge on an animal during transport were separated out either by experimental design (Bradshaw *et al.*, 1996b; Broom *et al.*, 1996; Cockram *et al.*, 1996) or by statistical analysis (Hall *et al.*, 1998c).

Each of these methods is of value because some are carefully controlled but less representative of commercial conditions, whilst others show what happens during commercial journeys but are less well controlled.

# Conclusions

Welfare is a characteristic of an individual animal at the time of observations or measurement. It can be assessed in an entirely objective way. Moral judgements can be made about the welfare of an animal when a person is given the scientific results and guidance on how to interpret them. The scientific assessment of welfare is separate from moral judgement. Welfare during transport and handling can be measured by behavioural or physiological measures. Measurements of injuries, bruises, mortality, morbidity and carcass quality can also be used as indicators of welfare during handling and transport. Experimental studies of the effects of transport are of value, but it is also important to monitor actual commercial journeys and procedures.

# References

Baldock, N.M. and Sibly, R.M. (1990) Effects of handling and transportation on heart rate and behaviour in sheep. *Applied Animal Behaviour Science* 28, 15–39.

Barton-Gade, P., Christenson, L., Brown, S.N. and Warriss, P.D. (1996) Effect of tier and ventilation during transport on blood parameters and meat quality in slaughter pigs. *Landbauforschung Volkenrode* 166 (Suppl.), 101–116.

Becker, B.A., Neinaber, J.A., Deshazer, J.A., and Hahn, G.L. (1985) Effect of transportation on cortisol concentrations and on the circadian rhythm of cortisol in gilts. *American Journal of Veterinary Research* 46, 1457–1459.

Becker, B.A., Mayes, H.F., Hahn, G.L., Neinaber, J.A., Jesse, G.W., Anderson, M.E., Heymann, H. and Hedrick, H.B. (1989) Effect of fasting and transportation on various physiological parameters and meat quality of slaughter hogs. *Journal of Animal Science* 67, 334.

Bradshaw, R.H., Hall, S.J.G. and Broom, D.M. (1996a) Behavioural and cortisol responses of pigs and sheep during transport. *Veterinary Record* 138, 233–234.

Bradshaw, R.H., Parrott, R.F., Goode, J.A., Lloyd, D.M., Rodway, R.G. and Broom, D.M. (1996b) Behavioural and hormonal responses of pigs during transport: effect of mixing and duration of journey. *Animal Science* 62, 547–554.

Bradshaw, R.H., Parrott, R.F., Forsling, M.L., Goode, J.A., Lloyd, D.M., Rodway, R.G. and Broom, D.M. (1996c) Stress and travel sickness in pigs: effects of road transport on plasma concentrations of cortisol, beta-endorphin and lysine vasopressin. *Animal Science* 63, 507–516.

Broom, D.M. (1986) Indicators of poor welfare. *British Veterinary Journal* 142, 524–526.

Broom, D.M. (1988) The scientific assessment of animal welfare. *Applied Animal Behaviour Science* 20, 5–19.

Broom, D.M. (1991a) Animal welfare: concepts and measurement. *Journal of Animal Science* 69, 4167–4175.

Broom, D.M. (1991b) Assessing welfare and suffering. *Behavioural Processes* 25, 117–123.

Broom, D.M. (1995) Quantifying pigs' welfare during transport using physiological measures. *Meat Focus International* 4, 457–460.

Broom, D.M. (1996) Animal welfare defined in terms of attempts to cope with the environment. *Acta Agriculturae Scandinavica Section A. Animal Science Supplement* 27, 22–28.

Broom, D.M. (1998) Welfare, stress and the evolution of feelings. *Advances in the Study Behaviour* 27, 371–403.

Broom, D.M. and Johnson, K.G. (1993) *Stress and Animal Welfare*. Chapman and Hall, London.

Broom, D.M. and Knowles, T.G. (1989) The assessment of welfare during the handling and transport of spent hens. In: Faure, J.M. and Mills, A.D. (eds) *Proceedings 3rd European Symposium Poultry Welfare*. World Poultry Science Association, Tours, pp. 79–91.

Broom, D.M., Knight, P.G. and Stansfield, S.C. (1986) Hen behaviour and hypothalamic–pituitary–adrenal response to handling and transport. *Applied Animal Behaviour Science* 16, 98.

Broom, D.M., Goode, J.A., Hall, S.J.G., Lloyd, D.M. and Parrott, R.F. (1996) Hormonal and physiological effects of a 15 hour journey in sheep: comparison with the responses to loading, handling and penning in the absence of transport. *British Veterinary Journal* 152, 593–604.

Cockram, M.S., Kent, J.E., Goddard, P.J., Waran, N.K., McGilp, I.M., Jackson, R.E., Muwanga, G.M. and Prytherch, S. (1996) Effect of space allowance during transport on the behavioural and physiological responses of lambs during and after transport. *Animal Science* 62, 461–477.

Dalin, A.M., Nyberg, L. and Eliasson, L. (1988) The effect of transportation/relocation on cortisol. CBG and induction of puberty in gilts with delayed puberty. *Acta Veterinaria Scandinavica* 29, 207–218.

Dalin, A.M., Magnusson, U., Haggendal, J. and Nyberg, L. (1993) The effect of transport stress on plasma levels of catecholamines, cortisol, corticosteroid-binding globulin, blood cell count and lymphocyte proliferation in pigs. *Acta Veterinaria Scandinavica* 34, 59–68.

Dämmrich, K. (1987) Growth and limb pathology in farm animals. In: Wiepkema, P.R. and van Adrichen, P.W.M. (eds) *Biology of Stress in Farm Animals*. Martinus Nijhoff, Dordrecht.

Esmay, M.L. (1978) *The Principles of Animal Environment*. AVI Publishing, Westport, Connecticut.

Fell, L.R., Shutt, D.A. and Bentley, C.J. (1985) Development of salivary cortisol method for detecting changes in plasma 'free' cortisol arising from acute stress in sheep. *Australian Veterinary Journal* 62, 403–406.

Fraser, A.F. and Broom, D.M. (1990) *Farm Animal Behaviour and Welfare*. CAB International, Wallingford, UK.

Freeman, B.M., Kettlewell, P.J., Manning, A.G.C. and Berry, P.S. (1984) Stress of transportation for broilers. *Veterinary Record* 114, 286–287.

Geverink, N.A. and Lambooij, E. (1994) Treatment of slaughter pigs during lairage in relation to behaviour and skin damage. In: *Proceedings of the 40th International Congress of Meat Science and Technology*. The Hague.

Geverink, N.A., Bradshaw, R.H., Lambooij, E., Wiegant, V.M. and Broom, D.M. (1998) Effects of simulated lairage conditions on the physiology and behaviour of pigs. *Veterinary Record* 143, 241–244.

Grandin, T. (1980) Observations of cattle behaviour applied to the design of cattle handling facilities. *Applied Animal Ethology* 6, 19–31.

Grandin, T. (1982) Pig behaviour studies applied to slaughter plant design. *Applied Animal Ethology* 9, 141–151.

Grandin, T. (1989) Behavioural principles of livestock handling. *Professional Animal Scientist* 5(2), 1–11.

Gregory, N.G. and Wilkins, L.J. (1989) Broken bones in domestic fowl: handling and processing damage in end-of-lay battery hens. *British Poultry Science* 30, 555–562.

Gregory, N.G., Wilkins, L.J., Eleperuma, S.D., Ballantyne, A.J. and Overfield, N.D. (1990) Broken bones in domestic fowls: effects of husbandry system and stunning method in end of lay hens. *British Poultry Science* 31, 59–69.

Guise, J. (1991) Humane animal management – the benefits of improved systems for pig production, transport and slaughter. In: Carruthers, S.P. (ed.) *Farm Animals: It Pays to be Humane.* CAS Paper 22, Centre for Agricultural Strategy, Reading, pp. 50–58.

Guise, J. and Penny, R.H.C. (1989) Factors affecting the welfare, carcass and meat quality of pigs. *Animal Production* 49, 517–521.

Hails, M.R. (1978) Transport stress in animals: a review. *Animal Regulation Studies* 1, 289–343.

Hall, S.J.G. (1995) Transport of sheep. *Proceedings of the Sheep Veterinary Society* 18, 117–119.

Hall, S.J.G. and Bradshaw, R.H. (1998). Welfare aspects of transport by road of sheep and pigs. *Journal Applied Animal Welfare Science* 1, 235–254.

Hall, S.J.G., Schmidt, B. and Broom, D.M. (1997) Feeding behaviour and the intake of food and water by sheep after a period of deprivation lasting 14 h. *Animal Science* 64, 105–110.

Hall, S.J.G., Broom, D.M. and Kiddy, G.N.S. (1998a) Effect of transportation on plasma cortisol and packed cell volume in different genotypes of sheep. *Small Ruminant Research* 29, 233–237.

Hall, S.J.G., Forsling, M.L. and Broom, D.M. (1998b) Stress responses of sheep to routine procedures: changes in plasma concentrations of vasopressin, oxytocin and cortisol. *Veterinary Record* 142, 91–93.

Hall, S.J.G., Kirkpatrick, S.M., Lloyd, D.M. and Broom, D.M. (1998c) Noise and vehicular motion as potential stressors during the transport of sheep. *Animal Science* 67, 467–473.

Hessing, M.J.C. (1994) Individual behavioural characteristics in pigs and their consequences for pig husbandry. PhD thesis, University of Utrecht.

Horton, G.M.J., Baldwin, J.A., Emanuele, S.M., Wohlt, J.E. and McDowell, L.R. (1996) Performance and blood chemistry in lambs following fasting and transport. *Animal Science* 62, 49–56.

Jensen, P. (1994) Fighting between unacquainted pigs: effects of age and of individual reaction pattern. *Applied Animal Behaviour Science* 41, 37.

Jensen, P., Rushen, J. and Forkman, B. (1995) Behavioural strategies or just individual variation in behaviour: a lack of evidence for active and passive piglets. *Applied Animal Behaviour Science* 43, 135–139.

Jones, A.R. and Price, S.E. (1992) Measuring the response of fallow deer to disturbance. In: Brown, R.D. (ed.) *The Biology of Deer*, Springer Verlag, Berlin.

Kelley, K.W. (1985) Immunological consequences of changing environmental stimuli. In: Moberg, G.P. (ed.) *Animal Stress.* American Physiological Association, pp. 193–223.

Kenny, F.J. and Tarrant, P.V. (1987) The reaction of young bulls to short-haul road transport. *Applied Animal Behaviour Science* 17, 209–227.

Kent, J.F. and Ewbank, R. (1986) The effect of road transportation on the blood constituents and behaviour of calves. III. Three months old. *British Veterinary Journal* 142, 326–335.

Knowles, T.G. and Broom, D.M. (1990a) Limb bone strength and movement in laying hens from different housing systems. *Veterinary Record* 126, 354–356.

Knowles, T.G. and Broom, D.M. (1990b) The handling and transport of broilers and spent hens. *Applied Animal Behaviour Science* 28, 75–91.

Knowles, T.G., Broom, D.M., Gregory, N.G. and Wilkins, L.J. (1993) Effects of bone strength on the frequency of broken bones in hens. *Research in Veterinary Science* 54, 15–19.

Knowles, T.G., Warriss, P.D., Brown, S.N. and Kestin, S.C. (1994) Long distance transport of export lambs. *Veterinary Record* 134, 107–110.

Knowles, T.G., Brown, S.N., Warriss, P.D., Phillips, A.J., Doland, S.K., Hunt, P., Ford, J.E., Edwards, J.E. and Watkins, P.E. (1995) Effects on sheep of transport by road for up to 24 hours. *Veterinary Record* 136, 431–438.

Lambooij, E. (1988) Road transport of pigs over a long distance: some aspects of behaviour, temperature and humidity during transport and some effects of the last two factors. *Animal Production* 46, 257–263.

Lambooij, E. and Engel, B. (1991) Transport of slaughter pigs by truck over a long distance: some aspects of loading density and ventilation. *Livestock Production Science* 28, 163–174.

Lambooij, E., Geverink, N., Broom, D.M. and Bradshaw, R.H. (1995) Quantification of pigs' welfare by behavioural parameters. *Meat Focus International* 4, 453–456.

Mendl, M., Zanella, A.J. and Broom, D.M. (1992) Physiological and reproductive correlates of behavioural strategies in female domestic pigs. *Animal Behaviour* 44, 1107–1121.

Nyberg, L., Lundstrom, K., Edfors-Lilja, I. and Rundgren, M. (1988) Effects of transport stress on concentrations of cortisol, corticosteroid-binding globulin and gluco-corticoid receptors in pigs with different halothane genotypes. *Journal of Animal Science* 66, 1201–1211.

Parrott, R.F., Misson, B.H. and Baldwin, B.A. (1989) Salivary cortisol in pigs following adrenocorticotrophic hormone stimulation: comparison with plasma levels. *British Veterinary Journal* 145, 362–366.

Parrott, R.F., Hall, S.J.G. and Lloyd, D.M. (1998a) Heart rate and stress hormone responses of sheep to road transport following two different loading responses. *Animal Welfare* 7, 257–267.

Parrott, R.F., Hall, S.J.G., Lloyd, D.M., Goode, J.A. and Broom, D.M. (1998b) Effects of a maximum permissible journey time (31 h) on physiological responses of fleeced and shorn sheep to transport, with observations on behaviour during a short (1 h) rest-stop. *Animal Science* 66, 197–207.

Parrott, R.F., Lloyd, D.M. and Brown, D. (1999) Transport stress and exercise hyperthermia recorded in sheep by radiotelemetry. *Animal Welfare* 8, 27–34.

Randall, J.M. and Bradshaw, R.H. (1998) Vehicle motion and motion sickness in pigs. *Animal Science* 66, 239–245.

Riad-Fahmy, D., Read, G.F., Walker, R.F. and Griffiths, K. (1982) Steroids in saliva for assessing endocrine function. *Endocrinology Review* 3, 367–395.

Rushen, J. (1986a) The validity of behavioural measure of aversion: a review. *Applied Animal Behaviour Science* 16, 309–323.

Rushen, J. (1986b) Aversion of sheep for handling treatments: paired choice experiments. *Applied Animal Behaviour Science* 16, 363–370.

Smart, D. Forhead, A.J., Smith, R.F. and Dobson, H. (1994) Transport stress delays the oestradiol-induced LH surge by a non-opioidergic mechanism in the early postpartum ewe. *Journal of Endocrinology* 142, 447–451.

Trunkfield, H.R. and Broom, D.M. (1990) The welfare of calves during handling and transport. *Applied Animal Behaviour Science* 28, 135–152.

Trunkfield, H.R. and Broom, D.M. (1991) The effects of the social environment on calf responses to handling and transport. *Applied Animal Behaviour Science,* 30, 177.

Trunkfield, H.R. and Broom, D.M., Maatje, K., Wierenga, H.K., Lambooy, E. and Kooijman, J. (1991) Effects of housing on responses of veal calves to handling and transport. In: Metz, J.H.M. and Groenestein, C.M. (eds) *New Trends in Veal Calf Production.* Pudoc, Wageningen, the Netherlands, pp. 40–43.

van Putten, G. (1982) Handling of slaughter pigs prior to loading and during loading on a lorry. In: Moss, R. (ed.) *Transport of Animals Intended for Breeding, Production and Slaughter.* Current Topics in Veterinary Medicine Animal Science 18, Martinus Nijhoff, The Hague, pp. 15–25

van Putten, G. and Elshof, W.J. (1978) Observations on the effect of transport on the well being and lean quality of slaughter pigs. *Animal Regulation Studies* 1, 247–271.

# Behavioural Principles of Handling Cattle and Other Grazing Animals Under Extensive Conditions

**5**

## Temple Grandin

*Department of Animal Sciences, Colorado State University, Fort Collins, CO 80523, USA*

## Introduction

In the late 1800s cowboys handled and trailed cattle quietly on the great cattle drives from Texas to Montana. In a cowboy's diary Andy Adams wrote: 'Boys, the secret of trailing cattle is to never let your herd know that they are under restraint. Let everything that is done be done voluntarily by the cattle' (Adams, 1903). Unfortunately, the quiet methods of the early 1900s were forgotten and some more modern cowboys were rough (Wyman, 1946; Hough, 1958; Burri, 1968). There is an excellent review of the history of herding in Smith (1998). Today, progressive producers of cattle know that reducing stress will improve both productivity and safety.

## Motivated by Fear

Cattle and other grazing, herding animals, such as horses, are a prey species. Fear motivates them to be constantly vigilant in order to escape from predators. Fear is a very strong stressor (Grandin, 1997). Fear stress can raise stress hormones higher than many physical stressors. When cattle become agitated during handling they are motivated by fear. The circuits in the brain that control fear-based behaviour have been studied and mapped (LeDoux, 1996; Rogan and LeDoux, 1996). Feedlot operators who handle thousands of extensively-raised cattle have found that quiet handling during vaccinating enabled cattle to go back on feed more quickly (Grandin, 1998a). Voisinet *et al.* (1997) reported that cattle which became highly agitated during restraint in a squeeze chute had lower weight gains than calm cattle which stood quietly in the chute.

©CAB *International* 2000. *Livestock Handling and Transport,* 2nd edn (ed. T. Grandin)

# Grazing-animal Perception

## Vision

To help them avoid predators, cattle have wide-angle (360°) panoramic vision (Prince, 1977) and vision has dominance over hearing (Uetake and Kudo, 1994). They can discriminate colours (Thines and Soffie, 1977; Darbrowska *et al.*, 1981; Gilbert and Arave, 1986; Arave, 1996). The latest research shows that cattle, sheep and goats are dichromats, with cones that are most sensitive to yellowish-green (552–555 nm) and blue-purple light (444–455 nm) (Jacobs *et al.*, 1998). Pick *et al.* (1994) tested a horse and found that it could discriminate red from grey and blue from grey, but could not discriminate green from grey. In another study, Smith and Goldman (1999) found that most horses could discriminate grey from red, blue, yellow and green, but one horse was not able to distinguish yellow from green. Dichromatic vision may provide better vision at night and aid in detecting motion (Miller and Murphy, 1995). The visual acuity of bulls may be worse than that of younger cattle or sheep (Rehkamper and Gorlach, 1998).

Grazing animals can see depth (Lemmon and Patterson, 1964). Horses are sensitive to visual depths cues in photographs (Keil, 1996). However, grazing animals may have to stop and put their heads down to see depth. This may explain why they baulk at shadows on the ground. Observations by Smith (1998) indicate that cattle do not perceive objects that are overhead unless they move. Smith (1998) also observed that, due to their horizontal pupil, cattle might see vertical lines better than horizontal lines. It is of interest that most grazing animals have horizontal pupils and most predators have round ones. Research with horses indicates that they have a horizontal band of sensitive retina, instead of a central fovea like a human (Saslow, 1999). This enables them to scan their surroundings while grazing. Grazing animals have a visual system that is very sensitive to motion and contrasts of light and dark. They are able to constantly scan the horizon while grazing and they may have difficulty in quickly focusing on nearby objects, due to weak eye muscles (Coulter and Schmidt, 1993). This may explain why grazing animals 'spook' at nearby objects that suddenly move.

Wild ungulates, domestic cattle and horses respect a solid fence and will seldom ram or try to run through a solid barrier. Sheets of opaque plastic can be used to corral wild ungulates (Fowler, 1995), whereas portable corrals constructed from canvas have been used to capture wild horses (Wyman, 1946; Amaral, 1977). Excited cattle will often run into a cable or chain-link fence because they cannot see it. A 30-cm-wide, solid, belly rail installed at eye height or ribbons attached to the fence will enable the animal to see the fence and prevent fence ramming (Ward, 1958). Cattle also have a strong tendency to move from a dimly illuminated area to a more brightly illuminated area (Grandin, 1980a,b). However, they will not approach a blinding light.

## Hearing

Grazing animals are very sensitive to high-frequency sounds. The human ear is most sensitive at 1000–3000 Hz, but cattle are most sensitive to 8000 Hz (Ames, 1974; Heffner and Heffner, 1983). Cattle can easily hear up to 21,000 Hz (Algers, 1984). Heffner and Heffner (1992) found that cattle and goats have a poorer ability to localize sound than most mammals. The authors speculate that, since prey species animals have their best vision directed to nearly the entire horizon, they may not need to locate sounds as accurately as an animal with a narrow visual field. Noise is stressful to grazing animals (Price *et al.*, 1993). The sounds of people yelling or whistling was more stressful to cattle than the sounds of gates clanging (Waynert *et al.*, 1999).

Lanier *et al.* (1999a, 2000) and Lanier (1999) found that cattle which became agitated in an auction ring were more likely to flinch or jump in response to sudden intermittent movement or sounds. Intermittent movements or sounds appear to be more frightening than steady stimuli. Talling *et al.* (1998) found that pigs were more reactive to intermittent sounds compared with steady sounds. High-pitched sounds increased a pig's heart rate more than low-pitched sounds (Talling *et al.*, 1996). Sudden movements have the greatest activating effect on the amygdala (LeDoux, 1996), the part of the brain that controls fearfulness (LeDoux, 1996; Rogan and LeDoux, 1996).

## Effects of Sudden Novelty

Cattle and other ungulates are frightened by novelty when they are suddenly confronted with it. Animals will baulk at a sudden change in fence construction or floor texture (Lynch and Alexander, 1973). Shadows, drain gates and puddles will also impede cattle movement (Grandin, 1980a). In areas where animals are handled, illumination should be uniform to prevent shadows and handling facilities should be painted all one colour to avoid contrasts. Contrasts have such an inhibiting effect on cattle movement that road maintenance departments have stopped cattle from crossing a road by painting a series of white lines across it (*Western Livestock Journal*, 1973).

Dairy cattle that are handled every day in the same facility will readily walk over a drain gate or a shadow because it is no longer novel. However, the same dairy cow will baulk and put her head down to investigate a strange piece of paper on the floor of a familiar alley. The paradoxical aspect of novelty is that it is both frightening and attractive (Grandin and Deesing, 1998). A clipboard on the ground will attract cattle when they can voluntarily approach it, but they will baulk and may refuse to step over it, if driven towards it.

A prey species must be wary of novelty because novelty can mean danger. For example, Nyala (antelope) in a zoo have little fear of people standing by their fence, but the novelty of people fixing a barn roof provoked an intense flight reaction. A review of the literature about cattle drives in the 1800s and early 1900s indicated that sudden novelty was the major cause of stampedes. Stampedes were caused by a hat flying in the wind, a horse bucking with a saddle under its belly, thunder, a cowboy stumbling or a flapping raincoat (Harger, 1928; Ward, 1958; Linford, 1977). Stampedes were also more likely to occur at night (Ward, 1958; Linford, 1977). Objects that move quickly are more likely to scare. Rapid motion has a greater activating effect on the amygdala (fear centre in the brain) than slow movement (LeDoux, 1996).

Dantzer and Mormede (1983) and Stephens and Toner (1975) both reported that novelty is a strong stressor. Placing a calf in an unfamiliar place is probably stressful (Johnston and Buckland, 1976). In tame beef cattle, throwing a novel-coloured ball into the pen caused a crouch–flinch reaction in half the animals (Miller *et al.*, 1986). Cattle which had previous handling experience in a livestock market settled down more quickly at the slaughter-plant stockyards because it appeared less novel and frightening to cattle which had been in a livestock market (Cockram, 1990). It is recommended to get cattle accustomed to being handled by people on foot, on horses and in vehicles, in order to prevent the animals from becoming excited by the novelty of handling at a feedlot, auction or slaughter plant. Zebu cattle reared in the Philippines are exposed to so much novelty that new experiences seldom alarm them. Halter-broken cows and their newborn calves are moved every day to new grazing locations along busy roads full of buses and cars.

## Studies of Handling Stress

There is an old saying: 'You can tell what kind of a stockman a person is by looking at his cattle.' Many cattlemen and women believe that early handling experiences have long-lasting effects (Hassal, 1974). Cattle with previous experiences with gentle handling will be calmer and easier to handle in the future than cattle that have been handled roughly (Grandin, 1981). Calves and cattle accustomed to gentle handling at the ranch of origin had fewer injuries at livestock markets because they had become accustomed to handling procedures (Wythes and Shorthose, 1984). Rough handling can be very stressful. In a review of many different studies, Grandin (1997) found that cortisol levels were two-thirds higher in animals subjected to rough treatment. Rough handling and sorting in poorly-designed facilities resulted in much greater increases in heart rate compared with handling in well-designed facilities (Stermer *et al.*, 1981). The severity and duration of a frightening handling procedure determine the length of time required for the heart rate to return to normal. Over 30 min is required for the heart rate to return to baseline levels after severe handling stress (Stermer *et al.*, 1981).

Measurement of cortisol levels has shown that animals can become accustomed to handling procedures. They will adapt to repeated non-painful procedures, such as moving through a race or having their blood samples taken through an indwelling catheter while they are held in a familiar tie stall

(Alam and Dobson, 1986; Fell and Shutt, 1986). Wild beef calves can adapt to a non-painful, relatively quick procedure, such as weighing. Peischel *et al.* (1980) reported that daily weighing did not affect weight gain. Cattle will not readily adapt to severe procedures which cause pain or to a series of rapidly repeated procedures where the animal does not have sufficient time to calm down between procedures. Fell and Shutt (1986) found that cortisol levels did not decrease after repeated trips in a truck where some animals fell down and lost their footing. Tame animals are likely to have a milder reaction to an aversive procedure than wild ones. Calves on an experiment station where they were petted by visitors had significantly lower cortisol levels after restraint and handling than calves which had less contact with people (Boandle *et al.*, 1989)

## Training Animals to Handling

Ried and Mills (1962) have suggested that animals can be trained to accept some irregularities in management, which would help reduce violent reactions to novelty. Exposing animals to reasonable levels of music or miscellaneous sounds will reduce fear reactions to sudden unexpected noises. When a radio is played in a pig barn, the pigs have a milder reaction to a sudden noise, such as a door slamming. Playing instrumental music or miscellaneous sounds at 75 dB improved weight gains in sheep (Ames, 1974). Louder sound reduced weight gains.

Binstead (1977), Fordyce *et al.* (1985) and Fordyce (1987) report that training young *Bos indicus* heifer calves produced calmer adult animals which were easier to handle. Training of weaner calves involves walking quietly among them in the corrals, working them through races and teaching them to follow a lead horseman (Fordyce, 1987). These procedures are carried out over a period of 10 days. Becker and Lobato (1997) also found that ten sessions of gentle handling in a race made zebu cross-bred calves calmer and less likely to attempt to escape or charge a person in a small pen. Training bongo antelope to voluntarily cooperate with injections and blood tests resulted in very low cortisol levels which were almost at baseline (Phillips *et al.*, 1998). All training procedures must be done gently. Burrows and Dillon (1997) suggest that training may provide the greatest benefit for cattle with excitable temperaments. There are great individual differences in how animals react to handling and restraint. Ray *et al.* (1972) found that cortisol levels varied greatly between individual cattle: in semi-tame beef cattle, one animal had almost no increase in cortisol levels during restraint and blood testing from the jugular, whereas the other five cattle in the experiments had substantial increases.

In extensively reared, untrained, wild 260 kg Gelbvieh × Simmental × Charolais cross cattle, behavioural traits were persistent over a series of four monthly handling and restraint sessions (Grandin, 1993). A small group of cattle (9% bulls and 3% steers) became extremely agitated and violently shook the squeeze chute (crush) every time they were handled. Another group (25% bulls and 40% steers) stood very calmly in the squeeze chute every time they were handled. There was also a large group of animals that was sometimes calm and sometimes agitated. The animals were handled carefully and gently during all the observed restraint sessions. These differences in temperament can probably be explained by a combination of genetic factors and handling experiences as young calves. The behaviour of the few extremely agitated animals did not improve over time. These observations illustrate that the behaviour patterns that are formed at a young age may be very persistent. There was also a tendency for the agitated animals to avoid coming through the race with the first bunches of cattle. Orihuelo and Solano (1994) found that animals first in line in a single file race moved more quickly through the race compared with animals last in line.

## Genetic and Species Differences

Genetic differences within a breed can affect stress responses during handling. Animals with flighty genetics are more likely to become extremely agitated when confronted with a sudden novel event such as seeing a waving flag for the first time than animals with a calmer temperament (Grandin and Deesing, 1998). A basic principle is that the animals with flighty, excitable genetics must have new experiences introduced more gradually than animals with calm genetics. One of the major differences between wild and domesticated animals is that the wild species have higher levels of fearfulness and a stronger reaction to environmental change (Price, 1998). Species such as American bison and antelope are so fearful that they often severely injure themselves when they are restrained. Whereas domestic cattle will tolerate being gently forced into a restraint device, bison and antelope are animals that need to be trained to voluntarily cooperate (Grandin, 1999). Jennifer Lanier in our research team has had some success in training American bison to move voluntarily through races for feed rewards (Lanier et al., 1999b). Bison, deer and other flighty species should be moved in small groups. They will remain calmer if each individual animal is brought from the forcing pen to the restraint device through a short single-file race. Whereas domestic cattle will stand quietly in a single-file race, many of the wild ungulates become stressed and agitated if they are made to wait in line. Even in domestic animals some individuals will habituate to a forced non-painful procedure and others may respond by getting increasingly stressed. Lanier et al. (1995) found that some pigs habituated quickly to swimming and their adrenaline levels dropped to baseline, whereas others remained frightened and their adrenaline levels remained high.

   In Holstein calves, the sire had an effect on cortisol response to trans-portation stress (Johnston and Buckland, 1976). The sire also has an effect on learning ability and activity levels in Holstein calves (Arave *et al.*, 1992). The breed of cattle has a definite effect on temperament. Brahman-cross cattle became more behaviourally agitated in a squeeze chute compared with shorthorns (Fordyce *et al.*, 1988). Both Fordyce *et al.* (1988) and Hearnshaw *et al.* (1979) report that temperament is heritable in *B. indicus* cattle. Stricklin *et al.* (1980) report that Herefords were the most docile British breed and Galloways the most excitable. The continental European breeds of *Bos taurus* were generally more excitable than British breeds. Within a breed the sire was found to have an effect on temperament scores.

LeNeindre *et al.* (1996) discuss problems associated with taking breeds which were developed for an intensive system and putting them out on an extensive range. For example, a bull can produce daughters that are gentle in an intensive system and aggressive towards handlers when raised on the range. The author has observed that these problems are most likely to occur in excitable, flighty cattle that panic in a new situation. Some genetic lines of Saler cattle are calm and easy to handle when they are with familiar people, but they panic and kick and charge people when confronted with the noise and novelty of an auction or slaughter plant. Grandin *et al.* (1995) and Randle (1998) found that cattle with small, spiral, hair whorls above the eyes had a larger flight distance and were more likely to become agitated during restraint than cattle with hair whorls below the eyes.

Different breeds of cattle also have different behavioural characteristics that affect handling. Pure-bred *B. indicus* cattle have a greater tendency to follow people or lead animals. It is sometimes easier to train these cattle to lead instead of driving them. Brahman and Brahman-cross cattle also tend to flock more tightly together when they are alarmed compared with British breeds. Brahman-type cattle are also more difficult to block at gates compared with British breeds (Tulloh, 1961).

Brahman and Brahman-cross cattle are more prone to display tonic immobility during restraint (Fraser, 1960; Grandin, 1980a). Brahman-cross cattle are more likely to lie down in a single-file race and refuse to move compared with British breeds (Grandin, 1980a). Excessive electric prodding of a submissive Brahman can kill it. It will usually get up if it is left alone. Fraser and Broom (1990) state that an uninjured downed cow will often get up if its environment is changed, such as moving it from inside to outside. Zavy *et al.* (1992) found that Hereford × Brahman crosses and Angus × Brahman crosses had higher cortisol levels during restraint in a squeeze chute compared with Hereford × Angus crosses. Brahman genetics increased cortisol levels and Angus × Brahman crosses had the highest levels.

## Flight Zone

The concept of flight distance was originally applied to wild ungulates. Hedigar (1968) states:

By intensive treatment, i.e. by means of intimate and skilled handling of the
wild animals, their flight distance can be made to disappear altogether, so
that eventually such animals allow themselves to be touched. This artificial
removal of flight distance between animals and man is the result of the process
of taming

This same principle applies to domestic cattle and wild ungulates.

Extensively raised wild cows on an Arizona ranch may have a 30 m flight
distance, whereas feedlot cattle may have a flight distance of 1.5–7.61 m
(Grandin, 1980a). Cattle with frequent contact with people will have a smaller
flight distance than cattle that seldom see people. Cattle subjected to gentle
handling will usually have a smaller flight zone than cattle subjected to
abusive handling. Excitement will enlarge the flight zone. Totally tame dairy
cattle may have no flight zone and people can touch them. The edge of the
flight zone can be determined by slowly walking up to a group of cattle. When
the animals turn and face the handler, he/she is outside the flight zone. When
a person enters the edge of the flight zone, the animals will turn around and
move away. When the flight zone of a group of bulls was invaded by a moving
mechanical trolley, the bulls moved away and maintained a constant distance
between themselves and the moving trolley (Kilgour, 1971). The flight
distance was determined by the size of a piece of cardboard attached to the
trolley. Cattle remain further away from a larger object (Smith, 1998). When a
person approaches full face, the flight zone will be larger than approaching
with a small sideways profile.

Cattle can be moved most efficiently if the handler works on the edge of the
flight zone (Grandin, 1980b, 1987). The animals will move away when the
flight zone is penetrated and stop when the handler retreats. Smith (1998)
explains that the edge of the flight zone is not distinct and that approaching an
animal quickly will enlarge the flight zone. Excited cattle will have a larger
flight zone and eye contact with the animals will also enlarge the flight zone.
To make an animal move forward, the handler should stand in the shaded area
marked A and B in Fig. 5.1 and stay out of the blind spot at the animal's rear.
To make an animal move forward the handler should stand behind the point of
balance at the shoulder and, to make the animal back up, the handler should
stand in front of the point of balance (Kilgour and Dalton, 1984). Another
principle is that grazing animals, either singly or in groups, will move forward
when a handler quickly passes the point of balance at the shoulder in the
opposite direction of desired movement (Grandin, 1998a; Fig. 5.2). Use of the
movements shown in Fig. 5.2 to induce cattle to enter a squeeze chute makes it
possible to greatly reduce or eliminate electric prod use.

When an animal is approached head on, it will turn right if the handler
moves left and vice versa (Kilgour and Dalton, 1984). Cattle in crowd pens and
other confined areas can be easily turned by shaking plastic strips on a stick
next to their heads (Fig. 5.3). For example, when a cow's vision is blocked on
the left side by the plastic strips, she will turn right. Handlers should avoid deep
penetration of the flight zone, because this may cause cattle to panic. If an

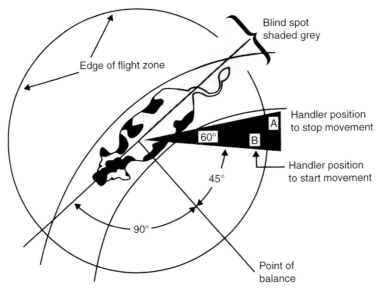

**Fig. 5.1.** Flight zone diagram showing the most effective handle positions for moving an animal forward.

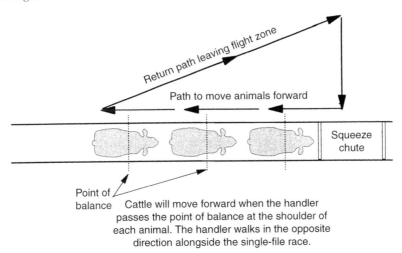

**Fig. 5.2.** Handler movement pattern to induce cattle to move forward in a race.

animal rears up in a race, handlers should back up to remove themselves from the animal's flight zone. Handlers should not attempt to push a rearing animal down, because deep penetration of flight zone causes increasing panic and attempts to escape. If cattle attempt to turn back in an alley, the handler should back up and remove him/herself from deep inside the flight zone.

The angle of approach and the size of enclosure an animal is in will also affect the size of the flight zone. Experiments with sheep indicated that animals

**Fig. 5.3.**    Plastic streamers on the end of a stick or whip are useful for turning cattle by shaking them alongside the animal's head.


confined in a narrow alley had a smaller flight zone than animals confined in a wider alley (Hutson, 1982). Cattle will have a larger flight zone when they are approached head on. Extremely tame cattle are often hard to drive because they no longer have a flight zone. These animals should be led. More information on flight zone can be found in Smith, 1998.


## Moving Large Groups

Ward (1958) described the methods used on the old cattle drives in the USA to move herds of 1000 cattle, in which many people were required to keep the cattle together. Bud Williams, a cowman in Canada, spent 30 years developing calm methods which enable one or two people to move hundreds of cattle. Unfortunately, he never published his methods, but I have had the opportunity to observe him and develop diagrams to help teach the principles.

The handler should spend time walking among the cattle so that they perceive him/her as a neutral entity, neither a predator (chasing them into the corrals) nor a feed source. All cattle movements are done at a slow walk with no yelling. Figure 5.4 shows the handler movement patterns which will keep a herd moving in an orderly manner. It will work both along a fence and on an open pasture. If a single person is moving cattle, position 2 on Fig. 5.4 shows the handler movement patterns which will keep a herd moving in an orderly manner. The principle is to alternately penetrate and withdraw from the flight zone. Continuous steady pressure will cause the herd to split. As the herd moves, the handler should keep repeating the movement pattern. For a more

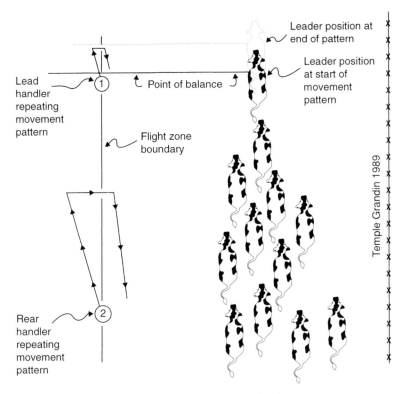

Leader position at
end of pattern

Leader position
at start of
movement
pattern

Lead
handler
repeating
movement
pattern

Point of balance

Flight zone
boundary

Temple Grandin 1989

Rear
handler
repeating
movement
pattern

**Fig. 5.4.** Handler movement patterns for moving a herd.

complete description, refer to Grandin (1990). Ward (1958) also showed a similar movement pattern. The principle is to move inside the flight zone in the direction opposite to the desired movement and to be outside the flight zone in the same direction as the desired movement.

Figure 5.5 illustrates how to bring the herd back together if it splits. The handler should not act like an attacking predator and run around behind the stragglers to chase them. He/she should move towards the stragglers while gradually impinging on the collective flight zones and stop at the point of balance of the last animal. After the herd closes up, he/she should walk forward at an angle to gradually decrease pressure on the collective flight zone.

## Gathering Cattle on Pasture

Wild and semi-wild cattle can be easily gathered on pasture by inducing their natural behaviour to loosely bunch. Figure 5.6 shows a 'windscreen-wiper' pattern, where the handler walks barely on the edge of the group's collective

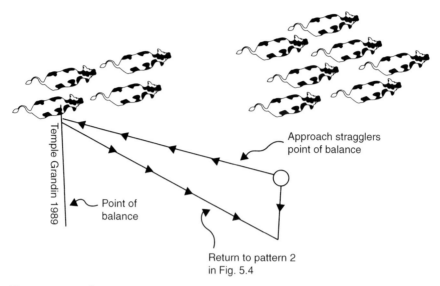

**Fig. 5.5.**    Handler movement pattern to bring a split herd back together.

flight zone. The handler moves at a slow walk and must be careful not to circle around the animals. The handler must also resist the urge to chase stragglers. When the bunching instinct is triggered, the herd will come together and the stragglers will join the other cattle. Care must be taken to be quiet and keep the animals moving at a walk. The principle is to induce bunching before any attempt is made to move the herd. The animals will move towards the pivot point of the 'windscreen wipers'. If too much pressure is applied to the collective flight zone prior to bunching, the herd will scatter. More information is in Grandin (1998b) and Smith (1998). This method will not work on completely tame animals with little or no flight zone.

## Why does it work?

The author speculates that the behavioural principles of moving cattle and other ungulates are based on innate instinctual antipredator behaviours (Grandin, 1998c). There appear to be four basic behaviours: (i) turn and orientate towards a novel stimulus, but keep a safe distance; (ii) point of balance; (iii) loose bunching; (iv) milling and circling. The study of many nature programmes on television indicated that the point-of-balance principle enables an animal to escape a chasing predator. Low-stress handling only activates a mild anxiety and the high stress of behaviour (iv) should be avoided. The least stressful handling procedure would be entirely voluntary. Smith (1998) states that there is no black and white dividing line between herding, leading and training. It is likely that cows gathered with the

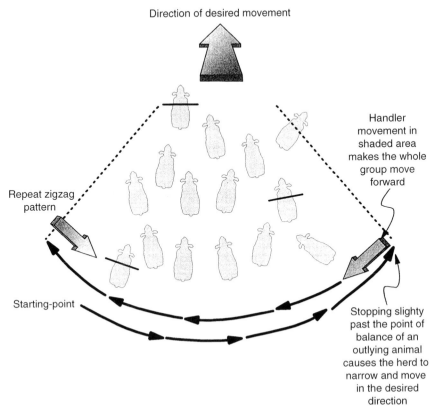

Direction of desired movement

Handler movement in shaded area makes the whole group move forward

Repeat zigzag pattern

Starting-point

Stopping slighty past the point of balance of an outlying animal causes the herd to narrow and move in the desired direction

**Fig. 5.6.**   Handler movement pattern to induce bunching for gathering cattle. The handler must zigzag back and forth to keep the herd going straight. Imagine that the leaders are the pivot point of a windscreen wiper and the handler is out on the end of the blade, sweeping back and forth. As the herd narrows and gets good forward movement, the width of the handler's zigzag narrows.

'windscreen-wiper' method may have slight anxiety at first and then become completely trained and have diminishing anxiety. Bud Williams, cattle-handling specialist in Canada, recommends using a straight zigzag motion, instead of the slight curve in the 'windscreen-wiper' pattern. The handler must not circle around the cattle. The arc should be very slight. Using these movement patterns probably triggers an instinct to bunch, similar to the behaviour of cattle in bear country, where they graze in tighter bunches.

## Working in Corrals

Figure 5.7 illustrates the correct positions to empty a pen and to sort cattle out through a gate (Smith, 1998). The diagram shows the movements to stop an

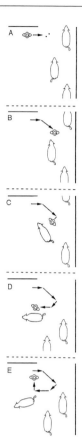

**Fig. 5.7.** Handler movement patterns for turning an animal back when sorting out of a gate (from Smith, 1998).

animal from going out of the gate. Eye contact can be used to hold back animals. The handler should avert his/her eyes away from animals sorted out of the gate. When a pen is emptied, the handler should avoid chasing the cattle out. They should move past the handler at a controlled rate, so that they learn that the handler controls their movements. Rancher Darol Dickinson states that you need to train cattle (McDonald, 1981). Additional methods for moving and loading cattle are shown in McDonald (1981). One of the most common mistakes is to place too many cattle in a crowd (forcing) pen which leads to a single-file race. An overloaded crowd pen causes problems because the cattle do not have room to turn. To utilize natural following behaviour, handlers should wait for the single-file race to become almost empty before refilling it (Grandin, 1980a).

Many handlers overuse and sometimes abuse electric prods and other persuaders. If tail twisting is used to move cattle through a race, pressure on the tail should be instantly released when the cow moves. Breeding cattle quickly learn that they can avoid having their tails twisted if they move promptly when the tail is touched.

Electric prods should be used as little as possible. On *B. indicus* cattle, electric prods should never be used on breeding animals. In slaughter plants and livestock auctions, electric-prod use should be confined to the single-file race. The handlers should wait until the tailgate of the squeeze chute is open before initiating the movement patterns in Fig. 5.2. If the movement pattern fails on the first attempt, walking past the point of balance a second time will often work. If one animal baulks, uncooperative behaviour will spread to other cattle. Harger (1928) discusses how one hysterical animal can have a negative influence on the rest of the group. Cattle are herd animals and they become stressed and upset when they are isolated from their herd mates (Ewbank, 1968). Isolated lone animals that have panicked cause many injuries to both people and cattle. To move a frantic animal some other cattle should be put in with it. Often the most difficult to handle animals are the last ones in a group to move through a race (Orihuela and Solano, 1994).

## Leaders

The natural following behaviour of cattle can be used to facilitate cattle movements. On the old cattle drives in the USA the value of leaders was recognized. The same leaders would lead a herd of 1000 cattle every day (Harger, 1928). A good leader is usually a sociable cow and is not the most dominant animal. Smith (1998) contains excellent information on the effect of social behaviour on handling.

Excitable, nervous animals that became leaders were usually destroyed and calm leaders were kept (Harger, 1928). If the cattle herd refused to cross a bridge or brook, a calf would be roped and dragged across to encourage the other cattle to follow (Ward, 1958). In Australia, a herd of tame 'coacher' cattle is used to assist in gathering wild feral cattle (Roche, 1988), and similar methods have also been used with wild horses (Amaral, 1977). Fordyce (1987) also recommends mixing a few quiet old steers in with *B. indicus* calves to facilitate training to handling procedures.

Cattle reared under extensive conditions can be easily trained to come when called. The animals learn to associate a vehicle horn with feed (Hasker and Hirst, 1987). In the northern USA when snow is on the ground cattle will come running when they see the hay truck. However, cattle can become a nuisance by chasing a truck for feed so the animals should be trained to associate the vehicle horn with feed, then the truck can be driven in a pasture without the animals running after it.

More and more ranchers are adopting intensive grazing systems where cattle are switched to new pasture every few days (Savory, 1978; Smith *et al.*, 1986). The cows quickly learn to make the switch, but calves are sometimes stressed when the cows rapidly run into the new pasture and leave them behind. To avoid calf stress, handlers should stand near the gate of the new pasture and make the cows walk by them at a controlled rate.

## Intensive Grazing Without Fences

There is increasing interest in practising intensive grazing methods without the expense of fences. Herding methods are being used to keep cattle bunched and to move them to different grazing areas. One of the big problems is that some cattle are 'bunch quitters' and do not want to stay with the herd (Nation, 1998). 'Bunch quitters' are most likely to be high-headed, nervous cows. Selling the cow is often the best option. The principle of herding without fences is to relieve pressure on the collective flight zone when cows stay where you want them and apply pressure when they go where you do not want them. Herders have to spend many hours with their herds and have lots of patience. Low-stress herding is most difficult with older cows from several different ranches that have had completely different previous experiences with herding and handling (Nation, 1999). More information on pasture herding can be found in Biggs and Biggs (1996, 1997), Herrmann (1998), Nation (1998), Smith (1998) and Williams (1998).

## Herding by Pastoral People

The herding methods described in the last section are a relearning of old pastoral herding methods that have been used for thousands of years. In all of these methods, a great deal of time is spent with the animals. Norwegian reindeer herders are in close contact with their animals and the animals associate the smells and noises of camp with serenity (Paine, 1994). The Fulani African tribesmen have no horses, ropes, halters or corrals (Lott and Hart, 1977). Their cattle are completely tame and have no flight zone. Instead of chasing the cattle the herdsman becomes a member of the herd and the cattle follow him (Lott and Hart, 1979). *B. indicus* cattle have a much stronger following instinct than *B. taurus*. Observations by the author indicate that tame pure-bred Brahmans are difficult to drive and they will often follow a person or a trained lead animal. In Australia they have been trained to follow lead dogs. The nomadic Fulani tribesmen use the animals' natural following, dominance, submission and grooming behaviour to control their behaviour. If a bull makes a broadside threat, the herdsman yells and raises a stick. The herdsman charges at the bull and hits it with a stick if it attempts a charge.

Similar methods have also been used successfully in other species. Raising a stick over the handler's head has been used to exert dominance over bull elk (B. Williams, personal communication; Smith, 1998). The stick is never used to hit the bull elk. The author has used similar methods to control aggressive pigs which exert dominance by biting and pushing against the other pig's neck (Houpt and Wolski, 1982). The aggressive behaviour was stopped by shoving a board against the pig's neck to simulate the bite of another pig. Using the animal's natural method of communication was more effective than slapping it on the rear. Exerting dominance is not beating an animal into submission:

the handler uses the animal's own behavioural patterns to become the 'boss' animal.

The aversion cattle have for manure can be used to keep them away from crops, by smearing the borders of a field with faeces (Lott and Hart, 1982). Manure is also smeared on the cow's udder to limit milk intake by the calf. The Fulani stroke their cattle in the same areas where a mother cow licks her calf (Lott and Hart, 1979), so adult cattle will approach and stretch out their necks to be stroked under the chin (Lott and Hart, 1982). Similar methods are used at the J.D. Hudgins Ranch in Hungerford, Texas, and the J. Carter Thomas Ranch in Cuero, Texas. Pure-bred Brahmans are led to the corrals and will eat out of the rancher's hand. Cows and bulls in the pasture will come up to Mr Thomas for stroking and brushing (Julian, 1978). Small herds of zebu cattle raised in the Philippines have no flight zone and they are easily led by small children. The author's observations indicate that taming cross-breeds of Brahman and *B. taurus* is more difficult. This may be partially due to a lower level of inquisitiveness, desire for stroking and following behaviour.

The cattle-herding methods of the Fulani are also practised by other African tribes, such as the Dinka (Deng, 1972; Schwabe and Gordon, 1988) and the Nuer (Evans-Pritchard, 1940). The less nomadic tribes do use corrals and tethers, but the cattle are still completely tame with no flight zone. Surplus bulls are castrated and kept as steers by all tribes. The cattle-handling practices of African tribes date back to before the great dynasties of Egypt (Schwabe, 1985; Schwabe and Gordon, 1988). It is also noteworthy that the religion of the Nuer and Dinka tribes centres around cattle (Seligman and Seligman, 1932; Evans-Pritchard, 1940). One factor which makes African tribal handling methods successful is that relatively small herds are handled and each tribe has many herdsmen. Therefore, each herdsman has time to develop an intimate relationship with each animal.

## Bull Behaviour

Dairy bulls have a bad reputation for attacking people, possibly due to the differences in the way beef bulls and dairy bulls are raised. Dairy bull calves are often removed from the cow shortly after birth and raised in individual pens, whereas beef bull calves are reared by the cow. Price and Wallach (1990) found that 75% of Hereford bulls reared in individual pens from 1 to 3 days of age threatened or attacked the handlers, whereas only 11% threatened handlers when they were hand-reared in groups. These authors also report that they have handled over 1000 dam-reared bulls and have experienced only one attack. Bull calves that are hand-reared in individual pens may fail to develop normal social relations with other animals and they possibly view humans as a sexual rival (Reinken, 1988). Similar aggression problems have also been reported in hand-reared male llamas (Tillman, 1981). Fortunately, hand-rearing does not cause aggression problems in females or castrated animals. It will make these animals easier to handle. More information on bulls can be found in Smith (1998).

## Conclusions

Cattle are animals that fear novelty and become accustomed to a routine. They have good memories and animals with previous experience of gentle handling will be easier to handle than animals with a history of rough handling. Both genetic factors and experience influence how cattle will react to handling. An understanding of natural behaviour patterns will facilitate handling. To reduce stress, progressive producers should work with their animals to habituate them to a variety of quiet handling methods such as people on foot, riders on horses and vehicles. Training animals to accept new experiences will reduce stress when the animals are moved to a new location.

## References

Adams A. (1903) *Log of a Cowboy*. Houghton Mifflin, New York. Reissued in 1964 by Bison Book, University of Nevada Press, Reno, Nevada.

Alam, M.G.S. and Dobson, H. (1986) Effect of various veterinary procedures on plasma concentrations of cortisol, luteinizing hormone and prostaglandin $F_2$ metabolite in the cow. *Veterinary Record* 118, 7–10.

Algers, B. (1984) A note on responses of farm animals to ultra sound. *Applied Animal Behaviour Science* 12, 387–391.

Amaral, A. (1977) *Mustang: Life and Legends of Nevada's Wild Horses*. University of Nevada Press, Reno, Nevada.

Ames, D.R. (1974) Sound stress and meat animals. In: *Proceedings of the International Livestock Environment Symposium*. SP-0174, American Society of Agricultural Engineers, St Joseph, Michigan. p. 324.

Arave, C.W. (1996) Assessing sensory capacity of animals using operant technology. *Journal of Animal Sciences* 74, 1996–2009.

Arave, C.W., Lamb, R.C., Arambel, M.J., Purcell, D. and Walters, J.L. (1992) Behaviour and maze learning ability of dairy calves are influenced by housing, sex and size. *Applied Animal Behaviour Science* 33, 149–163.

Becker, B.G. and Lobato, J.F.P. (1997) Effect of gentler handling on the reactivity of Zebu crossed calves to humans. *Applied Animal Behaviour Science* 53, 219–224.

Biggs, D. and Biggs, C. (1996) Using the magnetic nature of herd movement. *Stockman Grassfarmer*, September. Mississippi Valley Publishing, Ridgeland, Mississippi, pp. 15–17.

Biggs, D. and Biggs, C. (1997) Gatekeepers: low stress tactics for moving cattle. *Stockman Grassfarmer*, September. Mississippi Valley Publishing, Ridgeland, Mississippi, pp. 36–37.

Binstead, M. (1977) Handling cattle. *Queensland Agriculture Journal* 103, 293–295.

Boandle, K.E., Wohlt, J.E. and Carsia, R.V. (1989) Effect of handling, administration of a local anesthetic and electrical dehorning on plasma cortisol in Holstein calves. *Journal of Dairy Science* 72, 2193–2197.

Burri, R. (1968) *The Gaucho.* Crown Publishing, New York.

Burrows, H.M. and Dillon, R.D. (1997) Relationship between temperament and growth in a feedlot and commercial carcass traits of *Bos indicus* crossbreeds. *Australian Journal of Experimental Agriculture* 37, 407–411.

Cockram, M.S. (1990) Some factors influencing behavior of cattle in a slaughter house lairage. *Animal Production* 50, 475–481.

Coulter, D.B. and Schmidt, G.M. (1993) Special senses 1: vision. In: Swenson, M.J. and Reece, W.O. (eds) *Duke's Physiology of Domestic Animals,* 11th edn. Comstock Publishing, Ithaca, New York.

Dantzer, R. and Mormede, P. (1983) Stress in farm animals: a need for re-evaluation. *Journal of Animal Science* 57, 6–18.

Darbrowska, B., Harmata, W. and Lenkiewiez, Z. (1981) Color perception in cows. *Behaviour Process* 6, 1–10.

Deng, F.M. (1972) *The Dinka of Sudan.* Holt, Rinehart and Winston, New York.

Evans-Pritchard, E.E. (1940) *The Nuer.* Clarendon, Oxford, UK.

Ewbank, R. (1968) The behavior of animals in restraint. In: Fox, M.W. (ed.) *Abnormal Behavior in Animals.* W.B. Saunders, Philadelphia, Pennsylvania.

Fell, L.R. and Shutt, D.A. (1986) Adrenal response of calves to transport stress as measured by salivary cortisol. *Canadian Journal of Animal Science* 66, 637–641.

Fordyce, G. (1987) Weaner training. *Queensland Agricultural Journal* 113, 323–324.

Fordyce, G., Goddard, M.E., Tyler, R., Williams, C. and Toleman, M.A. (1985) Temperament and bruising in *Bos indicus* cross cattle. *Australian Journal of Experimental Agriculture* 25, 283–288.

Fordyce, G., Dodt, R.M. and Wythes, J.R. (1988) Cattle temperaments in extensive herds in northern Queensland. *Australian Journal of Experimental Agriculture* 28, 683–687.

Fowler, M.E. (1995) *Restraint and Handling of Wild and Domestic Animals,* 2nd edn. Iowa State University Press, Ames, Iowa.

Fraser, A.F. (1960) Spontaneously occurring forms of tonic immobility in farm animals. *Canadian Journal of Comparative Medicine* 24, 330–333.

Fraser, A.F. and Broom, D.M. (1990) *Farm Animal Behavior and Welfare.* Baillière Tindall, London.

Gilbert, B.J. and Arave, C.W. (1986) Ability of cattle to distinguish among different wavelengths of light. *Journal of Dairy Science* 69, 825–832.

Grandin, T. (1980a) Livestock behavior as related to handling facilities design. *International Journal for the Study of Animal Problems* 1, 33–52.

Grandin, T. (1980b) Observations of cattle behavior applied to the design of cattle handling facilities. *Applied Animal Ethology* 6, 19–31.

Grandin, T. (1981) Bruises on Southwestern feed lot cattle. *Journal of Animal Science* 53 (Suppl. 1), 213 (abstract).

Grandin, T. (1987) Animal handling. In: Price, E.O. (ed.) Farm Animal Behavior. *Veterinary Clinics of North America: Food Animal Practice* 3(2), 323–338.

Grandin, T. (1990) Forget that rodeo cowboy stuff. *Beef* (Intertec Publishing, Minneapolis, USA) 26(6), 14–18.

Grandin, T. (1993) Behavioral agitation during handling of cattle is persistent over time. *Applied Animal Behavior Science* 36, 1–9.

Grandin, T. (1997) Assessment of stress during handling and transport. *Journal of Animal Science* 75, 249–257.

Grandin, T. (1998a) Handling methods and facilities to reduce stress on cattle. *Veterinary Clinics of North America: Food Animal Practice* 14, 325–341.

Grandin, T. (1998b) Calm and collected. *Beef* (Minneapolis, Minnesota) March, 74–75.

Grandin, T. (1998c) Can acting like a predator produce low stress cattle handling? *Cattleman* (Winnipeg, Manitoba, Canada) October, 42.

Grandin, T. (1999) Safe handling of large animals. *Occupational Medicine: State of the Art Reviews* 14, 195–212.

Grandin, T. and Deesing, M.J. (1998) Behavioral genetics and animal science. In: Grandin, T. (ed.) *Genetics and the Behavior of Domestic Animals.* Academic Press, San Diego, California. pp. 1–30.

Grandin, T., Deesing, M.J., Struthers, J.J. and Swinker, A.M. (1995) Cattle with hair whorls above the eyes are more behaviorally agitated during restraint. *Applied Animal Behaviour Science* 46, 117–123.

Harger, C.M. (1928) *Frontier Days.* Macrae Smith, Philadelphia, Pennsylvania.

Hasker, P.J.S. and Hirst, D. (1987) Can cattle be mustered by audio conditioning? *Queensland Agriculture Journal* 113, 325–326.

Hassal, A.C. (1974) Behavior patterns in beef cattle in relation to production in the dry tropics. *Proceedings of the Australian Society of Animal Production* 10, 311–313.

Hearnshaw, H., Barlow, R. and Want, G. (1979) Development of a 'temperament' or handling difficulty score for cattle. *Proceedings Australian Association of Animal Breeding and Genetics* 1, 164–166.

Hedigar, H. (1968) *The Psychology and Behavior of Animals in Zoos and Circuses.* Dover Publications, New York.

Heffner, R.S. and Heffner, H.E. (1983) Hearing in large mammals: horse (*Equus caballus*) and cattle (*Bos taurus*). *Behavioral Neuroscience* 97(2), 299–309.

Heffner, R.S. and Heffner, H.E. (1992) Hearing in large mammals: sound-localization acuity in cattle (*Bos taurus*) and goats (*Capra hircus*). *Journal of Comparative Psychology* 106, 107–113.

Herrman, J. (1998) Patience and silence are the major virtues in successful stockherders. *Stockman Grass Farmer* (Mississippi Valley Publishing, Ridgeland, Mississippi) October, 55(10), 1, 8.

Hough, E. (1958) The roundup. In: Neider, C. (ed.) *The Great West.* Coward-McCann, New York.

Houpt, K. and Wolski, T.R. (1982) *Domestic Animal Behavior for Veterinarians and Animal Scientists.* Iowa State University Press, Ames, Iowa.

Hutson, G.D. (1982) Flight distance in Merino sheep. *Animal Production* 35, 231–235.

Jacobs, G.H., Deegan, J.F. and Neitz, J. (1998) Photopigment basis for dichromatic colour vision in cows, goats and sheep. *Visual Neuroscience* 15, 581–584.

Johnston, J.D. and Buckland, R.B. (1976) Response of male Holstein calves from seven sires to four management stresses as measured by plasma corticoid levels. *Canadian Journal of Animal Science* 56, 727–732.

Julian, S. (1978) Gentle on the range. *Western Livestock Journal* (Denver, Colorado) November, 24–30.

Kilgour, R. (1971) Animal handling in works: pertinent behavior studies. In: *Proceedings of the 13th Meat Industry Research Conference.* Hamilton, New Zealand, pp. 9–12.

Kilgour, R. and Dalton, C. (1984) *Livestock Behavior: a Practical Guide.* Granada Publishing, Frogmore, St Albans, UK.

Lanier, E.K., Friend, T.H., Bushong, D.M., Knabe, D.A., Champney, T.H. and Lay, D.G. (1995) Swim habituation as a model for eustress and distress in the pig. *Journal of Animal Science* 73 (Suppl. 1), 126 (abstract).

Lanier, J. (1999) The effect of gender, breed and hair whorl position on the reaction of cattle to stimulus associated with an auction ring. Masters thesis, Colorado State University.

Lanier, J.L., Grandin, T., Green, R.D., Avery, D and McGee, K. (1999a) The effect on cattle of sudden intermittent movements and sounds associated with an auction ring. In: *1999 Beef Program Report* Colorado State University, Fort Collins, Colorado, pp. 225–230.

Lanier, J.L., Grandin, T., Chaffin, A. and Chaffin, T. (1999b) Training American bison calves 1999. *Bison World* 24 (4), 94–99.

Lanier, J.L., Grandin, T., Green, R.D., Avery, D. and McGee, K. (2000) The relationship between reaction to sudden intermittent movements and sounds to temperament. *Journal of Animal Science* (in press).

LeDoux (1996) *The Emotional Brain.* Simon and Schuster, New York.

Lemmon, W.B. and Patterson, G.H. (1964) Depth perception in sheep: effects of interrrupting the mother–neonate bond. *Science* 145, 835–836.

LeNeindre P., Boivin, X. and Boissy, A. (1996) Handling extensively kept animals. *Applied Animal Behavior Science* 49, 73–81.

Linford, L. (1977) Stampede. *New Mexico Stockman* 42(11), 48–49.

Lott, D.F. and Hart, B. (1977) Aggressive domination of cattle by Fulani herdsmen and its relation to aggression in Fulani culture and personality. *Ethos* 5, 172–186.

Lott, D.F. and Hart, B. (1979) Applied ethology in a nomadic cattle culture. *Applied Animal Ehtology* 5, 309–319.

Lott, D.F. and Hart, B.L. (1982) The Fulani and their cattle: applied behavioral technology in a nomadic cattle culture and its psychological consequences. *National Geographic Research Reports* 14, 425–430.

Lynch, J.J. and Alexander, G. (1973) *The Pastoral Industries of Australia.* University Press, Sydney, Australia, pp. 371–400.

McDonald, P. (1981) Thinking like a cow makes the job easier. *Beef* (Intertec Publishing, Minneapolis, Minnesota) October pp. 81–86.

Miller, C.P., Wood-Gush, D.G.M. and Martin, P. (1986) The effect of rearing systems on the responses of calves to novelty. *Biology of Behaviour* 11, 50–60.

Miller P.E. and Murphy, C.J. (1995) Vision in dogs. *Journal of the American Veterinary Medical Association* 12, 1623–1634.

Nation, A. (1998) Western grazers practice management intensive grazing without fences. *Stockman Grass Farmer* September. Mississippi Valley Publishing, Ridgeland, Mississippi, pp. 1–12.

Nation, A. (1999) Herding is not as easy as it looks. *Stockman Grassfarmer* (Mississippi Valley Publishing, Ridgeland, Mississippi) September, pp. 1–13.

Orihuelo, J.A. and Solano, J.J. (1994) Relationship between order of entry in slaughterhouse raceway and time to traverse raceway. *Applied Animal Behaviour Science* 40, 313–317.

Paine, R. (1994) *Herds of the Tundra: a Portrait of Saami Reindeer Pastoralism.* Smithsonian Institute Press, Washington, DC.

Peischel, P.A., Schalles, R.R. and Owenby, C.E. (1980) Effect of stress on calves grazing Kansas Hills range. *Journal of Animal Science* Suppl. 1, 24–25 (abstract).

Phillips, M., Grandin, T., Graffam, W., Irlbeck, N.A. and Cambre, R.C. (1998) Crate conditioning of Bongo (*Tragelephus eurycerus*) for veterinary and husbandry procedures at Denver Zoological Gardens. *Zoo Biology* 17, 25–32.

Pick, D.F., Lovell, G., Brown, S. and Dail, D. (1994) Equine color vision revisited. *Applied Animal Behavior Science* 42, 61–65.

Price, E.O. (1998) Behavioral genetics and the process of animal domestication. In: Grandin, T. (ed.) *Genetics and the Behavior of Domestic Animals*. Academic Press, San Diego, California, pp. 31–65.

Price, E.O. and Wallach, S.J.R. (1990) Physical isolation of hand reared Hereford bulls increases their aggressiveness towards humans. *Applied Animal Behaviour Science* 27, 263–267.

Price, S., Sibley, R.M. and Davies, M.H. (1993) Effects of behavior and handling on heart rate in farmed red deer. *Applied Animal Behavior Science* 37, 111–123.

Prince, J.H. (1977) The eye and vision. In: Swenson, M.J. (ed.) *Dukes Physiology of Domestic Animals*. Cornell University Press, New York, pp. 696–712.

Randle, H.D. (1998) Facial hair whorl position and temperament in cattle. *Applied Animal Behavior Science* 56, 139–147.

Ray, D.E., Hansen, W.J., Theurer, B. and Stott, G.H. (1972) Physical stress and corticoid levels in steers. *Proceedings Western Section, American Society of Animal Science* 23, 255–259.

Rehkamper, G. and Gorlach, A. (1998) Visual identification of small sizes by adult dairy bulls. *Journal of Dairy Science* 81, 1574–1580.

Reinken, G. (1988) General and economic aspects of deer farming. In: Reid, H.W. (ed.) *The Management and Health of Farmed Deer*. London, pp. 53–59.

Ried, R.L. and Mills, S.C. (1962) Studies of carbohydrate metabolism of sheep. XVI The adrenal response to physiological stress. *Australian Journal of Agricultural Research* 13, 282–294.

Roche, B.W. (1988) Coacher mustering. *Queensland Agricultural Journal* 114, 215–216.

Rogan, M.T. and LeDoux, J.E. (1996) Emotion: systems, cells and synaptic plasticity. *Cell* (Cambridge, Massachusetts) 83, 369–475.

Saslow, C.A. (1999) Factors affecting stimulus visibility in horses. *Applied Animal Behaviour Science* 61, 273–284.

Savory, A. (1978) Ranch and range management using short duration grazing. In: *Beef Cattle Science Handbook*. Agriservices Foundation, Clovis, California, pp. 376–379.

Schwabe, C.W. (1985) A unique surgical operation on the horns of African bulls in ancient and modern times. *Agricultural History* 58, 138–156.

Schwabe, C.W. and Gordon, A.H. (1988) The Egyptian w3s-sceptor and its modern analogue: uses in animal husbandry, agriculture and surveying. *Agricultural History* 62, 61–89.

Seligman, C.G. and Seligman, B.Z. (1932) *Pagan Tribes of the Nilotic Sudan.* Routledge and Kegan Paul, London.

Smith, B. (1998) *Moving 'Em: a Guide to Low Stress Animal Handling.* Graziers Hui, Kamuela, Hawaii.

Smith, B., Leung, P. and Love, G. (1986) *Intensive Grazing Management: Forage, Animals, Men, Profits.* Graziers Hui, Kamuela, Hawaii.

Smith, S. and Goldman, L. (1999) Color discrimination in horses. *Applied Animal Behaviour Science* 62, 13–25.

Stephens, D.B. and Toner, J.N. (1975) Husbandry influences on some physiological parameters of emotional responses in calves. *Applied Animal Ethology* 1, 233–243.

Stermer, R., Camp, T.H. and Stevens, D.G. (1981) *Feeder Cattle Stress During Transportation*. Paper No. 81–6001, American Society of Agricultural Engineers, St Joseph, Michigan.

Stricklin, W.R., Heisler, C.E. and Wilson, L.L. (1980) Heritability of temperament in beef cattle. *Journal of Animal Science* 51 (Suppl. 1), 109 (abstract).

Talling, J.C., Waran, N.K. and Wathes, C.M. (1996) Behavioral and physiological responses of pigs to sound. *Applied Animal Behavior Science* 48, 187–202.

Talling, J.C., Waran, N.K., Wathers, C.M. and Lines, J.A. (1998) Sound avoidance by domestic pigs depends upon characteristics of the signal. *Applied Animal Behavior Science* 58(3–4), 255–266.

Thines, G. and Soffie, M. (1977) Preliminary experiments on color vision in cattle. *British Veterinary Journal* 133, 97–98.

Tillman, A. (1981) *Speechless Brothers: the History and Care of Llamas*. Early Winters Press, Seattle, Washington.

Tulloh, N.M. (1961) Behavior of cattle in yards. II. A study of temperament. *Animal Behavior* 9, 25–30.

Uetake, K. and Kudo, Y. (1994) Visual dominance over hearing in feed acquition procedure of cattle. *Applied Animal Behavior Science* 42, 1–9.

Voisinet, B.D., Grandin, T., Tatum, J.D., O'Connor, S.F. and Struthers, J.J. (1997) Feedlot cattle with calm temperaments have higher average daily gains than cattle with excitable temperaments. *Journal of Animal Science* 75, 892–896.

Ward, F. (1958) *Cowboy at Work*. Hasting House Publishers, New York.

Waynert, D.E., Stookey, J.M., Schwartzkopf-Gerwein, J.M., Watts, C.S. and Waltz, C.S. (1999) Response of beef cattle to noise during handling. *Applied Animal Behavior Science* 62, 27–42.

Western Livestock Journal (1973) 'Put on' foils escape minded cattle. *Western Livestock Journal* (Denver, Colorado) July, 65–66.

Williams, E. (1998) How to place livestock and have them stay where you want them. *Stockman Grass Farmer* (Mississippi Valley Publishing, Ridgeland, Mississippi) September, pp. 6–7.

Wyman, W.D. (1946) *The Wild Horse of the West*. Caxton Printers, Caldwell, Idaho.

Wythes, J.R. and Shorthose, W.R. (1984) Marketing Cattle: its Effect on Live Weight and Carcass Meat Quality. Australian Meat Research Committee Review no. 46 Australian Meat and Research Corporation, Sydney, Australia.

Zavy, M.T., Juniewicz, P.E., Williams, A.P and Von Tungeln, D.L. (1992) Effect of initial restraint, weaning and transport stress on baseline and ACTH stimulated cortisol responses in beef calves of different genotypes. *American Journal of Veterinary Research* 53, 552–557.

# Handling Cattle in Intensive Systems | 6

## Roger Ewbank

*Formerly Director of Universities Federation for Animal Welfare, 19 Woodfield Road, Ealing, London, UK*

## Introduction

Cattle have been domesticated by humans for several thousands of years and they are currently used in very large numbers as sources of meat and/or milk and for draught purposes. As a result of this long and close contact, there is great practical knowledge within the agricultural community as to their management, care and handling.

Information on the behaviour of cattle – an important component of any understanding of how they can be rationally handled – is to be found in Hafez and Bouissou (1975), Phillips (1993) and Albright and Arave (1997). Ewbank (1968), Grandin (1987, 1998) and Lawrence (1991) give accounts of the behaviour of animals in restraint, and the general behaviour and welfare of farm animals, with much attention to bovines, is discussed in Fraser and Broom (1990).

## Effect of Genetics and Experience

Many different breeds of cattle have been developed (Epstein and Mason, 1984) and it is generally accepted that these can be grouped into two main types: the humped (zebu) (*Bos indicus*) animal and the non-humped (*Bos taurus*) cattle of European origin. There are some behavioural differences between these cattle types (Hafez and Bouissou, 1975) and it is believed, without much real evidence, that they react differently to handling.

The husbandry systems under which they have been reared is probably more significant than the effects of any genetic differences between the breeds or types themselves (Murphey *et al.*, 1980; Kabuga and Appiah, 1992). It is

©CAB *International* 2000. *Livestock Handling and Transport,* 2nd edn
(ed. T. Grandin)

held as a general rule, however, that within European (*B. taurus*) cattle, females of the dairy breeds are quieter to handle than females of the beef breeds, but that the converse is true for the males.

The most important factor which influences the ease, or otherwise, with which cattle (or any other creature) can be handled is the extent and severity of their previous experiences of contact with humans. For a general account of the effects of this human–animal relationship in stock farming, see Hemsworth and Coleman (1998). It is widely believed, and there is much field evidence to support the view, that cattle which have had frequent, gentle and early contact with humans are usually tame and easy to handle. This early contact may have to be over an extended period if it is to have a long-term effect. Handling restricted to the first month of the calf's life (Sato *et al.*, 1984) seemed to have little influence on its subsequent behaviour; the regular handling of heifers up to the age of 9 months (Bouissou and Boissy, 1988), however, was shown permanently to reduce their fear of humans. A detailed discussion on the effects of early experience on farm livestock is to be found in Creel and Albright (1987).

Young cattle are inquisitive, active and playful, while older animals are more placid, but both types can be readily trained (Albright, 1981; Lemenager and Moeller, 1981; Dickfos, 1991). For general accounts of learning and training in the domesticated species, see Kilgour (1987) and Universities Federation for Animal Welfare (1990).

## Types of Intensive Systems

In the intensive, traditional, family-farm systems of stock-keeping, small numbers of cattle are reared and kept in close contact with humans. They are grazed under careful supervision in small fields or on tethers and/or are kept indoors in tie-up cowsheds or in covered yards. Their calves are often hand-reared by humans or, if kept on their dams, are habituated to the near presence of humans from an early age. As a general rule, cattle husbanded in these ways are tame, cooperative and easily handled. It seems that they associate handling with particular locations (Rushen *et al.*, 1998) and that they may, at times, differentiate between different human handlers (Taylor and Davis, 1998).

In the extensive systems of stock farming, cattle are kept in considerable numbers in large fields or on free range; their welfare is supervised by humans but the animals are relatively untamed and only occasionally handled, and then mainly in groups in gathering pens (corrals), races and crushes.

The modern moves towards large-scale intensification in cattle-keeping methods have meant that on many big farms, nowadays, large numbers of animals are looked after by small numbers of attendants. Many of the regular stock tasks, such as feeding, milking and dung clearing, are at least partly carried out by machines; calves and young stock are not often as closely handled as they are on the traditional family farm and there is little time for individual attention to any one animal. As a result of the low need for close regular physical contact between humans and cattle, there has, perhaps, been a tolerance of uncooperative behaviour and less incentive to select for quietness in the stock. Whatever the causes, it is widely held that many of the cattle kept in the large-scale intensive systems are increasingly seen to be nervous, uncooperative and difficult to control.

Cattle in the intensive systems can be handled in four different ways:

1. While being confined closely together as a group in a small space.
2. In groups via forcing pens, races and crushes (squeeze chutes).
3. While being held in their own tie-up stall, etc.
4. Individually by means of ropes and pieces of specially designed equipment.

## Confined closely together as a group in a small pen

In this method of handling, the animals, which can be young stock, polled/dehorned dairy cows or intensively housed beef cattle, are put into a restricted area (pen) such that they are in close contact with each other and have little space for movement. Once the cattle have settled down, the human handler(s) can enter the pen, push between the animals and catch and/or handle them at will. The animals usually remain quiet. This is possible because: (i) they have little scope for individual movement; and/or (ii) they remain relaxed while in close physical contact with their peers. An animal separated from its group soon becomes agitated and difficult to handle.

## In groups via forcing pens, races and crushes

In many intensive dairy- and beef-cattle enterprises, the animals are regularly passed through specially designed handling units, which are made up of various combinations of funnels, forcing (gathering) pens, races, crushes (squeeze chutes), shedding gates and automatic cow traps (Fig. 6.1). Some of the layouts incorporate foot-baths, weighing crates and/or loading ramps.

For dairy cattle these facilities are usually placed so that, each time the animals leave the milking area, they traverse all or part of the handling unit on their way back to the feeding/resting areas or the grazing fields. Beef cattle will often be regularly passed – once a week, say – through the foot-bath and/or weigher. In both types of enterprise, the animals are used to being routinely handled while still being in contact with other members of their social (pen) group.

If the facilities have been well designed and constructed – there are numerous advisory booklets and pamphlets on these matters (e.g. Shepherd, 1972; Graves, 1983; Ministry of Agriculture, Fisheries and Food, 1984;

**Fig. 6.1.** Automatic cow traps.

Gilbert, 1991) – and the stockmen and women are skilled and patient, the handling can often be efficient, relatively stress-free to both human and animal and humane.

The terminal part of the handling unit is usually some form of so-called cattle crush (restraint device). These devices are essentially a solid or bar-sided box designed and constructed to hold one animal. There is a vertically pivoted release gate on the front, and a sliding back gate (or bars) which can be pushed across the race at the rear, to stop the animal backing away. The animal's neck can be either caught in a yoke (stanchion) which is built into the front gate or held between two vertically pivoted side panels or bars set so that, when they swing out, they trap the neck of the animal between them as it stands just short of the front of the crush. The yoke can only be used if the animal volunteers – or can be persuaded – to put its head through the gap in the front gate; successful use of the side panel/bar neck restrainer does not depend on this kind of cooperation.

Sometimes cattle restrained in crushes become very stubborn, put their heads down and refuse to move and even adopt a so-called kneeling submissive position (Ewbank, 1961) – a possible form of tonic immobility (Fraser, 1960; see also later).

Once the head is held, various hinged panels (gates) in the sides of the crush can be swung open to gain access to the body of the animal (Fig. 6.2). Some of the more complex handling crates have one of the sides so pivoted

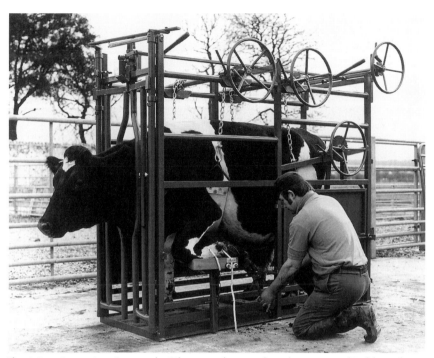

**Fig. 6.2.** Foot-trimming crush. (Photograph reproduced courtesy of Poldenvale Ltd, Taunton, UK.)

(Fig. 6.3) that it can be moved in to squeeze the animal and hold it tight for restraint, and some are even equipped with leg winders (see Fig. 6.2), so that the feet of the animal can be readily restrained for examination and treatment.

At times it is necessary to take a cow out of the crush and handle it in the adjacent open area. To do this a halter should be put on its head, the front gate opened and the animal led forward. This manoeuvre is easy with a side-panel/bar head-restraint crush and with a front-gate yoke in which the neck of the animal is being held between the gate and the frame of the crush (Fig. 6.4). In both these cases, the head can be released while the animal is standing still. In the central yoke gate design (Fig. 6.5), however, the cow has first to be driven back clear of the gate and then the gate swung open.

Cattle, especially those which routinely traverse a group housing system, will often readily follow each other through the funnels, forcing pens and races. Once an individual animal is held in the cattle crush, however, the forward movement of the group stops and can only start again when the 'crushed' animal is released. It may then be difficult to get the next animal in line to move forward and enter the crush. This is especially true if the previous handled animal had become in any way distressed. To overcome this start–stop–start problem, it is common practice for the handlers, who may be

**Fig. 6.3.** Modified arm-over-side-of-the-face headgrip being used on an adult cow held in a pivoted side-panel/bar yoke cattle crush. (Photograph reproduced courtesy of Industrial and Agricultural Engineers, Leek, UK.)

standing on a raised catwalk running alongside the run-up race, to attempt to examine or treat animals by reaching over the top of the sides of the crush and the run-up race. In this technique, only the heads and the top-lines of the cattle are directly accessible to the stockmen and women. If necessary, the heads of the animals can be held by hand; the animals usually stand still as they have little space for body movement and they feel relatively secure, being in close contact with other members of their group. Handling under these circumstances is really a linear form of 'being confined closely together as a group in a small space' – with the advantage that the human handler is safely outside the race (pen).

Once the cattle are released from the crush via the front gate, they can be diverted via sorting gates into side pens, which may contain automatic cow traps or neck yokes set at the feed troughs, or they can be allowed to return to the grazing fields or their housed accommodation.

### Handled while being held in their own tie-up stalls

The advantages for the stockperson in handling cattle while they are being held in their own tie-up stalls (stanchions) are as follows:

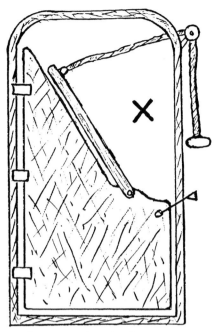

**Fig. 6.4.** Diagrammatic front gate of crush. The cow is expected here to put its head through the gap at X, that is, between the pivoted neck bar and the frame.

**1.** The animal is already caught, i.e. it does not have to be chased and captured.
**2.** The head of the animal is being held in either the chain tie or the neck yoke.
**3.** The animal is in its own living place and is between known companions in the adjacent stalls, i.e. it is usually relaxed and confident.
**4.** The husbandry system is generally such that the animal is used to frequent physical contact with humans while it is being fed/milked/having its bedding replaced, etc.
**5.** It is surrounded by bars and pipes to which the various ropes, etc. used in restraint (see later) can be attached.

There are also a number of disadvantages:

**1.** There may not be much space around the animal. Cattle in adjacent stalls can be moved away, but this often distresses the one left isolated.
**2.** The floor is probably hard (concrete) and the cow may damage itself if it falls.
**3.** It may begin to associate its own stanchion with unpleasantness and become reluctant to enter it again. This is especially true if painful operations, e.g. foot dressing, are repeatedly carried out in the home stall. It may be better to remove the animal to a 'neutral' place for potentially distressing procedures.

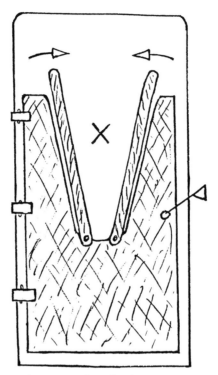

**Fig. 6.5.** Diagrammatic front gate of crush. The cow is expected to put its head through the gap at X, that is, into the V notch in the front gate.

### Handled individually by means of ropes and specially designed equipment

There are many descriptions of the way ropes and specially designed pieces of handling equipment can be used to control and manipulate cattle (Miller and Robertson, 1959; Battaglia and Mayrose, 1981; McNitt, 1983; Holmes, 1991; Battaglia 1998). There are also a number of veterinary texts (e.g. Leahy and Barrow, 1953; Stober, 1979a; Aanes, 1980; Kennedy, 1988; Fowler, 1995) which give details of the numerous specialized handling and surgical restraint techniques which can be used, as appropriate, on calves, young stock and adult cattle.

## Common Factors

Whatever technique is used, there are a number of common factors which may come into play.

## The place

The cattle should be handled in a place where they are likely to suffer little damage if they escape or go down. The area should, if possible, be enclosed, i.e. escape-proof, the fences/walls free of sharp projections and the floor non-slip and, if made of concrete or other hard material, well bedded with straw or other litter. Ideally, the animal should be in the centre of a pen or large loose box and beneath overhead attachment points for ropes or slings, although in practice it may, of necessity, have to be in a tie-up stall, in a cow cubicle, in the space alongside the cattle crush or in the area immediately adjacent to a tubular-metal pen side or wooden fence line.

## The equipment

The necessary equipment should be ready at hand and should have been well maintained and recently tested. Ropes, ideally, should be of cotton; they should have a fairly wide diameter and be soft-surfaced and free of knots. Webbing should be clean, pliable and not frayed. All leather parts should be soft (oiled) and not perished in any way. All metal components should be free of rust and dirt and have smooth surfaces.

## The human handlers

The persons handling the cattle should have both stock skills and stock sense. Workers on the intensive traditional family farms often have both these attributes; some of the animal attendants on the larger, more factory-like, intensive units can, at times, be good stock-system managers but poor animal handlers. The whole subject of stockmanship is complex – for details, see Seabrook (1987), English *et al.* (1992) and Hemsworth and Coleman (1998).

Regardless of the talents (or otherwise) of the handling team, its members should have been instructed and rehearsed in the particular techniques to be used. They should also have been made aware that one person – the team leader – makes the decisions as to when each step of the technique is or is not to be carried out. The efficiency of the handling and the safety of both the humans and the animals depend on the attitudes and skills of the handlers.

# Keeping the Animals as Calm/Quiet as Possible

Whenever possible, the cattle should be kept calm and relaxed. If this can be done, the handling will usually be quick and safe to both the humans and the animals.

Calmness in the animals and, indeed, in the human handlers will be encouraged by having them both, if possible, well trained. The animals should be handled in familiar surroundings. The equipment should be reliable and ready at hand.

Calmness/control in cattle can sometimes be encouraged by a number of factors.

### The use of blinds

Cattle can often be quieted for handling purposes by blindfolding them. The blindfold should be made of a soft, fully opaque material and be free from any foreign matter, e.g. sawdust or straw, which might enter and irritate the eyes. The arm-over-the-side-of-the-face head grip, as used on younger (smaller) stock is reputed to calm by a blindfold effect, as one of the animal's eyes is covered by the handler's arm and the other is pressed against the handler's body. When polled cattle are held in a neck stanchion in a squeeze chute, they can be easily restrained for a blood sampling or intravenous (IV) injection by a person leaning against their heads so that the person's rear covers their eyes. The head can be gently turned sideways (the person follows this movement) and the animal will remain calm because its eyes are covered still (T. Grandin, 1997, personal communication).

Bulls running free at pasture were sometimes equipped with specially constructed masks (Fig. 6.6), made so that they had to raise their heads in order to look forward. The mask was supposed to have a general quieting effect

**Fig. 6.6.**  Example of mask on dehorned bull used for calming the animal.

and to make it difficult for the bull to attack, in that, when it put its head down in the attack position, it could no longer see its potential victim. There is, however, much dispute as to whether any bull normally uses sight to correct its line of charge once its head is down and it is moving quickly forward.

## The use of chemical agents

Chemical agents are occasionally used to quieten cattle before they are handled. It would be very convenient if some drug could be put into the food or water of, say, an aggressive bull so that it was calm and cooperative by the time it was to be handled. However, there does not seem to be such a material. Chloral hydrate has been tried but it is not readily consumed by most cattle and, even when taken in, its action is somewhat uncertain. Many injectable agents have been used (Stober, 1979b; Heath, 1980; Thurman, 1986; Trim, 1987; Hall and Clarke, 1991). The current drug of choice is still probably xylazine (Rompun, Bayer) given by intramuscular injection. This agent should always be used under veterinary supervision. Low doses produce sedation; higher doses can sometimes cause the cattle to go down.

## The use of sound

Many stockmen and women claim that the human voice can be used to calm cattle and, indeed, this may be true if the animals and the humans are well acquainted and trust each other. Recorded sound (music) has sometimes been broadcast to cattle entering milking parlours in the belief that they can be more readily handled during the milking process. The benefits, if any, are probably due to the broadcast sound making the routine bangs and clatters produced by the parlour machinery less noticeable to the stock. It should always be remembered that, while cattle readily become habituated to sounds in their environment, they are easily startled by sudden, unexpected noises. See Chapter 5 for more information on the effects of sound. Recent research shows that loud shouting at cattle is very aversive (Pajor *et al.*, 1999).

## The use of pressure/force on the animal's body

### Applied to the whole body

It has long been recognized that the tightening of single or multiple rope loops, which have been placed round the bodies of cattle, e.g. udder kinches, chest twitches and Reuff's method of casting, cause the animals to stand still and, if tightened further, to appear partly paralysed and go down. If the rope loops are kept tight, the animals stay down. The reason as to why this works is largely unknown.

It may be that this partial paralysis is related to the so-called tonic immobility (fear paralysis) seen in some animals when pressure is applied to their bodies and/or when they are exposed to fear-inducing situations. This phenomenon has been studied under experimental conditions in chickens, rodents and rabbits and there is now some understanding (Carli, 1992) of the behavioural and physiological mechanisms involved. It is known that in rodents sudden and/or aversive stimuli can, at times, lead to tonic immobility and that this is sometimes accompanied by a drop in the animal's heart rate (Hofer, 1970; Steenbergen *et al.*, 1989). It has been noted in cattle (Leigh, 1937, p. 177) that the tightening of a rope round the body slows down the movement of the heart.

The quietness shown by animals in the squeeze crush (squeeze chute), when being restrained in close physical contact with others of their kind and while being held close together in handling pens, raceways or loose boxes may be at least partly a result of this 'pressure on the animal's body' phenomenon.

### Applied to part of the body

A rope noose tightened round the hind leg of a cow just above the hock joint or the application of a specially designed pincer-clip device to each side of the tendon that runs down to the point of the hock can result in the leg seemingly becoming temporarily paralysed. The animal itself is not really calmed and is often somewhat distressed, but the leg is easily handled. Occasionally the animal will go down, but this seems to be more a result of a loss of balance on the three non-affected legs, rather than any general calming/immobilization effect. The removal of the rope noose or the pincer clip soon results in the restoration of the animal's ability to move and use the leg, although there is sometimes a degree of swelling at the application site and an associated, temporary, slight lameness.

The holding of a cow or calf's tail in a near vertical position seems to partially restrict the movement of the hind legs and allows a relatively safe examination of the groin or udder region of animals that are otherwise unhandlable.

The grasping of the nose of the animal by placing a finger and thumb each side of the septum nasi or by the application of blunt-ended pincer-like tongs (bulldogs) to the same site usually results in the animal standing still and tolerating a certain amount of interference to the rest of its body. The impression is given that the animal's whole attention is concentrated on its nose. It has been suggested that it will only continue to stand still as long as the distress/pain of the interference is equal to or less than the distress/pain of its nose; once the interference distress is greater than the nose distress, control will be lost.

A major disadvantage of using a pincer is that the cattle remember the pain and they will resist application of the pincer in the future. In most instances, a halter or head collar can be used to restrain the head. The twitch used on the horse is a more subtle device, in that the loop of cord round the horse's upper lip can be relaxed and tightened at will and, in effect, the stimulus to the lip can be 'titrated' against the stimulus of the interference being carried out on other parts of the animal. In the horse, it has been suggested that the twitch has a somewhat similar action to acupuncture (Lagerweiz *et al.*, 1984), with an increase of circulatory endorphins causing possible changes in pain susceptibility and/or mood of the animal. It is not known whether there are increases in endorphin levels in the blood of cattle subjected to pressure/force being applied to all or part of their bodies.

The metal rings placed through the soft anterior parts of the septum nasi in bulls, to which various staffs, chains, etc., can be attached, probably help to control the animals by causing them discomfort/pain if they do not respond to the pushes and pulls exerted on the rings.

## Restraint of the head and neck

Most handling of cattle demands that the head/neck is restrained in some way. Animals which have their necks held in yokes or tie-chains in their individual standings (stanchions) are still able to move their heads about, and are usually capable of some limited movement of their whole bodies. This is also true of animals detained within the various designs of cattle crushes. Additional physical control of these animals is generally achieved by the grasping of the heads. This is usually done by a human handler, who should be standing to one side of the animal's head, taking hold of the horn or ear closest to him/her and, at the same time, seizing the septum nasi of the nose between a finger and thumb of the other hand. When this is done the cattle will usually keep their heads and bodies still. If the head is pulled sideways and back towards the animal's flank, i.e. 'rotated' around the chain tie/bar of the yoke, the cow will tend to swing itself away from the handler and, in many instances, steady itself by pushing the side of its body against the adjacent pen division, stall side or wall.

The arm-over-the-side-of-the-face head-grip technique should only really be used on young, relatively small, polled or dehorned stock. The handler, standing to one side of the animal's head, reaches over the head so that his/her arm covers the animal's far eye and cups his/her hand round the anterior part of the animal's upper lip. The handler's other hand goes to the side of the face and either grasps the lower jaw, with a finger or thumb pressing on the gum line just behind the incisor teeth, or seizes the septum nasi between his/her finger and thumb. The animal is restrained by a mixture of force, calming through the use of the arm as a blind on the smallest animals and distraction via the action of the hands/fingers inside the animal's mouth. A modified version of the arm-over-the-side-of-the-face head grip can be used on larger animals safely held by the neck in a stall stanchion (neck yoke) or a cattle crush head gate (see Fig. 6.3).

## Conclusions

Cattle being managed under intensive systems can be restrained by a whole variety of techniques and devices (see the standard texts recommended earlier). The efficiency and humaneness of the restraint depend on the stockpersons building up a routine which meets the purpose of the handling, utilizes the facilities available and is within the capacity of the handlers. An understanding of the advantages/disadvantages and the rationale of the various components common to most handling techniques is essential if success is to be assured.

Cattle kept under intensive systems are dependent on humans for the state of their health and well-being. Efficient and humane handling – a procedure very much under the control of the stockperson – can play an important part in ensuring that their welfare needs are met.

## References

Aanes, W.A. (1980) Physical restraint. In: Amstutz, H.E. (ed.) *Bovine Medicine and Surgery*, 2nd edn. vol. 2. American Veterinary Publications, Santa Barbara, California, pp. 1128–1148.

Albright, J.L. (1981) Training dairy cattle. In: *Dairy Science Handbook*, vol. 14, Agriservices Foundation, Clowis, USA, pp. 363–370.

Albright, J.L. and Arave, C.W. (1997) *The Behaviour of Cattle*. CAB International, Wallingford, UK.

Battaglia, R.A. (1998) *Handbook of Livestock Management Techniques*. Prentice Hall, Englewood Cliffs, New Jersey.

Battaglia, R.A. and Mayrose, V.B. (eds) (1981) *Handbook of Livestock Management Techniques*. Burgess, Minneapolis, Minnesota.

Bouissou, M.F. and Boissy, A. (1988) Effects of early handling on heifers' subsequent reactivity to humans and to unfamiliar situations. In: Unshelm, J., van Putten, S., Zeeb, K. and Ekesbo, I. (eds) *Proceedings of the International Congress on Applied Ethology in Farm Animals, Skara, 1988*. KTLB, Darmstadt, pp. 21–38.

Carli, G. (1992) Immobility responses and their behavioral and physiological aspects, including pain: a review. In: Short, C.E. and Van Poznak, A. (eds) *Animal Pain*. Churchill Livingstone, New York, pp. 543–554.

Creel, S.R. and Albright, J.L. (1987) Early experience. *Veterinary Clinics of North America – Food Animal Practice* 3, 251–268.

Dickfos, J.A. (1991) *Training Cattle for Scientific Experiments*. CSIRO, Rockhampton, Queensland, Australia.

English, P., Burgess, G., Segundo, R. and Dunne J. (1992) *Stockmanship: Improving the Care of the Pig and Other Livestock*. Farming Press, Ipswich, UK.

Epstein, H. and Mason, I.L. (1984) Cattle. In: Mason, I.L. (ed.) *Evolution of Domesticated Animals*. Longman, London, pp. 6–27.

Ewbank, R. (1961) The behaviour of cattle in crushes. *Veterinary Record* 73, 853–856.

Ewbank, R. (1968) The behavior of animals in restraint. In: Fox, M.W. (ed.) *Abnormal Behavior in Animals*. Saunders, Philadelphia, Pennsylvania, pp. 159–178.

Fowler, M.E. (1995) *Restraint and Handling of Wild and Domestic Animals,* 2nd Ed, Iowa State University Press, Ames, Iowa.

Fraser, A.F. (1960) Spontaneously occurring forms of 'tonic immobility' in farm animals. *Canadian Journal of Comparative Medicine* 24, 330–333.

Fraser, A.F. and Broom, D.M. (1990) *Farm Animal Behaviour and Welfare,* 3rd edn. Baillière Tindall, London.

Gilbert, J.E. (convenor) (1991) *Farm Cattle Handling.* Standing Committee on Agriculture Report No. 35, CSIRO, East Melbourne, Australia.

Grandin, T. (1987) Animal handling. *Veterinary Clinics of North America – Food Animal Practice* 3, 323–338.

Grandin, T. (1998) Safe handling of large animals. In: Langley, R.L. (ed.) *Occupational Medicine: State of the Art Reviews* 14, 195–212.

Graves, R.E. (1983) *Restraint and Treatment Facilities for Dairy Animals.* Special Circular 289, Pennsylvania State University Co-operative Extension Service, University Park, Pennsylvania.

Hafez, E.S.E. and Bouissou, M.F. (1975) The behaviour of cattle. In: Hafez, E.S.E. (ed.) *The Behaviour of Domestic Animals,* 3rd Edn. Baillière Tindall, London, pp. 203–245.

Hall, L.W. and Clarke, K.W. (1991) *Veterinary Anesthesia,* 8th edn. Baillière Tindall, London.

Heath, R.B. (1980) Chemical restraint and anesthesia. In: Amstutz, H.E. (ed.) *Bovine Medicine and Surgery,* 2nd edn, Vol. 2. American Veterinary Publications, Santa Barbara, California, pp. 1125–1128.

Hemsworth, P.H. and Coleman, G.J. (1998) *Human–Livestock Interactions. The Stockperson and the Productivity and Welfare of Intensively Farmed Animals.* CAB International, Wallingford, UK.

Hofer, M.A. (1970) Cardiac and respiratory function during prolonged immobility in wild rodents. *Psychosomatic Medicine* 32, 633–647.

Holmes, R.J. (1991) Cattle. In: Anderson, R.S. and Edney, A.T.B. (eds) *Practical Animal Handling.* Pergamon Press, Oxford, pp. 15–38.

Kabuga, J.D. and Appiah, P. (1992) A note on the ease of handling and the flight distance of *Bos indicus, Bos taurus* and their crossbreeds. *Animal Production* 54, 309–311.

Kennedy, G.A. (1988) Surgical restraint: cattle and sheep. In: Oehme, F.W. (ed.) *Textbook of Large Animal Surgery,* 2nd edn. Williams and Wilkins, Baltimore, Maryland, pp. 67–73.

Kilgour, R. (1987) Learning and training of farm animals. *Veterinary Clinics of North America – Food Animal Practice* 3, 323–338.

Lagerweiz, E., Nelis, P.C., Wiegart, V.M. and Van Ree, J.L. (1984) The twitch in horses: a variant of acupuncture. *Science* 225, 1172–1174.

Lawrence, A.B. (1991) Introduction: the biological basis of handling animals. In: Anderson, R.S. and Edney, A.T.B. (eds) *Practical Animal Handling.* Pergamon Press, Oxford, UK, pp. 1–13.

Leahy, J.R. and Barrow, P. (1953) *Restraint of Animals,* 2nd edn. Cornell Campus Store, Ithaca, New York

Leigh, M.M. (1937) *Highland Homespun.* Bell, London.

Lemenager, R.P. and Moeller, N.J. (1981) Cattle management techniques. In: Battaglia, R.A. and Mayrose, V.B. (eds) *Handbook of Livestock Management Techniques.* Burgess, Minneapolis, Minnesota, pp. 105–181.

McNitt, J.I. (1983) *Livestock Husbandry Techniques.* Granada, London.

Miller, W.C. and Robertson, E.D.S. (eds) (1959) *Practical Animal Husbandry,* 7th edn. Oliver and Boyd, Edinburgh.

Ministry of Agriculture, Fisheries and Food (1984) *Cattle Handling.* Booklet 2495, MAFF Publications, Alnwick, UK.

Murphey, R.M., Duarte, F.A.M. and Penedo, M.C.T. (1980) Approachability of bovine cattle in pastures: breed comparison and a breed × treatment analysis. *Behavior Genetics* 10, 171–181.

Pajor, E.A., Rushen, J. and de Passille, A.M.B. (1999) Aversion learning techniques to evaluate dairy cow handling practices. *Journal of Animal Science* 77 (Suppl. 1), 149 (abstract).

Phillips, C.J.C. (1993) *Cattle Behaviour.* Farming Press, Ipswich, UK.

Rushen, J., Munksgaard, L., de Passille, A.M.B., Jensen, M.B. and Thodberg, K. (1998) Location of handling and dairy cows' responses to people. *Applied Animal Behaviour Science* 55, 259–267.

Sato, S., Shiki, H. and Yamasaki, F. (1984) The effects of early caressing on later tractability of calves. *Japanese Journal of Zootechnical Science* 55, 322–338.

Seabrook, M. (ed.) (1987) *The Role of the Stockmen in Livestock Productivity and Management: Proceedings of a Seminar.* Report EUR 10982 EN, Commission of the European Communities, Luxemburg.

Shepherd, C.S. (1972) *Layout and Equipment for Handling Dairy and Suckler Cows.* Bulletin No. 157, Farm Buildings Departments, West of Scotland Agricultural College, Glasgow.

Steenbergen, J.M., Koolhaas, J.M. Stoubbe, J.M. and Bohus, B. (1989) Behavioural and cardiac responses to a sudden change in environmental stimuli: effect of forced shift in food intake. *Physiology and Behaviour* 45, 729–733.

Stöber, M. (1979a) Handling cattle – calming by mechanical means of restraint. In: Rosenberger, G. (ed.) *Clinical Examination of Cattle,* translated by Roy Mack from the German 2nd edn, 1977. Verlag Paul Parey, Berlin, pp. 1–28.

Stöber, M. (1979b) Handling cattle – sedation, immobilization and analgesia by the use of drugs. In: Rosenberger, G. (ed.) *Clinical Examination of Cattle,* translated by Roy Mack from the German 2nd edn, 1977. Verlag Paul Parey, Berlin, pp. 28–30.

Taylor, A.A. and Davis, H. (1998) Individual humans as discriminative stimuli for cattle (*Bos taurus*). *Applied Animal Behavior Science* 58, 13–21.

Thurman, J.C. (ed.) (1986) Anesthesia. *Veterinary Clinics of North America – Food Animal Practice* 2(3), 507–785.

Trim, C.M. (1987) Special anesthesia considerations in ruminants. In: Short, C.E. (ed.) *Principles and Practice of Veterinary Anesthesia.* Williams and Wilkins, Baltimore, Maryland, pp. 285–300.

Universities Federation for Animal Welfare (eds) (1990) *Animal Training: Proceedings of a Symposium.* UFAW, Potters Bar, UK.

# Handling Facilities and Restraint of Range Cattle

**7**

## Temple Grandin

*Department of Animal Sciences, Colorado State University, Fort Collins, CO 80523, USA*

## Introduction

Even though extensively reared cattle have a large flight zone and are not completely tame they will become calmer and easier to handle if they are trained to seeing people on foot, on horseback and in vehicles. Doing this will have the added advantage of reducing agitation and stress when the animals are transported to a feedlot or a slaughter plant. Cattle which have been handled only by people on horses may become highly agitated when they first see a person on foot.

It is important that the first experience with a mounted rider or a person on foot is a good first experience. First experiences make a big impression on animals (Grandin, 1997a). Handling in a new set of corrals will be easier if the animal's first experiences with the facility are positive. The first time they enter the corral, they should be walked through it and fed. Ideally, this should be done several times before any actual work is done.

Facilities that are suitable for intensively raised tame cattle are not suitable for cattle reared on large ranches or properties in the USA, Australia or South America. Facilities that take advantage of the natural behavioural characteristics of cattle will reduce stress on the animals and improve labour efficiency.

## Corral Design

Round pens for gathering herding animals were invented over 11,000 years ago in the Middle East. The Syrians herded wild ungulates into a round pen to slaughter them (Legge and Conway, 1987). A round pen is efficient because

©CAB *International* 2000. *Livestock Handling and Transport,* 2nd edn
(ed. T. Grandin)

there are no sharp corners for animals to bunch up in. To prevent the animals from running back out of the entrance, the pen was shaped like a heart with the entrance between the shoulders. It is interesting that designs which are really effective keep being reinvented. The same design was used by cowboys to capture wild horses (Ward, 1958), and fishing nets are laid out in a similar manner. When gathering pens for cattle are being designed, sufficient space must be allocated. Minimum space is 2.3 m² for every cow (Daly, 1970) or 3.3 m² for every cow and calf pair. These space recommendations are for short-term holding of less than 24 h.

In rough country, cattle can be difficult to gather, and chasing them with helicopters, horses or vehicles is stressful and labour intensive. Cattle can be easily gathered by building corrals with trap gates around water sources (Cheffins, 1987). A trap gate acts as a valve, as the cattle can move through a closed trap gate in only one direction. Trap-gate designs are described by Howard (1953), Ward (1958) and Anderson and Smith (1980). Several weeks before gathering, the previously open trap gate is closed a little more each day. On the last day, the space between the ends of the two gates are so close together that a cow has to push them apart to get through. After she has gone through, she is unable to return. Cheffins (1987) describes an improved self-gathering system which has separate entrance and exit trap gates. Training the cattle is easier, because the animals become accustomed to moving in both directions through the trap gates. To make a self-gathering system work, all water sources must be enclosed with corrals equipped with trap gates. In areas with numerous water sources, corrals baited with molasses or palatable hay can be used to trap cattle. The animals will be easier to gather if the pasture conditions are poor. When the cattle eat the feed, they trip a trigger wire that closes the corral gate (Adcock *et al.*, 1986; Webber, 1987). A more modern version of this design would be to use a radio signal to close the gate.

## Sorting Facilities

Sorting cattle into age, sex or condition categories is an important handling procedure. Single-file races with sorting gates are used by ranchers in the south-western USA (Ward, 1958). These systems are similar to races used for sorting sheep. Another design utilizes a triangular gate to direct the cattle (Murphy, 1987). A person standing on a platform over the race can operate the gate. The triangular design facilitates stopping an animal that attempts to push its way past the gate. The disadvantage of both types of single-file sorting races is that it is difficult visually to appraise cattle as they rapidly move by. Many American ranchers hold cattle in a 3.5 m wide alley during sorting. Individual animals are separated from the group and directed to sorting pens by a person on the ground (Grandin, 1980a). This system works well with European breeds and makes visual appraisal easier, but it may work poorly with Brahman and zebu (*Bos indicus*) cattle, due to their greater tendency to bunch together. *Bos indicus* cattle are more reluctant to separate from a group.

The Australians developed pound yards (Daly, 1970), which are small round pens, 3–6 m in diameter with four to eight gates around the perimeter (Powell, 1986). A single person standing on an elevated platform over the pen can operate the gates. This system works well with zebu- or Brahman-cross cattle and visual appraisal is easy because the person has more time to look at the cattle. Other sorting systems are the hub sort (Oklahoma State University, 1973) and the circular alley sort (Arbuthnot, 1979).

## Corral Layouts

With the advent of truck transport, squeeze chutes (crushes for cattle restraint) and large feedlots, corral designs became more complex. Modern curved races and round crowd pens evolved independently in Australia (Daly, 1970), New Zealand (Kilgour, 1971; Diack, 1974) and the USA (Oklahoma State University, 1973). During the early to middle 1960s, the construction of large feedlots in Texas stimulated design of truly modern systems with curved single-file races, round crowd pens and long narrow diagonal pens (Paine *et al.*, undated). Prior to this time, ranchers used corrals with square pens, with little regard to behavioural principles.

Grandin (1980a) combined the best features of Texas feedlot designs with the round gathering pens from Ward (1958). Figure 7.1 illustrates a general-purpose corral. The wide curved lane serves two functions: (i) to hold cattle going to the loading ramp or squeeze; or (ii) as a reservoir for cattle which were being sorted back into the diagonal pens by a person on the ground. The wide curved lane facilitates moving cattle into the crowd pen. Each diagonal pen in Fig. 7.1 holds one truck-load of 45 mature cattle. The curved and diagonal layout avoids sharp 90° corners for cattle to bunch up in. For larger herds, additional diagonal pens can be added and the gathering pen can be enlarged. The corral is easy to build because the curved single-file race, round crowd pen and wide curved lane consist of three half circles with three radius points on a layout line (Fig. 7.2).

The Weean yard is used on many ranches in Australia (Thompson, 1987). It incorporates a curved single-file race and a pound yard for sorting in an economical, easy-to-build system (Fig. 7.3; Powell, 1986; Thompson, 1987). Powell (1986) contains many corral designs that are especially adapted to Australian conditions.

## Individual Animal Identification

Designs where cattle can be sorted after they leave the squeeze chute will become increasingly popular, as more and more people switch to individual

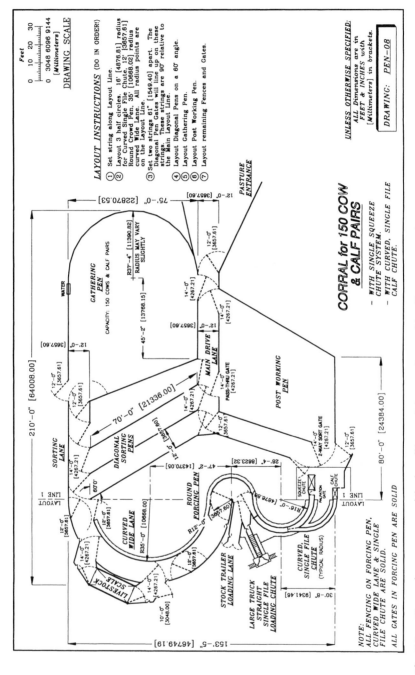

**Fig. 7.1.** General-purpose corral for handling weights, sorting and loading range cattle.

**Fig. 7.2.**   Curved single-file race, round crowd pen and curved lane.

animal identification (Fig. 7.4). Each animal can be weighed in a squeeze chute mounted on load cells, evaluated with ultrasound or other technology and then sorted. Facilities where cattle can be easily sorted when they leave the squeeze chute will be required when ranchers and feedlot operators sell cattle under contract, which will have strict specifications for fat thickness, frame score, weight and other specifications. The system in Fig. 7.4 also has the advantage of training cattle to go through the squeeze chute.

## Race, Crowd-pen and Loading-ramp Design

Single-file races, crowd pens (forcing pens) and loading ramps should have solid sides (Fig. 7.5; Rider *et al.*, 1974; Grandin, 1980a, 1997b). Solid sides in these areas help to keep cattle calmer and facilitate movement, because they prevent the cattle from seeing distractions outside the fence, such as people and vehicles. Solid sides are especially important when handling wild animals that are unaccustomed to close contact with people. The crowd gate that is closed behind the cattle must also be solid. Circular crowd pens and curved single-file races can reduce the time required to move cattle by up to 50% (Vowles and Hollier, 1982). A curved single-file race is more efficient because cattle cannot see people and motion up ahead when they enter the race. Another reason why a curved race is more efficient is that the cattle think they are going back to where they came from. Cattle also back up less in a curved race. Observations indicate that moving people are more threatening to cattle than people who stand completely still. A curved race provides the greatest

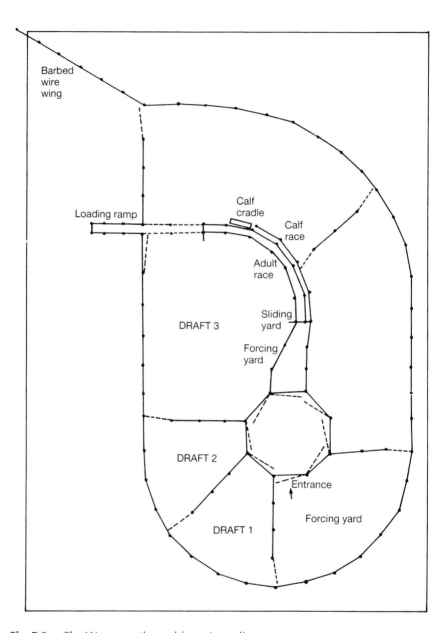

**Fig. 7.3.**   The Weean cattle yard from Australia.

advantage when cattle have to wait in line for vaccinations or other procedures. It provides no advantage when cattle run freely through the race without having to be detained for handling (Vowles *et al.*, 1984). The recommended inside radius for a curved single-file chute is 3.5–5 m and the

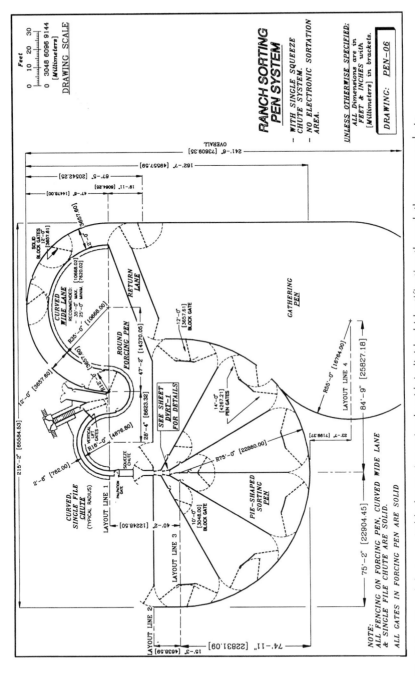

**Fig. 7.4.** Corral design suitable for sorting animals with individual identification through the squeeze chute.

**Fig. 7.5.** Curved single file race and round crowd pen with solid fences.

radius of the crowd pen should not exceed 3.5 m. Cattle bunch up if the crowd pen is too big. On funnel-shaped crowd pens one side should be straight and the other should be on a 30° angle. To reduce baulking, a curved single-file race must not be bent too sharply at the junction between the single-file race and the crowd pen. A cow standing in the crowd pen must be able to see two to three body lengths up the race. If the race appears to be a dead end the cattle will refuse to enter. Bending the single-file race too sharply where it joins the crowd pen is the most serious design mistake. Recommended dimensions for single-file chutes can be found in Grandin (1983a) and Midwest Plan Service (1987). If one side of the curved single-file race has to have open bars for vaccinating, the outer fence should be solid and the inner fence should be solid up to the 60 cm level. To avoid problems with cattle refusing to enter sheds or buildings, the wall of a building should never be located at the junction between the single-file race and the crowd pen (Grandin, 1980a, 1987).

Following behaviour can be used to facilitate cattle movement through a system. Cattle in adjacent single-file races will move when they see an animal in an adjacent race move. The outer walls are solid but cattle can see through the bars of the inner partitions. This principle was first used at the Swift Meat Plant in Arizona in 1974 and has also been used successfully by Grandin (1982, 1990a) for moving pigs and Syme (1981) for moving sheep. John Kerston in New Zealand moves cattle into two squeeze chutes (crushes) with two parallel races. When a cow is leaving one squeeze chute, the next cow is entering the empty squeeze chute on the other side (Andre, 1991). Andre (1991) also describes a system, designed by Roy Atherton, which has three, parallel, single-file races leading to a single squeeze chute. This system will work best if cattle are kept calm.

A crowd pen should be level and never be built on a ramp. Groups of animals that are standing still on a ramp will often pile up against the crowd gate and get trampled. When animals are handled on a wide ramp, they should be kept moving. However, cattle can stand safely in a single-file race which is on a ramp. Loading ramps must not be made too steep. The maximum recommended angle is 20–25°. On concrete ramps, stair steps are more efficient. A 10 cm rise and 30 cm tread is recommended. If cleats or ridges are used the spacing should be 20 cm to match the stride length of cattle (Mayes, 1978). Further information on loading ramps is given in Grandin (1990a) and Powell (1986).

## Things That Impede Cattle Movement

Corrals and races must be free of distractions that make animals baulk (see Chapter 20). Distractions such as a small piece of chain dangling in a race will make cattle baulk (Grandin, 1998). Distractions and lighting problems, such as a race entrance being too dark, can ruin the efficiency of a well-designed system. The author has observed that cattle often move more easily through outdoor corrals. If a building is built over a race, it should be equipped with either skylights or plastic side panels to let in light. White translucent panels are best because they let in lots of light but eliminate shadows. The ideal lighting inside a building looks like a bright cloudy day. Cattle tend to approach light. In one facility, cattle refused to enter the race unless a large door was left open to admit light. When the door was closed, cattle baulked and refused to enter the dark race entrance. Lamps can be used to attract cattle into buildings at night. The lighting must be indirect. The author has found that, in most facilities, cattle can be moved into squeeze chutes without electric prods. However, a lighting problem can make it almost impossible to move cattle quietly because they constantly baulk.

## Restraint Devices

Before the invention of squeeze chutes (crushes), range cattle were caught and restrained with a lariat (Ward, 1958). The invention of mechanized restraint devices both improved animal welfare and reduced labour requirements. They also required less skill to operate than a lariat. One of the first mechanical devices for restraining cattle was patented by Reck and Reck (1903). It had squeeze sides, which pressed against the animal, and a stanchion to hold the head. During the 1920s, Thompson (1931) developed a head-catching gate, designed for wild, horned cattle, which was installed on the end of a single-file race.

A squeeze chute restrains the animal with two devices. A stanchion is closed around the neck and side panels press against the animal's body to control movement (Fig. 7.6). They are available with either manual or hydraulic controls. The best squeeze chutes have two side panels which close in evenly on both sides. This design enables the animal to stand in a balanced position. An animal will struggle and resist being restrained if pressure is applied to only one side of the body. Today, most adult range cattle are restrained in a squeeze chute, but lariats are still used on some large ranches for restraining calves for branding and vaccinations. One of the first mechanical devices for holding calves was invented by Thompson and Gill (1949).

There are five different types of head gates (head bail or stanchions) for restraining cattle. They can be used either alone on the end of a single-file race or in conjunction with a complete squeeze chute. The five types are self-catch (Pearson, 1965), positive (Thompson, 1931; Heldenbrand, 1955), scissors-stanchion, pivoting and rotating head gates (Moly Manufacturing, Lorraine, Kansas, USA, 1995; Cummings and Son Equipment Company, Garden City, Kansas, USA, 1997). The advantage of pivoting or rotating head gates is that they open up to the full width of the squeeze and animals can exit more easily. These new designs may also help reduce shoulder injuries. A picture of the rotating head gate is in Grandin (1998). The Cummings and Son (1997) head gate requires very little force to restrain the animals compared with scissors-stanchion head gates. All types are available with either straight, vertical neck bars or curved neck bars. Stanchions with straight neck bars are

**Fig. 7.6.** Hydraulically operated squeeze chute (crush). Covering the open barred sides with cardboard will keep cattle calmer because it will prevent them from seeing people in their flight zone.

recommended for general-purpose uses, as they are less likely to choke an animal. Pressure on the carotid arteries exerted by a neck stanchion will quickly kill cattle (White, 1961; Fowler, 1995). Stanchions with straight, vertical neck bars are the safest because cattle can lie down while their neck is in the stanchion; there is no pressure on the carotid arteries. Curved bar stanchions provide a good compromise between control of head movement and safety for the animal. Positive head gates (Thompson, 1931), which clamp tightly around the neck, provide better head control but have an increased hazard of choking the cattle (Grandin, 1975, 1980b). Both positive-type head gates and a curved-bar stanchion must be used with squeeze sides or some other apparatus to prevent the animal from lying down. On chutes where the squeeze sides are hinged at the bottom, the width at the bottom must be narrow enough so that the V formed by the sides supports the animal's body in a standing position. A lift plate under the belly can also be used to support an animal (Marshall *et al.*, 1963). The squeeze sides must be designed so that there is no tendency to throw the animal off balance.

Hydraulically activated squeeze chutes have become increasingly popular. A properly adjusted hydraulic chute is safer for both people and cattle. Operator safety is improved because long, protruding, lever arms are eliminated. The pressure-relief valve must be properly set to prevent severe injury from excessive pressure. Additional information on proper adjustments can be found in Grandin (1980b, 1983b, 1990b). If an animal vocalizes (moos or bellows) when it is squeezed in a hydraulic chute, the pressure setting must be reduced. Newer models have a quiet pump and motor, whereas with older, noisy, hydraulic squeeze chutes the pump and motor should be removed and located away from the animal. Some new squeeze chutes have plastic inserts to reduce noise and prevent metal-to-metal contact when gates open and close.

Carelessness and rough handling are the major cause of injuries to cattle in squeeze chutes (Grandin, 1980a), but there is still a need to develop better restraint devices. Even under the best of conditions, bruises directly attributable to the squeeze chute occur in 2–4% of cattle. In one study, bruises occurred in five out of seven feedlots; 1.6–7.8% of the cattle had increased bruises compared with animals which were not handled in the squeeze chute (Brown *et al.*, 1981). Observations also indicate that cattle can be injured when the head gate is suddenly closed around the neck of a running animal. Cattle should be handled quietly so that they walk into a squeeze chute and walk out. Hitting the head gate too hard can cause haematomas and bruises. Shoulder meat may still be damaged when cattle are slaughtered.

## Electronic Measurement of Restraint-device Use

Progressive managers have found that quiet handling in squeeze chutes and reduction of electric prod use will enable cattle to go back on feed more quickly. As stated in Chapters 1 and 20, continuous evaluation and measurement of handling performance are essential to prevent workers from becoming rough. The technology is now available to electronically evaluate handling of cattle in a squeeze chute. Australian researchers Burrow and Dillon (1997) used a radar unit to measure the speed at which cattle left a squeeze chute. Animals that ran out at a high speed grew more slowly. Voisinet *et al.* (1997) found that cattle that became agitated in a squeeze chute had lower weight gains. Canadian researchers have developed ways to measure how hard cattle hit the head gate and how much animals jiggled the squeeze chute (Schwartzkopf-Genswein *et al.*, 1998). They recorded jiggling by recording signals from the load cells in the electronic scale, which was already under the squeeze. Electronic measurements of speed and jiggling could be easily correlated with weight gain, health records and feed conversion (Grandin, 1998). As more and more operations use electronic identification of cattle, electronic measurement of animal behaviour both in the squeeze chute and as the animal exits, will be easy to do. Vocalization scoring, described in Chapters 1 and 20, can also be used to evaluate handling in squeeze chutes.

## Behavioural Reactions to Restraint

One of the reasons for cattle becoming agitated in a squeeze chute is due to the operator being deep inside their flight zone. They can see him/her through the open bar sides. Cattle will remain calmer in a restraining device which has solid sides and a solid barrier around the head gate to block the animal's vision (Grandin, 1992). Cattle struggled less in a restraint device if their vision was blocked until they were completely restrained (Grandin, 1992). Cattle are less likely to attempt to lunge through the head opening if there is a solid barrier in front of the head opening which prevents them from seeing a pathway of escape. Restraint device designs that have been successfully used in slaughter plants could be adapted for handling on the ranch and feedlot (Marshall *et al.*, 1963; Grandin, 1992; see Fig. 20.10). Most cattle stood quietly when the Marshall *et al.* (1963) restraint device was slowly tightened against their bodies (Grandin, 1993). Solid sides prevented the animals from seeing the operator or other people inside their flight zone. Observations also indicate that cattle which are not accustomed to head restraint will remain calmer if body restraint is used in conjunction with head restraint. Head restraint without body restraint can cause stress (Ewbank *et al.*, 1992). More information on the design and operation of restraint devices can be found in Chapter 20.

Breeders of American bison prevent injuries and agitation by covering the open-barred sides of squeeze chutes and installing a solid gate (crash barrier) about 1–1.5 m in front of the head gate. Covering the sides of a squeeze chute so that the animal does not see the operator standing beside it will keep animals with a large flight zone calmer. When bison are handled, the top must also be covered to prevent rearing. Commercially available squeeze chutes are now available with rubber louvres on the sides to block the animal's vision

(Fig. 7.7). The louvres are mounted on a 45° angle and the drop bars on the side of the chute can still be opened. A cow's-eye view of a squeeze chute equipped with louvres is in Grandin (1998). Covering the open barred sides of a squeeze chute with cardboard will also result in calmer cattle.

Cattle will remain calmer if they 'feel restrained'. Sufficient pressure must be applied to hold the animal snugly, but excessive pressure will cause struggling due to pain. There is an optimal amount of pressure. If an animal struggles due to excessive pressure, the pressure should be slowly and smoothly reduced. Many people mistakenly believe that the only way to stop animal movement is to greatly increase the pressure. Sudden jerky movements of the apparatus will cause agitation, and smooth, steady movements help keep the animal calm (Grandin, 1992). Fumbling a restraining procedure will also cause excitement (Ewbank, 1968). It is important to properly restrain the animal on the first attempt. See Chapter 20 for more information on restraint.

## Dark-box Restraint

For artificial insemination and pregnancy testing, mechanical holding devices can be eliminated by using a dark-box race (Parsons and Helphinstine, 1969; Swan, 1975). It consists of a narrow stall with solid sides, a solid front and a solid top. Very wild cattle will stand still in the darkened enclosure. A cloth can be hung over the cow's rump to darken the chamber completely. Comparisons between a dark box and a regular squeeze chute with open-bar sides indicated

**Fig. 7.7.** Rubber louvres installed on the squeeze slide to prevent incoming cattle from seeing people.

that cows in the dark box were less stressed (Hale *et al.*, 1987). Further experiments indicated that cortisol levels were lower in the dark box, but heart-rate data were highly variable due to the novelty of the box (Lay *et al.*, 1992c). Blindfolding of both poultry and cattle reduces heart and respiration rates (Douglas *et al.*, 1984; Jones *et al.*, 1998; Don Kinsman, personal communication). Observations by Jennifer Lanier in our lab showed that blindfolds on American bison had to be opaque to have the greatest calming effect. Installation of a solid top on a squeeze chute also kept the bison calmer. If large numbers of cattle are inseminated, two or three dark boxes can be constructed in a herring-bone configuration (McFarlane, 1976; Canada Plan Service, 1984; Fig. 7.8). The outer walls are solid, with open-barred partitions between the cows. Side-by-side bodily contact helps to keep cows calmer (Ewbank, 1968). To prevent cattle from being frightened by a novel dark box, the animals should be handled in the box prior to insemination.

The effectiveness of dark-box restraint is probably due to a combination of factors, such as blocking the view of an escape route and preventing the animal from seeing people that are inside its flight zone. Darkness, however,

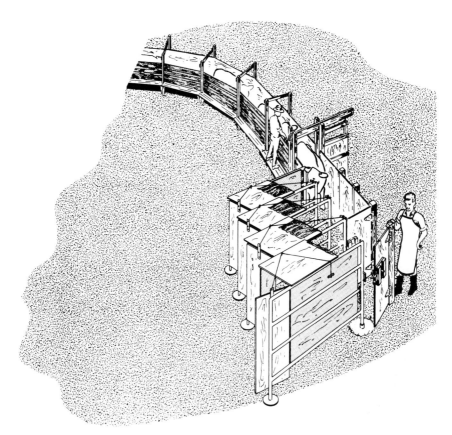

**Fig. 7.8.**   Dark box restraint races in a herring-bone configuration.

has a strong calming effect. Wild ungulates remain much calmer in a totally dark box. Small light leaks sometimes cause animals to become agitated. A well-designed dark box for domestic cattle has small slits in the front to admit light (Fig. 7.9). Cattle will enter easily because they are attracted to the light. For wild ungulates, it may be desirable to block the slits after the animal is in the box.

## Adaptation to Restraint

Cattle remember painful or frightening experiences for many months (Pascoe, 1986) so the use of aversive methods of restraint should be avoided. Restraint devices should be designed so that an animal is held securely in a comfortable upright position. If a restraint device causes pain, cattle will often become agitated and refuse to enter the device the next time it is used. For example, cattle restrained with nose tongs will toss their heads and be more difficult to restrain in the future compared with cattle restrained with a halter (Grandin, 1987, 1989a). When cattle are restrained with a halter for repeated blood testing, they will often learn to turn their heads voluntarily. Proper use of the halter is described in Holmes (1991).

Tame animals can be trained within 1 day to voluntarily enter a relatively comfortable restraint device for a feed reward (Grandin, 1984, 1989b). The restraint device should be introduced gradually and care must be taken to avoid hurting the animal. If the animal resists restraint, it must not be released until it has stopped struggling (Grandin, 1989a).

**Fig. 7.9.** A cow's-eye view into a well-designed dark box. Light slits attract the cow in. In some facilities the slits are covered after the cow is in the box.

Tame animals can be trained more quickly than wild animals. To reduce stress a wild animal must be trained and tamed over a period of days or weeks. The 1-day method (Grandin, 1989b) is recommended for tame animals. Wilder, more excitable cattle became increasingly agitated when they were repeatedly run through the squeeze chute in one afternoon. The wilder animals need time to calm down before the next training treatment. Grandin *et al.* (1995) and Phillips *et al.* (1998) found that many weeks were required to train nyala and bongo antelope to voluntarily enter a crate for injections and blood sampling. Each new sight or sound had to be introduced very gradually in tiny increments to avoid frightening the animals. Very low, almost baseline, cortisol levels of 4.4–8.5 ng ml$^{-1}$ were obtained (Phillips *et al.*, 1998). Their study indicates that training an animal to voluntarily cooperate greatly reduces stress.

Extensively raised beef cattle were restrained and had nine blood samples taken from the vena cava during a 16-day period. These animals had large reductions in cortisol (stress hormone) levels, and they became less excited as the days progressed (Crookshank *et al.*, 1979). They were blood-tested every day for the first 4 days and then 2–3 days apart for the remaining 12 days. The largest reduction in cortisol occurred after the first five tests (Crookshank *et al.*, 1979). It appears that four or five restraint sessions were required for the animals to become accustomed to the procedure. After a 5-day training period, which included only three restraint sessions, wild cattle still reacted to 20 min of restraint with steadily increasing cortisol levels up to 30 ng ml$^{-1}$ (Lay *et al.*, 1992a,b). Tame animals, such as dairy cattle, which have become accustomed to restraint devices will have a lower cortisol reaction. Possibly the wild cattle in Lay *et al.*'s (1992a) experiment had experienced some adaptation because their cortisol levels did not rise to the high of 63 ng ml$^{-1}$ which was recorded in a poorly designed slaughter facility (Cockram and Corley, 1991).

Stahringer *et al.* (1989) found that, in Brahman heifers, the excitable animals had higher levels of serum cortisol than calm heifers. Crookshank *et al.* (1979) also found that extensively reared calves that had been subjected to 12 h of trucking and weaning shortly before blood testing responded with increasing cortisol levels up to 46 ng ml$^{-1}$ for the first four blood tests and then levels dropped back down for the last five blood tests. Cattle that were not subjected to the added stress of weaning and transport had a peak level of only 24 ng ml$^{-1}$ during the entire experiment. One can tentatively conclude that subjecting cattle to closely spaced, multiple, stressful procedures will delay adaptation to handling.

## Learning and Restraint

The use of highly aversive methods of restraint such as electroimmobilization, is not recommended (Lambooy, 1985). An electronic immobilizer restrains an animal by tetanizing the muscles with electricity. There is no analgesic or anaesthetic effect (Lambooy, 1985). Application of the immobilizer to my arm felt like a disagreeable electric shock. Cows which had been immobilized had elevated heart rates 6 months later when they approached the chute where they had received the shock (Pascoe, 1986). A choice test in a Y-shaped race indicated that sheep preferred the tilt squeeze table to electroimmobilization. After one or two experiences, sheep avoided the race which led to the immobilizer (Grandin *et al.*, 1986). When a choice test is used to test aversiveness of restraint methods, naïve animals which have never been in the testing facility should be used. New cattle should be used for each test. Cattle which have developed a strong preference for one of the races will often refuse to switch races to avoid mildly aversive treatment, such as being gently restrained in a squeeze chute (Grandin *et al.*, 1994). Initially, they quickly learn to avoid the aversive side, but they often refuse to switch when the aversive treatment is switched to the other side. When the treatments are switched, the animal's brain registers the switch because the amount of looking back and forth at the decision point increases. However, cattle that had been accidentally banged on the head with the head gate were more likely to avoid the squeeze chute in the choice test (Grandin *et al.*, 1994). From a species survival standpoint, it makes sense to keep going down the previously learned safe path if something mildly aversive happens, such as being restrained gently by the head gate, but when something really aversive happens, such as being banged on the head or electrically immobilized, the animal will immediately switch paths to avoid the head gate. Deer show a similar reluctance to change. After 18 training sessions with no aversive treatment, deer still quickly entered a race after the first aversive treatment (Pollard *et al.*, 1992).

Cattle will learn to differentiate between a head stanchion that hits them on the head and a scale which causes no discomfort. Cattle that were handled five times became progressively more willing to enter a single-animal scale and somewhat less willing to enter a squeeze chute (Grandin, 1993). However, many of the animals that refused to enter the squeeze chute entered the squeeze section willingly but refused to place their head in the head-gate stanchion. They had learned that pressure on the body does not cause discomfort but the head-gate stanchion hurts when it slams shut. Cattle were also more likely to become agitated in the squeeze chute: 2% of the animals became agitated on the scale but 13% became agitated in the squeeze.

Research by Virginia Littlefield in our lab showed that cattle will habituate to repeated daily restraint in a squeeze chute if they are handled gently. The animals baulked less on each successive day and became less and less agitated in the squeeze chute. However, they will become harder and harder to drive into a squeeze chute if electric prods are used (Gooneswarden *et al.*, 1999). The animals in our study were all restrained in a chute with a stanchion-scissors head gate and care was taken to avoid banging them on the head. Some head-gate designs are more likely to be aversive to cattle than others. Poorly adjusted self-catching head gates may hurt the shoulders or put excessive pressure on the animal's neck. Pollard *et al.* (1992) made a similar observation in deer. Heart rate increased when the deer approached the second treatment for antler removal, but it decreased when they approached the second restraint only treatment.

The associations that animals make appear to be very specific. Tame sheep approached a familiar person more quickly than wild sheep. Practical experience has shown that cattle can recognize a familiar person's voice. Research with pigs indicates that they can recognize people by the colour of their clothing (Koba and Tanida, 1999). Behavioural measurements of struggling indicated that taming did not generalize to other procedures such as shearing or handling in a race (Mateo *et al.*, 1991). However, an animal's learning will generalize to another similar situation. Cattle which had been handled four times in the same squeeze chute and single-animal scale were able to recognize these items when they were handled at a different location with a slightly different scale and squeeze chute. Taming may reduce stress, even though the animal struggles during restraint. Free-ranging deer had cortisol levels that were more than double the levels in hand-reared deer during restraint (Hastings *et al.*, 1992). Both groups vocalized and actively resisted restraint. To reduce stress and improve welfare, livestock should be both acclimatized to people and specific procedures.

## Dip Vats

Pharmaceutical products, such as ivermectin, have replaced dipping for external parasites in most places. If dipping ever becomes required due to resistance to pharmaceutical agents, design information can be found in (Hewes (1975); Texas Agricultural Extension Service (1979); Grandin, (1980a,c); Sweeten (1980); Sweeten *et al.* (1982); Kearnan *et al.* (1982); Midwest Plan Service (1987).

## Conclusions

Curved races and round crowd-pen and corral layouts which eliminate square corners facilitate efficient and humane handling of cattle under extensive conditions. These systems utilize behavioural principles and work well. Proper layout is essential, as layout mistakes can ruin efficiency. Wild cattle will be calmer and less stressed if their vision is blocked by solid walls on races and restraint devices. Stress can be reduced by gentle handling, training animals and the use of relatively comfortable methods of restraint.

## References

Adcock, D., Kingstone, T. and Lewis, M. (1986) Trapping cattle with hay. *Queensland Agricultural Journal* 112, 243–246.

Anderson, D.M. and Smith, J.N. (1980) A single bayonet gate for trapping range cattle. *Journal Range Management* 33, 316–317.

Andre, P.D. (1991) Keep 'em coming. *Beef* July, 28–35.

Arbuthnot, N.J. (1979) *Livestock Marketing, Foundation for Technical Advancement*. Shire Engineer, Shire of Mildura, Victoria, Australia.

Brown, H., Elliston, N.G., McAskill, J.W. and Tonkinson, L.V. (1981) The effect of restraining fat cattle prior to slaughter on the incidence and severity of injuries resulting in carcass bruises. *Western Section, American Society of Animal Science* 33, 363–365.

Burrow, H.M. and Dillon, R.D. (1997) Relationships between temperament and growth in a feedlot and commercial carcass traits of *Bos indicus* crossbreds. *Australian Journal of Experimental Agriculture* 37, 407–411.

Canada Plan Service (1984) *Herringbone, A.I.* Breeding Chute Plan 1819, Agriculture Canada, Ottawa, Ontario, Canada.

Cheffins, R.I. (1987) Self mustering cattle. *Queensland Agricultural Journal* 113, 329–336.

Cockram, M.S. and Corley, K.T.T. (1991) Effect of pre-slaughter handling on the behavior and blood composition of beef cattle. *British Veterinary Journal* 147, 444–454.

Crookshank, H.R., Elissalde, M.H., White, R.G., Clanton, D.C. and Smolley, H.E. (1979) Effect of handling and transportation of calves upon blood serum composition. *Journal of Animal Science* 48, 430–435.

Daly, J.J. (1970) Circular cattle yards. *Queensland Agricultural Journal* 96, 290–295.

Diack, A. (1974) Design of round cattle yards. *New Zealand Farmer* 26 September, 25–27.

Douglas, A.G., Darre, M.D. and Kinsman, D.M. (1984) Sight restriction as a means of reducing stress during slaughter. In: *Proceedings 30th European Meeting of Meat Research Workers*. Bristol, UK, pp. 10–11.

Ewbank, R. (1968) The behavior of animals in restraint. In: Fox, M.W. (ed.) *Abnormal Behavior in Animals*. Saunders, Philadelphia, Pennsylvania. pp. 159–178.

Ewbank, R., Parker, M.J. and Mason, C.W. (1992) Reactions of cattle to head restraint at stunning: a practical dilemma. *Animal Welfare* 1, 55–63.

Fowler, M.E. (1995) *Restraint and Handling of Wild and Domestic Animals*, 2nd Edn. Iowa State University Press, Ames, Iowa.

Gooneswarden, L.A., Price, M.A., Okine E. and Berg, R.T. (1999). Behavioral responses to handling and restraint on dehorned and polled cattle. *Applied Animal Behavior Science* 64, 159–167.

Grandin, T. (1975) Survey of behavioral and physical events which occur in restraining chutes for cattle, Master's thesis, Arizona State University, Tempe, Arizona.

Grandin, T. (1980a) Observations of cattle behavior applied to the design of cattle handling facilities. *Applied Animal Ethology* 6, 19–31.

Grandin, T. (1980b) Good cattle restraining equipment is essential. *Veterinary Medicine and Small Animal Clinician* 75, 1291–1296.

Grandin, T. (1980c) *Safe Design and Management of Cattle Dipping Vats*. American Society of Agricultural Engineers, Paper No. 80–5518, St Joseph, Michigan.

Grandin, T. (1982) Pig behavior studies applied to slaughter plant design. *Applied Animal Ethology* 9, 141–151.

Grandin, T. (1983a) Design of ranch corrals and squeeze chute for cattle. In: *Great Plains Beef Cattle Handbook*. Cooperative Extension Service, Oklahoma State University, Stillwater, Oklahoma, pp. 5251.1–5251.6.

Grandin, T. (1983b) Handling and processing feedlot cattle. In: Thompson, G.B. and O'Mary, C.C. (eds) *The Feedlot*. Lea and Febiger, Philadelphia, Pennsylvania, pp. 213–236.

Grandin, T. (1984) Reduce stress of handling to improve productivity of livestock. *Veterinary Medicine* 79, 827–831.

Grandin, T. (1987) Animal handling. In: Price, E.O. (ed.) *Farm Animal Behavior, Veterinary Clinics of North America, Food Animal Practice* 3(2), 323–338.

Grandin, T. (1989a) Behavioral principles of livestock handling. *Professional Animal Scientist, American Registry of Professional Animal Scientists* 5(2), 1–11.

Grandin, T. (1989b) Voluntary acceptance of restraint by sheep. *Applied Animal Behavior Science* 23, 257–261.

Grandin, T. (1990a) Design of loading facilities and holding pens. *Applied Animal Behavior Science* 28, 187–201.

Grandin, T. (1990b) Chuting to win. *Beef* 27(4), 54–57.

Grandin, T. (1992) Observations of cattle restraint device for stunning and slaughtering. *Animal Welfare* (Universities Federation for Animal Welfare, Potters Bar, UK) 1, 85–91

Grandin, T. (1993) Previous experiences affect behavior during handling. *Agri-Practice* 14(4), 15–20.

Grandin, T. (1997a) Assessment of stress during handling and transport. *Journal of Animal Science* 75:249–257.

Grandin, T. (1997b) The design and construction of facilities for handling cattle. *Livestock Production Science* 49, 103–119.

Grandin, T. (1998) Handling methods and facilities to reduce stress on cattle. *Veterinary Clinics of North America (Food Animal Practice)* 14, 325–341.

Grandin, T., Curtis, S.E., Widowski, T.M. and Thurman, J.C. (1986) Electro-immobilization versus mechanical restraint in an avoid choice test. *Journal of Animal Science* 62, 1469–1480.

Grandin, T., Odde, K.G., Schutz, D.N. and Behrns, L.M. (1994) The reluctance of cattle to change a learned choice may confound preference tests. *Applied Animal Behavior Science* 39, 21–28.

Grandin, T., Rooney, M.B., Phillips, M., Irlbeck, N.A. and Grafham, W. (1995) Conditioning of nyala (*Tragelaphus angasi*) to blood sampling in a crate using positive reinforcement. *Zoo Biology* 14, 261–273.

Hale, R.H., Friend, T.H. and Macaulay, A.S. (1987) Effect of method of restraint of cattle on heart rate, cortisol and thyroid hormones. *Journal of Animal Science* Suppl. 1 (abstract).

Hastings, B.E., Abbott, D.E., George, L.M. and Staler, S.G. (1992) Stress factors influencing plasma cortisol levels and adrenal weights in Chinese Water deer. *Research in Veterinary Science* 53, 375–380.

Heldenbrand, L.E. (1955) Cattle headgate. US Patent Nos 2,714,872 and 2,466,102, US Patent Office, Washington, DC.

Hewes, F.W. (1975) Stock dipping apparatus. US Patent No. 3,916,839. US Patent Office, Washington, DC.

Holmes, R.J. (1991) Cattle. In: Anderson, R.S. and Edney, A.T.B. (eds) *Practical Animal Handling*. Pergamon Press, Oxford, pp. 15–38.

Howard, K.F. (1953) *A Trap Gate for Cattle*. Advisory Leaflet No. 58, Queensland Department of Agriculture and Stock, reprinted from *Queensland Agricultural Journal* July, 1953.

Jones, R.B., Satterlee, D.G. and Cadd, G.G. (1998) Struggling responses of broiler chickens shackled in groups on a moving line: effects of light intensity hoods and curtains. *Applied Animal Behavior Science* 38, 341–352.

Kearnan, A.F., McEwan, T. and Ried, T.J. (1982) Dip vat construction and management. *Queensland Journal of Agriculture* 108, 25–47.

Kilgour, R. (1971) Animal handling in works: pertinent behavior studies. In: *13th Meat Industry Conference*. MIRINZ, Hamilton, New Zealand.

Koba, Y. and Tanida, H. (1999) How do miniature pigs discriminate between people. The effect of exchanging cues between a non-handler and their familiar handler on discrimination. *Applied Animal Behaviour*, 61, 239–252.

Lambooy, E. (1985) Electro-anesthesia or electro-immobilization of calves, sheep and pigs by Feenix Stockstill. *Veterinary Quarterly* 7, 120–126.

Lay, D.C., Friend, T.H., Randel, R.D., Bowers, C.C., Grissom, K.K. and Jenkins, O.C. (1992a) Behavioral and physiological effects of freeze and hot iron branding on crossbred cattle. *Journal of Animal Science* 70, 330–336.

Lay, D.C., Friend, T.H., Bowers, C.C., Grissom, K.K. and Jenkins, O.C. (1992b) A comparative physiological and behavioral study of freeze and hot iron branding using dairy cows. *Journal of Animal Science* 70, 1121–1125.

Lay, D.C., Friend, T.H., Grissom, K.K., Hale, R.L. and Bowers, C.C. (1992c) Novel breeding box has variable effect on heartrate and cortisol response of cattle. *Journal of Animal Science* 70, 1–10.

Legge, A.J. and Conway, R. (1987) Gazelle killing in stone age Syria. *Scientific American* 257, 88–95.

McFarlane, I. (1976) Rationale in the design of housing and handling facilities. In: Engminger, M.E. (ed.) *Beef Cattle Science Handbook 13*. Agriservices Foundation, Clovis, California, USA, p. 233.

Marshall, M., Milberg, E.E. and Shultz, E.W. (1963) Apparatus for holding cattle in position for humane slaughtering. US Patent 3,092,871, US Patent Office, Washington, DC.

Mateo, J.M., Estep, D.Q. and McCann, J.S. (1991) Effects of differential handling on the behavior of domestic ewes. *Applied Animal Behaviour Science* 32, 45–54.

Mayes, H.F. (1978) *Design Criteria for Livestock Loading Chutes*. American Society of Agricultural Engineers, Paper No. 78–6014, St Joseph, Michigan.

Midwest Plan Service (1987) *Beef Housing and Equipment Handbook*, 4th edn, Midwest Plan Service, Iowa State University, Ames, Iowa.

Murphy, K. (1987) Overhead drafting facilities. *Queensland Agricultural Journal* 113, 353–354.

Oklahoma State University (1973) *Expansible Corral with Pie Shaped Pens*. Plan No. Ex. OK-724–25, Cooperative Extension Service, Stillwater, Oklahoma.

Paine, M., Teter, N. and Guyer, P. (undated) Feedlot layout. In: *Great Plains Beef Cattle Handbook*. Cooperative Extension Service, Oklahoma State University, Stillwater, Oklahoma, pp. 5201.1–5201.6.

Parsons, R.A. and Helphinstine, W.M. (1969) *Rambo A.I. Breeding Chute for Beef Cattle. One Sheet Answers*, University of California Agricultural Extension Service, Davis, California.

Pascoe, P.J. (1986) Humaneness of electro-immobilization unit for cattle. *American Journal of Veterinary Research* 10, 2252–2256.

Pearson, L.B. (1965) Automatic livestock headgate. US Patent No. 3,221,707, US Patent Office, Washington, DC.

Phillips, M., Grandin, T., Graffam, W., Irlbeck, N.A. and Cambre, R.C. (1998) Crate conditioning of bongo (*Tragelaphus eurycerus*) for veterinary and husbandry procedures at Denver Zoological Garden. *Zoo Biology* 17, 25–32.

Pollard, J.C., Littlejohn, R.P., Johnstone, P., Laas, F.J., Carson, I.D. and Suttie, J.M. (1992) Behavioral and heart rate response to velvet antler removal in red deer. *New Zealand Veterinary Journal* 40, 56–61.

Powell, E. (1986) *Cattle Yards*. Information Service Q 186014, Queensland Department of Primary Industries, Brisbane, Australia.

Reck, E. and Reck, J.P. (1903) Cattle stanchion. US Patent No. 733,874 US Patent Office, Washington, DC.

Rider, A., Butchbaker, A.F. and Harp, S. (1974) *Beef Working, Sorting and Loading Facilities*. American Society of Agricultural Engineers, Paper No. 74–4523, St Joseph, Michigan, USA.

Schwartzkopf-Genswein, K.S., Stookey, J.M., Crowe, T.G. and Genswein, B.M. (1998) Comparison of image analysis, exertion force and behavior measurements for assessment of beef cattle response to hot iron and freeze branding. *Journal of Animal Science* 76, 972–979.

Stahringer, R.C., Randel, R.D. and Nevenforff, D.A. (1989) Effect of naloxone on serum luteinizing hormone and cortisol concentration in seasonally anestrous Brahman heifers. *Journal of Animal Science* 67 (Suppl. 7), 359 (abstract).

Swan, R. (1975) About A.I. facilities. *New Mexico Stockman*, 11 February, 24–25.

Sweeten, J.M. (1980) Static screening of feedlot dipping vat solution. *Transactions American Society Agricultural Engineers* 23(2), 403–408.

Sweeten, J.M., Winslow, R.B. and Cochran, J.S. (1982) *Solids Removal from Cattle Dip Pesticide Solution with Sedimentation Tanks*. American Society Agricultural Engineers, Paper No. 82–4083, St Joseph, Michigan.

Syme, L.A. (1981) *Improved Sheep Handling Design for a Self Feeding Race*. CSIRO, Division of Land Resources Management, Perth, Western Australia.

Texas Agricultural Extension Service (1979) *Portable Cattle Dipping Vat*. Drawing No. 600, Texas A&M University, College Station, Texas.

Thompson, A.C. (1931) Cattle holding and dehorning gate. US Patent No. 1,799,073, US Patent Office, Washington, DC.

Thompson, A.C. and Gill, C. (1949) Restraining chute for animals. US Patent No. 2,477,888, US Patent Office, Washington, DC.

Thompson, R.J. (1987) Radical new yard proven popular. *Queensland Agricultural Journal* 113, 347–348.

Voisinet, B.D., Grandin, T., Tatum, J.D., O'Connor, S.F. and Struthers, J.J. (1997) Feedlot cattle with calm temperaments have higher average daily gains than cattle with excitable temperaments. *Journal of Animal Science* 75, 892–896.

Vowles, W.J. and Hollier, T.J. (1982) The influence of yard design on the movement of animals. *Proceedings of the Australian Society of Animal Production* 14, 597.

Vowles, W.J., Eldridge, C.A. and Hollier, T.J. (1984) The behavior and movement of cattle through single file handling races. *Proceedings of the Australian Society of Animal Production* 15, 767.

Ward, F. (1958) *Cowboy at Work*. Hastings Press, New York.

Webber, R.J. (1987) Molasses as a lure for spear trap mustering. *Queensland Agricultural Journal* 113, 336–337.

White, J.B. (1961) Letter to the editor. *Veterinary Record* 73, 935.

# Dairy Cattle Behaviour, Facilities, Handling and Husbandry

**8**

## J.L. Albright

*Purdue University, West Lafayette, IN 47907, USA*

## Introduction

Early Indiana research over 100 years ago showed an economic advantage in providing housing for dairy cows during the winter months instead of leaving them outside (Plumb, 1893). During good weather, to enrich their environment and to improve overall health and well-being, cows should be moved from indoor stalls into the barnyard, where they can groom themselves and one another (Wood, 1977; Bolinger *et al.*, 1997), stretch, sun themselves, exhibit oestrous behaviour and exercise (Albright *et al.*, 1999). Exercise decreases the incidence of leg problems, mastitis, bloat and calving-related disorders (Gustafson, 1993).

## Housing and Facilities

Housing systems vary widely, from fenced pastures, corrals and exercise yards with shelters to insulated and ventilated barns with special equipment to restrain, isolate and treat cattle. Generally, self-locking stanchions/headlocks (one per cow), corrals and sunshades are used in warm, semi-arid regions. Pastures and shelters are common in warm, humid areas, naturally ventilated barns with free stalls are used widely in cool, humid climatic regions and insulated and ventilated barns with tie stalls are common in colder climates (Albright *et al.*, 1999). Free-stall housing with open sides (or no side walls) is common in hot, humid areas with rainfall over 64 cm or 25–30 cm in a 6-month period, e.g. San Joaquin Valley in California (D.V. Armstrong, Arizona, 1999, personal communication). The range of effective dimensions for pens and stalls for calves, heifers, dry cows, maternity or isolation, special needs, milking cows and mature bulls have been published in the 'Ag Guide' (Albright *et al.*, 1999). Recommended sizes of free stalls and tie stalls as related to weights of Holstein female dairy cows were updated (Albright *et al.*, 1999).

Keeping cows out of the mud increases their productivity and reduces endoparasitic and foot infections. Current trends and recommendations favour keeping dairy cows on unpaved dirt lots in south-west USA and on concrete or pasture in north USA throughout their reproductive lifetimes. Concrete floors should be grooved to provide good footing and to reduce injury (Albright, 1994, 1995; Jarrett, 1995). The concrete surface should be rough but not abrasive, and the microsurface should be smooth enough to avoid abrading the feet of cattle.

Data are limited on the long-term effects of intensive production systems; however, concern has been expressed about the comfort, well-being, behaviour, reproduction and udder, foot and leg health of cows kept continuously on concrete. As a safeguard, most cows are moved from concrete to dirt lots or pasture, at least during the dry period. Also, the rate of detection and the duration of oestrus are higher for cows on dirt lots or pastures than for cows on concrete (Britt *et al.*, 1986).

Physical accommodation for dairy cattle should provide a relatively clean, dry area where the animals can lie down and be comfortable (Jarrett, 1995). It should be conducive to cows lying for as many hours of the day as cows desire. Recent work indicates that blood flow to the udder, which is related to the level of milk production, is substantially higher (28%) when a cow is lying than when it is standing (Metcalf *et al.*, 1992; Jarrett, 1995).

Criteria for a satisfactory environment for dairy cattle include thermal comfort (effective environmental temperature), physical comfort (injury-free space and contact surfaces), disease control (good ventilation and clean surroundings) and freedom from fear. Cattle can thrive in almost any region of the world if they are given ample shelter from excessive wind, solar radiation and precipitation.

Heat stress affects the comfort of cattle more than cold stress does. Milk production can be increased during hot weather by the use of sunshades, sprinklers and other methods of cooling (Roman-Ponce *et al.*, 1977; Armstrong *et al.*, 1984, 1985; Schultz *et al.*, 1985; Buchlin *et al.*, 1991; Armstrong, 1994; Armstrong and Welchert, 1994), as well as by dietary alterations. Temperatures that are consistently higher than body temperature can cause heat prostration of lactating cows, but additional energy intake ($+1\%$ $°C^{-1}$) and higher heat production by the cow can compensate for lower temperatures, even extremely low ones. Consideration also needs to be given to humidity levels and wind-chill factors in determining effective environmental temperatures. Adaptation to cold results in a thicker hair coat and more subcutaneous fat, which also reduces cold stress (Curtis, 1983; Holmes and Graves, 1994).

## Bedding

Resting dairy cattle should have a dry bed. Ordinarily, stalls should have bedding to allow for cow comfort and to insulate the udder against cold temperatures. Of all the factors that discourage cows from using free stalls, the condition of the bed is likely to be the most important (Bickert and Smith, 1998). (On the other hand, what is in front of the cow has as much to do with cow comfort as what is under her (S.D. Young, Ontario, Canada, 1999, personal communication). If the neck rails are placed too low, the cow feels cramped and may be reluctant to use the stall(s).) Avoid beds that are too hard (concrete, concrete with solid rubber mats and compacted earth). Swollen hocks and knees result from a bed that does not provide sufficient cushion. Also, avoid beds with mounds, lumps or holes. Such conditions reduce cow comfort and cause difficulties when cows get up. The lack of comfort and difficulty in rising both discourage free-stall usage (Bickert and Smith, 1998). When handled properly, many fibrous and granular bedding materials may be used (Midwest Plan Service, 1997), including long or chopped straw, poor quality hay, sand, wood chips, sawdust, shavings, bark, shredded newspapers, composted manure, maize stalks or peanut or rice hulls. Inorganic bedding materials (sand or ground limestone) provide an environment that is less conducive to the growth of mastitis pathogens. Sand bedding may also keep cows cooler than straw or sawdust. Regional climatic differences and diversity of bedding options should be considered when bedding materials are being selected. Bedding should be absorbent, free of toxic chemicals or residues that could injure animals or humans and of a type not readily eaten by the animals. Bedding rate should be sufficient to keep the animals dry between additions or changes. Any permanent stall surface, including rubber mats, should be cushioned with dry bedding (Albright, 1983; Albright *et al.*, 1999). Bedding material added on top of the base absorbs moisture and collects manure tracked into the stall, adds resiliency, makes the stall more comfortable and reduces the potential for injuries (Midwest Plan Service, 1997).

Bedding mattresses over hard stall bases, such as concrete or well-compacted earth, can provide a satisfactory cushion. This relatively new innovation cuts bedding use in stall (tie stall and stanchion) and free-stall barns, providing cushion and traction, with less bedding required, and reduces stall maintenance (House, 1999). A bedding mattress consists of bedding material compacted to 8–10 cm (3–4 in) and enclosed in a fabric (heavy-weight polypropylene or other similar material). Shredded rubber may be used and is recommended as a mattress filler (Underwood *et al.*, 1995). Rubber should be packed firmly to prevent shifting and settling. Small amounts of bedding (chopped straw) on top of the mattresses keep the surface dry and the cows clean (Midwest Plan Service, 1997). There are estimates that as many as 1,200,000 commercial mattresses have been produced and installed (House, 1999). The general public and consumers of dairy products, who know very little about dairy housing, have heard that cows are sleeping on mattresses.

This enhances their opinion that farmers are concerned about the welfare of their animals. Also, there is the fact that rubber tyres are being recycled rather than going into landfills (House, 1999).

Two bedding methods have emerged as top candidates (Bickert and Smith, 1998): (i) mattresses with bedding on top; and (ii) a deep layer of sand. According to Bickert and Smith (1998), sand can be considered to be the gold standard for a free-stall base and bedding. If other materials are to be considered as alternatives or evaluated on the basis of cow comfort, sand is the basis for comparison. The only logical choice for not using sand has little to do with cow comfort and udder health, but with the difficulty it adds to the manure system or with the availability of high-quality sand (no rocks or pebbles, as they can cause hoof damage or lameness). Furthermore, loose sand conforms to the shape of body components – knees, hocks, etc. This reduces pressure on projecting bones and body part contacts by distributing force or weight over a larger lying surface area (Bickert and Smith, 1998).

## Automation

Housing and herd-management developments have important effects on the well-being of dairy cattle, and the cattle enterprise is well suited for the application of electronics and automation (Albright, 1987; Smith *et al.*, 1998). Automation, considered by some to be detrimental to the husbandry and welfare of animals in intensive units, needs to be reconsidered. The time saved, together with reduced work and drudgery, could free people for more human–animal interactions, thus allowing better care. Automation can increase the time given to the training of handlers for observational or 'eyes-on' rather than manipulative 'hands-on' skills. Automation should take the pressure off existing workers, which, in turn, should reduce accidents and injury. If stress on the worker is reduced, animals in large intensive units may be treated more humanely (Kilgour, 1985).

Over time, capital investments for cow comfort and sanitary requirements have increased markedly. Labour-saving practices have been developed to reduce the drudgery of dairy farming. Many of the top-producing cows continue to be housed and milked in labour-intensive stall barns. For these stall barns there are now silo unloaders, gutter cleaners, battery-operated silage carts, portable straw choppers, automatic detaching milking machines with low milk lines, and mechanized manure handling.

## Milking-centre design

Until the advent of centralized milking centres, most cows were milked in their stalls. A disadvantage of this method is that it is labour intensive. The idea of milking cows on an elevated herring-bone platform originated in Australia

(O'Callaghan, 1916; Albright and Fryman, 1964). Early US designs enabled a single person to milk two cows while seated on a swivel chair or to use elevated side-opening parlours (Albright and Fryman, 1964). Due to labour shortages and high wages, New Zealand dairy farmers were motivated to develop rotary centres (Gooding, 1971). One large rotary moved 64 cows and milking machines on a rotating platform. Other rotary designs have been developed (Anon., 1971; Fellows, 1971). At the time, these systems were a great innovation, but had high maintenance costs. Simple layouts with automated gates and milking-machine detachers became popular. Possibly due to shorter walking distances (Smith *et al.*, 1998) and greater efficiency and automation at the entry and exit points, there is currently a new wave of rotary parlours for larger herds in the USA (D.V. Armstrong, Arizona, 1999, personal communication). Quaife (1999) claims that today's rotary parlours will remain a viable option for some larger producers and not fade away as they did in the 1970s.

Extensive time-and-motion studies have been conducted on different milking-centre designs (Armstrong and Quick, 1986; Armstrong, 1992; Smith *et al.*, 1999). The addition of automation, such as powered gates, enabled simple designs, such as herring-bones, trigons and polygons, to achieve greater labour efficiency than the early smaller rotaries. Good reviews illustrating different milking-centre layouts can be found in Bickert (1977), Midwest Plan Service (1985, 1997) and Armstrong (1992). The most commonly used design is the herring-bone where two rows of four to 20 cows are milked from a central pit (Fig. 8.1). One design that has grown in popularity is the parallel milking centre (Fig. 8.2). The milking machines are attached from the rear between the cow's legs instead of from the traditional side position. During milking, the cows stand at 90° relative to the pit, with

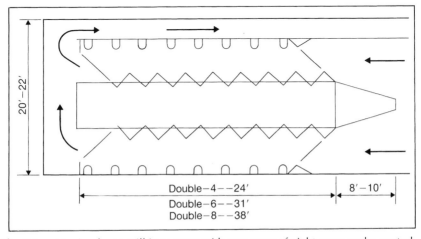

**Fig. 8.1.** Herring-bone milking centre with two rows of eight cows and a central pit for the milker.

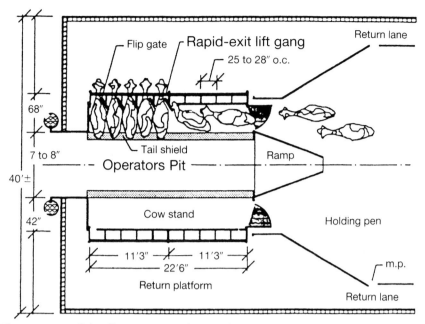

**Fig. 8.2.**   Parallel milking centre with a rapid exit.

their heads restrained in self-locking stanchions. This design is more efficient than a herring-bone (Armstrong *et al.*, 1989, 1990), especially with rapid-exit stalls (Fig. 8.3).

The use of a powered crowd gate to make the holding pen smaller induces cows to enter the milking centre voluntarily. Crowd gates should not be used to forcibly push cows or apply electric shocks. Proper training of cows and of milking-centre operators will also improve the efficiency of cow movement through the facility. Cows should be encouraged to enter voluntarily without prodding. The milker should avoid leaving the pit to chase cows, as this conditions the cows to wait for the milker to come after and chase them. Cows also have individual preferences for music, weather, certain people and the side of the milking centre they will enter (Albright *et al.*, 1992). Since cows are creatures of habit, it is imperative to be consistent from one milking to the next. Recommendations for milking machine and udder sanitation can be found in the 'Ag Guide' (Albright *et al.*, 1999).

## Behaviour and Management

The dairy cow has been called 'the foster mother of the human race' (Rankin, 1925). A relationship develops between the milker and the cow, which is a vital part of the milk-extraction process, and as machine-milking took over from hand-milking this relationship was considered by many to have

**Fig. 8.3.** Cows set for release in a parallel milking centre with a rapid-exit system.

diminished. After her calf is removed, the cow is milked with a minimum of manual stimulation in highly automated surroundings.

Caretakers in high-producing herds are aware of the importance of such changes. For as long as cows have been milked, there has been an art of cow care, which results in more milk from healthier, contented cows. It has been recognized that the dairy cow's productivity can be adversely affected by discomfort or maltreatment. Alert handlers have the perception and ability to read 'body language' in animals. For example, healthy calves and cows will exhibit a good stretch after they get up and then relax to a normal posture. Increased standing of cattle is now often taken as a sign of discomfort or discontent in studies of cow and calf confinement (Albright, 1987).

Cattle under duress show signs by bellowing, butting or kicking. Behavioural indications of adjustments to the environment are always useful signs of whether the environment needs to be improved. In some cases, the way animals behave is the only clue that stress is present (Stephens, 1980; Albright, 1983).

Clues to a cow's mood and condition can be obtained by observing the animal's tail. When the tail is hanging straight down, the cow is relaxed, grazing or walking, but, when the tail is tucked between the cow's legs, it means the animal is cold, sick or frightened. During mating, threat or investigation, the tail hangs away from the body. When galloping, the tail is held straight out, and a kink can be observed in the tail when the animal is in a bucking, playful mood (Kiley-Worthington, 1976; Albright, 1986a; Albright and Grandin, 1993; Albright and Arave, 1997).

According to Kiley-Worthington (1976), when studying the cause and function of tail movement it is necessary to consider the whole posture of the animal, as well as the contexts that give rise to it. In cattle (and horses) the immediate association one makes with lateral tail movements is with cutaneous irritation. In these species there are morphological changes of the tail which point to its use as a fly switch.

## Tail-docking

According to the 'Ag Guide' with references on this subject (Albright *et al.*, 1999), docking of tails is a controversial and yet common practice performed on cows that are milked from the rear or that have filthy switches. Tail-docking has been prohibited in the UK and some other European countries. Under conditions of high fly numbers, tail-docked heifers tail-flick more often and are forced to use alternative behaviours such as rear leg stomps and head turning to try to rid themselves of flies. More flies settle on tail-docked cows than on intact cows; the proportion of flies settling on the rear of the cow increases as tail length decreases. Grazing and rumination are disturbed when fly attacks are intense, and substantial losses to the US cattle industry have been attributed to flies causing interference with grazing. Excellent fly control is therefore especially important for tail-docked cattle. A study of tail-docking in New Zealand (Matthews *et al.*, 1995) found no difference in cortisol concentrations between docked and intact cows, but there were also no differences in milk yields, body-weights, somatic cell counts, frequency of mastitis or milker comfort among the treatments studied (intact tails, trimmed tails, docked tails). Trimming switches with clippers or fastening the switch out of the way are preferred as alternatives to tail-docking in research or teaching herds. Further research is needed on the short- and long-term consequences of tail-docking in calves and cows (Albright *et al.*, 1999).

## Stray voltage

Stray electrical voltages from malfunctioning electrical equipment can cause discomfort to dairy cows and lower milk production. Numerous research studies have quantified the physiological and behavioural responses of dairy cattle to electrical currents (Lefcourt, 1991; Aneshansley *et al.*, 1992). The electrical currents required for perception, behavioural change or physiological effects to occur are widely variable. Dairy cows can feel very low voltages of only 1.0 V when they occur between a water bowl and the rear hooves (Gorewit *et al.*, 1989). Furthermore, symptoms associated with problems of stray voltage or electrical current are not unique, and many factors other than stray voltage and electrical current can cause similar problems in behaviour, health or milk production (Gorewit *et al.*, 1992).

The sources of relatively small amounts of electrical currents passing through animals are often very difficult to locate. Stray voltage or electrical currents may arise because of poor electrical connections, corrosion of switches, frayed insulation, faulty equipment or heavily loaded power lines.

Information on how to detect and correct stray-voltage problems is contained in Appleman (1991). Periodic evaluation of facilities for stray voltage is suggested. Solutions include voltage reduction, control of sources of voltage leakage, gradient control by use of equipotential planes and transition zones and isolation of a portion of the grounding or grounded neutral system from the animals. Proper installation of electrical equipment and complete grounding of stalls and milking-centre equipment should help prevent stray-voltage problems. Although stray voltages and electrical currents cannot be totally eliminated, they can be reduced (Albright *et al.*, 1991, 1999; Lefcourt, 1991; Gorewit *et al.*, 1992).

## Social Environment

Dairy cattle are social animals that function within a herd structure and follow a leader to and from the pasture or milking centre. Cows exhibit wide differences in temperament, and their behaviour is determined by inheritance, instinct, physiology, hormones, prior experience and training. Cows are normally quiet and thrive on gentle treatment by handlers. Handling procedures are more stressful for isolated animals; therefore, attempts should be made to keep several cows together during medical treatment, artificial insemination or when moving cows from one group to another (Whittlestone *et al.*, 1970; Arave *et al.*, 1974). Cattle should have visual contact with each other and with their caretakers (Albright *et al.*, 1999).

Many dairymen and women allow their cows to develop their own individual personalities as long as no special care or treatment is required. Mass handling of cows dictates that individual cows fit into the system rather than the system conforming to the habits of the cow. The slow milker, the kicker, the boss cow, the timid cow, the explorer and the finicky eater are usually removed from larger herds, regardless of pedigree.

Although concern is expressed from time to time about temperament and behavioural problems, most attempts at reinforcing correct behaviour and disciplining improper behaviour have been successful. One dairy study showed that behaviour as a reason for disposal was less than 1% of cases. Other categories included: reproductive disorders and diseases, 36%; udder problems and mastitis, 23%; anatomical problems (feet, legs and skeleton), 11%; digestive problems, 11%; metabolic problems, 7%; and low yield, 4%. The cows culled for behaviour mainly represent the truly wild ones which would not conform to training and management (Albright and Beecher, 1981; Albright, 1986b).

Although creatures of habit, gentle dairy animals may be excited into rebellion by the use of unnecessarily severe methods of handling (e.g. shouting and shock prods) and restraint. Attempts to force an animal to do something it does not want to do often end in failure and can cause the animal to become confused, disorientated, frightened or upset. Handling livestock requires that they be 'outsmarted' rather than outfought and that they be 'outwaited' rather than hurried (Battaglia, 1998b). Most tests of will between the handler and the cow are won by the cow.

Considerable self-stimulation and 'inwardness' occur in cattle due to the rumination process. During rumination, cows appear relaxed with their heads down and their eyelids lowered. Resting cows prefer to lie on their chest, facing slightly uphill. Also, through cud-chewing as well as mutual and self-grooming, aggression is reduced and there is little or no boredom (Albright, 1986b).

Management developments which have improved the comfort and well-being of dairy cattle include raising calves in individual pens or hutches (Baker, 1981), providing exercise prior to calving (Lamb et al., 1979), grooving or roughening polished, slick, concrete flooring (Albright, 1983, 1994, 1995), making use of pasture or earthen exercise lots and removing slatted floors (Albright, 1983), and eliminating stray voltage (Appleman, 1991). Individual stalls (cubicles/free stalls) have resulted in cleaner cows and fewer teat injuries than loose housing. Dairy cattle thrive best when they are kept cool and free from flies and pests and provided with a dry, comfortable bed to lie down on (Albright, 1986b).

Dairy cattle have traditionally been kept in groups of 40–100 cows. In commercial dairy herds in Arizona, New Mexico and Texas, variation in group size – small (50–99), medium (100–199) and large (200 or more) – does not cause a problem per se. Large herd size, however, can affect management decisions, because overcrowding, with an insufficient number of headlocks or inadequate manger space per cow, irregular or infrequent feeding and excessive walking distance to and from the milking parlour have a greater impact on behaviour and well-being than does group size (Albright et al., 1999).

In order to evaluate the effects of restraint using self-locking stanchions, 64 Holstein cows from peak to late lactation were restrained at feeding time for 4 h per day for four weekly periods. Milk production, somatic cell counts, mastitis or other health concerns, plasma cortisol concentrations and total daily feed intake were unaffected by restraint. For the cows locked in stanchions, their eating frequency over 24 h was significantly reduced, but dry-matter intake was not affected. Total rumination frequency over 24 h was not significantly different for cows that were restrained; however, restrained cows ruminated less during the day following release. Behaviourally, cows that were locked in the stanchions spent significantly more time lying in free stalls after release from restraint. Grooming was also one of the first behaviours performed following release. Grooming was considered to be a behavioural need and was significantly increased during all times when cows were not locked up. Acts of aggression were elevated during all periods following restraint. The use of self-locking stanchions did not appear to affect substantially the overall well-being of the cow (Bolinger *et al.*, 1997).

Other Purdue work with detailed observations, using intact and cannulated cows, suggests a behavioural need for the cow to rest and to ruminate on her left side (Grant *et al.*, 1990; Albright, 1993).

## Cow and Calf Handling

Milk production is a by-product of the reproductive process. Therefore, an essential part of the onset of lactation is the birth of a calf. Unfortunately, newborn calves are sometimes cast aside, especially during economic downturns. Calves require special handling and care from the time they are born. The most important point to remember is to feed the newborn calf colostrum soon after birth and within the first 6 h. A calf should be given 8–10% of its body-weight in fresh colostrum by bottle, bucket or tube feeder; twice within 24 h following birth. Colostrum is nutrient-rich and provides the calf with vital immunoglobulins. Good nutrition, along with proper handling, starts a calf on its way toward a healthy life. If young calves are to be marketed, the following three procedures should be used:

**1.** Provide individual care and colostrum for 2–3 days after birth.
**2.** Calves should always have a dry hair coat, have a dry navel cord and walk easily before being transported. A day-old calf can stand, but it is unsteady and wobbly and is not ready for market (Albright and Grandin, 1993). In England and Canada, the sale of calves under 1 week old is forbidden. Calves should not be brought to a livestock market until they are strong enough to walk without assistance. To reach adequate strength and vigour, calves need to be a minimum of 5 days old (Grandin, 1990).
**3.** Handle calves in transit carefully, protecting them from the sun and heat stress in the summer and from the cold and wind chill in winter.

By observing behaviour and carefully following these recommendations, healthier and contented dairy cattle are assured:

**1.** Always keep hooves trimmed to prevent lameness. A cow with properly trimmed hooves and healthy feet and legs will stand quietly and occasionally shift her weight. Cows with feet and leg problems are more restless, crampy and uncomfortable; they appear to walk in place. Foot lameness is probably the single greatest insult to the welfare of the modern high-producing dairy cow (Albright *et al.*, 1999).
**2.** Breed first-calf heifers to bulls with a reputation for easy calving.
**3.** Use caution with calf pullers to prevent internal injuries.
**4.** If internal injury happens during calving, lift the cow into a standing position for rehabilitation. Non-ambulatory or downed animals must not be

dragged. Provisions should exist for lifting downer cows. Devices to aid and promote standing include hip lifters (hip clamps), slings (wide belt and hoist), inflatable bags and warm-water flotation systems (Albright *et al.*, 1999).

**5.** To prevent downed cattle from getting milk fever and other metabolic disorders, obtain the services of a competent veterinarian or dairy specialist.

**6.** To prevent mastitis, keep the udder dry and dip teats before and after milking.

**7.** When loading dairy animals for shipping, allow plenty of handling space. Cattle need ample room to turn, the leaders will then move into the chute, with other animals following. This is an example of leadership–followership.

**8.** Stair steps are recommended for loading ramps. Each step should be 10 cm high, with a 30 cm tread width.

**9.** Loading ramps for young stock and animals that are not completely tame should have solid sides.

**10.** Never attempt to transport cows which become emaciated or too weak to stand. If rehabilitation does not occur within a reasonable time, the animal should be humanely killed on the farm (Livestock Conservation Institute, 1992).

**11.** When transporting young dairy animals or producing cows, always handle them gently. Since cows are curious, allow them to quietly investigate their new environment and ease into it without outside distractions.

**12.** Try to ship dairy animals under favourable weather conditions. Avoid extremely hot or extremely cold temperatures, which create undue stress and may cause sickness.

Dairy producers have much to gain when cows and young stock are properly handled and cared for (Albright and Grandin, 1993; Albright *et al.*, 1999).

The Canadian guide (Agriculture Canada, 1990) contains a complete transportation section, including definitions, general information, vehicles, containers, space requirements, protecting cattle, food, water and rest for cattle in transit, unfit cattle, pregnant cattle, precautions in cold or hot, humid weather and transportation stress. There is also a section on assembly yards, sale yards and processing facilities, which includes facilities, unfit cattle, holding and handling, education of personnel, slaughter and emergency procedures.

## Bull Handling

The safety of humans and animals is the chief concern underlying management practices. By virtue of their size and disposition, bulls may be considered as one of the most dangerous domestic animals. Management procedures should be designed to protect human safety and to provide for bull welfare (Albright *et al.*, 1999).

## Threat postures

There are certain major behavioural activities related to bulls. These are threat displays, challenges, territorial activities, female seeking and directing (nudging) and female tending. These behavioural activities tend to flow from one to another (Fraser, 1980). Threat displays in bulls and ungulates (e.g. antelope, bison) are a broadside view (Fig. 8.4) when a person or a conspecific invades its flight zone. The threat display of the bull puts him in a physiological state of fight or flight. The threat display often begins with a broadside view with back arched to show the greatest profile, followed by the head down, sometimes shaking the head rapidly from side to side, protrusion of the eyeballs and piloerection of the hair along the back. The direct threat is head-on, with head lowered, shoulders hunched and neck curved to the side toward the potential object of the aggression. Pawing the ground with the forefeet, sending the earth flying behind or over the back as well as rubbing or horning the earth are often components of the threat display. If in response to the threat display the recipient animal advances with head down in a fight mode, a short fight with butting of horns or heads ensues. If the recipient of the threat has been previously subdued by that animal, he is likely to withdraw with no further interaction (Albright and Arave, 1997).

While a bull is showing a threat display, if an opponent, such as another bull (or a person), withdraws to about 6 m, the encounter will subside and the bull will turn away. If not, the bull will circle, drop into the cinch (flank) body

**Fig. 8.4.** Broadside threat display as a warning that a human has invaded his flight zone. A direct threat is head-on with head lowered, shoulders hunched and neck curved to the side. (Reproduced from Albright and Arave (1997) with permission).

position or start with a head-to-head or head-to-body pushing. At the first sign of any of the above behaviours, humans should avoid the bull and exit rapidly and hopefully via a predetermined route.

Many people lack the background, attitude and precaution of dealing with dangerous bulls and parturient cows; therefore additional training and bull/cow behaviour information are needed. It is wise to respect and be wary of all bulls, especially dairy bulls, as they are not to be trusted. Any bull is potentially dangerous. He may seem to be a tame animal, but on any given day he may turn and severely injure or perhaps kill a person, young or old, inexperienced or experienced. This is especially true when a cow is in oestrus and needs to be removed from 'his' group or when the group is moved to the holding pen for milking. Never handle the bull alone and never turn your back on a bull. To move cattle or to appear larger and to protect yourself carry a cane, stick, handle, plastic pole with flap or baseball bat. For further information about bull behaviour and handler safety, refer to Albright and Arave (1997).

## Other dairy animals

In addition to bulls, humans must be careful around certain steers, heifers and recently calved cows protecting their calves. Some animals are different and do not follow the threat-display behaviour previously mentioned. Be careful of following behaviour, walking the fence, bellowing, a cow in oestrus and the bull which protects the cow, thereby attacking the handler. An animal's first attack should be its last and it should be sent to the slaughterhouse (Wilson, 1998).

The system of management under which dairy cattle are raised and kept has a profound effect on their temperament, and this is not always taken into consideration. For example, bull calves should never be teased, played with as a calf, treated roughly or rubbed vigorously on the forehead and area of the horns. The Fulani herdsmen stroke under the chin (rather than on top of the head) as an appeasement, i.e. taming, grooming-like behaviour. This is essentially the way cows groom each other (Hart, 1985; Albright and Grandin, 1993; Albright and Arave, 1997).

# Transport Developments

Transportation was reviewed by Albright and Arave (1997). Knowledge and utilization of the flight zone (see Chapter 5) are important during the movement of dairy cattle. Cows should be moved at a slow walk, particularly if the weather is hot and humid or if the flooring is slippery.

Heart-rate transmitters were implanted in lactating Holstein cows prior to travel (Ahn et al., 1992). Cows were transported 402 km in about 6 h over various road surface conditions in an 8.2 m long livestock trailer. The two-way journeys started in the morning and ended late in the afternoon. Cows stayed overnight and were brought back in the late afternoon. This 2-day journey was repeated 1 week later. Feed and water were provided during the interim between travel, with cows receiving their normal ration for that period. Cows were milked by portable machine according to their regular schedule and confined to a fenced corral of approximately 0.4 ha. The heart rate taken as travel commenced averaged 89.7 beats per minute (b.p.m.) and differed significantly ($P < 0.01$) from all hourly readings. The heart rate for hours 2, 3, 4, 5 and 6 averaged 77.0, 74.8, 71.3, 74.4 and 72.9 b.p.m., respectively (which are all similar to a resting heart rate of 76.5 b.p.m.). Heart rates differed significantly ($P < 0.01$) depending on road surface, averaging 83.3 b.p.m. on a dirt road, 81.2 b.p.m. on a paved rural three- and four-lane road, 76.1 b.p.m. on a paved two-lane desert road and 73.6 b.p.m. on the paved motorway. Heart rates observed gave evidence of habituation on the day of travel and also from week 1 to week 2.

Transport is particularly stressful for young calves, who experience mortality rates greater than 20% and bruised stifles at an incidence of 50% or more (Hemsworth *et al.*, 1995).

## Human–Animal Interactions

Recently, this subject has been reviewed by Albright and Arave (1997) and Hemsworth and Coleman (1998).

### The behaviour of the cowman/woman

Studies on homogeneous dairy herds, as defined by similar feeding policy, feeding levels, breed and genetic potential, grazing management and climate, demonstrate the effect of the cowman's behaviour and personality (Seabrook, 1972, 1977, 1991, 1994). The highest-performance cowmen, in terms of milk yield for a given level of input, have the following traits: self-reliant; considerate; patient; independent; persevering; difficult to get on with; forceful; confident; suspicious of change; not easygoing; unadaptable; not neat; not modest; not a worrier; not talkative (quiet); uncooperative; and non-sociable ('grumpy'). In summary, they are confident introverts. Some of these traits may seem to be socially undesirable, but it is the cow's and not another human's reaction which is critical. The cowmen with these traits were more stable and had an air of confidence, enabling them to develop a relationship with their cows which positively influenced the animal's performance. Cows under the care of such a person easily outproduced a person lacking confidence or a confident extrovert ('cheerful Charlie'), who tended to have only average production achievement from their cows.

Building on this work, Reid's (1977) study on high-producing herds both in North America (Canada and the USA) and in England yielded some important results. Reid concluded that the high-production cowman was able to minimize output of adrenaline by the cow and obtain a higher percentage of the milk yield which her genetic capacity permits than others would obtain from the same cow under similar conditions. The high-production cowman achieves this by constant attention to the behaviour patterns or performance of each individual cow in the herd. Other interests of Reid's 'confident introverts' included vegetable growing, but the most startling fact was that they also grew either roses, gladioli or chrysanthemums, species that have different varieties requiring specific treatment and which respond to feeding at specific times of the year. The best cowmen were also attuned to instant recognition of each animal in the herd and the individuality of their cows, plus a close identification with the herd. In many cases, it was difficult to define whether the herd was regarded as an extension of the family or the reverse.

## The behaviour of the cow

Albright (1978) and Seabrook (1980) have shown that animal behaviour differs among dairy herds. One factor which varies both within and between groups of cows is flight distance (basically, how close one can approach an individual animal without it moving away). In some dairy herds, this distance may be almost zero, whereas in others it may be as high as 6 m (20 ft). For individual animals in these herds there will be ranges of values, but they may be lower for one herd than the lowest for another herd. Why do these differences exist and how do they arise? Some variation could be attributed to conditional learning, e.g. the 'memory' of being struck by a handler, but there is little evidence to account for all of the differences. Seabrook (1994) has shown that animals are effective discriminators and perceive by experience and learning. Cows made the greatest number of approaches under test conditions to the familiar person and fewest to the stranger. Cowman behaviour in the milking parlour showed 2.1 vocal interactions per cow min$^{-1}$ for higher-yielding dairy units as compared with 0.5 vocal interactions per cow min$^{-1}$ for lower-yielding units. Likewise, cowman behaviour in the milking centre talking 'with' and 'to' cows were 2.1 times per cow min$^{-1}$ and 9.1 words per cow min$^{-1}$ in higher-yielding dairy units, while in the lower yielding dairy units they were 0.3 times per cow min$^{-1}$ and 2.1 words per cow min$^{-1}$, respectively. Table 8.1 summarizes responses of dairy cows to using pleasant or aversive handling.

Observations of identical one-person units show behaviour differences in terms of how long it takes cows to enter the milking centre. In some herds the cows are keen to enter while in others they are reluctant to do so. Studies showed the milking centres and their identically sized and shaped

**Table 8.1.** General response of animals under different handling treatments (Seabrook, 1991).

| Action of cow | Pleasant handling | Aversive handling |
|---|---|---|
| Mean entry time to parlour | Shorter (9.9 s per cow) | Longer (16.1 s per cow) |
| Flight distance (nervousness) | Low (0.5 m) | High (2.5 m) |
| Dunging in parlour | Less (3.0 h$^{-1}$) | More (18.2 h$^{-1}$) |
| Free approaches to humans | More (10.2 min$^{-1}$) | Less (3.0 min$^{-1}$) |

collecting yards to be in excellent condition. It is the relationship between the cowmen and women and the cows which seems to explain differences in entry time. It is fallacious to talk about the behaviour of dairy cows in isolation; the actual pattern is a reflection of the relationship between human and cow. This connection was realized in the 1940s by Rex Patterson, the pioneer of large-scale dairy farming in England, when he publicly stated that the biggest effect on herd yield and cow behaviour on his one-person dairy units was exerted by the cowman (Seabrook, 1972, 1977, 1980).

Research (Munksgaard *et al.*, 1995; de Passille *et al.*, 1996) with cows and calves show clearly that cattle learned to discriminate between humans based on their previous experience and cues based on the colour of clothing worn – approaching pleasant handlers positively and avoiding those who handled them aversively. Aversive handling can result in a generalized fear of people, making handling more difficult, and increases the chances of injury to both animal and handler. This fear can be overcome by positive handling. Discrimination was generalized to other locations and cattle appear to be more fearful of humans in an unfamiliar location (de Passille *et al.*, 1996).

In order to determine if an aversion corridor could be used to evaluate various handling practices, 60 cows were randomly assigned to five different treatments: electric prod, shouting, hitting, tail twisting and control. Cows walked down a corridor and treatments were applied at the end of the corridor. Preliminary results suggest that cows found the electric prod most aversive, followed by shouting, hitting, tail twisting and control (Pajor *et al.*, 1998). In a follow-up experiment, 54 cows were randomly assigned to four treatments (hit/shout, brushing, control and food). The time and force required for cows to walk down the corridor were measured. Cows on the hit/shout treatment took more time and required more force to walk through the corridor than cows on other treatments ($P < 0.001$). In addition, brushed cows took longer to move through than cows given food ($P < 0.05$) (Pajor *et al.*, 1999). Aversion learning methods show promise as an effective method to determine which handling procedures cows find more aversive or friendly.

## Establishing the relationship

In higher-performance herds, where cowman/woman and cow enjoy a good relationship, the animals have a short flight distance, tend to move quickly into the milking centre and are comfortable in the cowman's presence. The cowmen establish and maintain the relationship by frequently touching and communicating with the animals, treating them with special care at critical points, such as calving and first milking after calving, and assuming the roles of both boss animal and caring mother substitute. This close relationship enables the cowman to spot changes in the cows' behaviour quickly and thus to prevent situations from developing which could adversely affect performance. In addition, the atmosphere created by this kind of psychological environment seems to be more conducive to rest, which means that the cows may be able to reserve more energy for milk production.

## The implications for animal welfare

The animals in the herd where there is a good relationship between the cowman/woman and cow produce more milk, as they release less adrenaline to block milk let-down. The cows are less nervous, more settled and steady in an environment created by a confident cowman. The pertinent point, from an animal-welfare point of view, is that these are not necessarily the best equipped herds technically, e.g. in milking-centre design. In other words, cows can be under stress in a well-designed system if they cannot develop a good relationship with their cowman. Similarly, they may be in a poor system technically, but may be content and under little stress if they have confidence in and a good relationship with the person who tends them.

Efficient dairy management and animal welfare would both be served by selecting people who have the correct traits and then further training them to develop a relationship with their animals and so ensure that the animals are able to live in an environment where stress is reduced to a minimum. Design of a system from a welfare perspective is only part of the solution. The most important factor in determining stress is the behaviour and attitude of the cowman (Seabrook, 1980).

## Husbandry Procedures

Certain dairy cattle behaviour (e.g. aggression and kicking) put at risk the health and well-being of herdmates, as well as the humans handling the cattle. Several devices and procedures can reduce or modify these behaviours. Certain identification procedures, clipping milk cows, training them with a halter and milking procedures must be done properly to minimize negative effects on cattle health (Agriculture Canada, 1990; Battaglia, 1998a; Albright *et al.*,

1999). Step-by-step, learn-by-doing techniques for identification, milking, mastitis treatment, downer cows, weaning and training to eat and drink, and body-condition scoring of dairy cows have been summarized (Battaglia, 1998a). Castration may be performed on bull calves, except those being raised as veal calves. The same is true early in life for dehorning (Battaglia, 1998a). Many dairy calves are born with more than the usual four teats. These supernumerary teats can grow and develop much like a normal teat. They detract from the general appearance of the animal and have the potential to disrupt the milking process later on and to become infected. For these reasons, it is a good practice to remove these extra teats as early as possible in the calf's life. If it is done immediately following birth at the same time as the navel is treated, the calf is easy to handle and one qualified person can accomplish the task (Battaglia, 1998a). Because the calf is very young, the cut bleeds only slightly. Older calves and heifers close to calving should have extra teats removed under local anaesthetic by a qualified person (Albright *et al.*, 1999). Removal can be performed in the first 3 months of life with sharp scissors or a scalpel.

## Conclusions

Observation of dairy cattle has been going on for centuries and helps to increase knowledge and to improve husbandry techniques. A more logical approach to the study of cow behaviour and training is now advocated, linking it with commercial operations. Time saved through automation should be invested in observing animals. A knowledge of normal behaviour patterns provides an understanding about cattle and results in improved management and handling, which will achieve and maintain higher milk yields and animal comfort and well-being (Albright and Grandin, 1993). Dairy cattle must fit in well with their herdmates as well as their handlers. For those who like to work with dairy cattle, the proper mental attitude of handlers must blend in with skilful management and humane care in today's highly competitive, technological, urban-based society.

## References

Agriculture Canada (1990) *Recommended Code of Practice for the Care and Handling of Dairy Cattle*. Publication 18531E, Ottawa, Ontario, 41 pp.

Ahn, H.M., Arave, C.W., Bunch, T.D. and Walters, J.L. (1992) Heart rate changes during transportation of lactating dairy cows. *Journal of Dairy Science* 75 (Suppl. 1), 166 (abstract).

Albright, J.L. (1978) The behaviour and management of high yielding dairy cows. In: *Proceedings British Oil and Cake Mills–Silcock Conference*. BOCM-Silcock, London, UK, 31 pp.

Albright, J.L. (1983) Putting together the facility, the worker and the cow. In: *Proceedings of the 2nd National Dairy Housing Congress.* American Society of Agricultural Engineers. St Joseph, Michigan, pp. 15–22.

Albright, J.L. (1986a) Bovine body language. In: Broom, D.M., Sambraus, H.H. and Albright, J.L. (eds) *Farmed Animals.* Torstar Books, New York, pp. 36–37.

Albright, J.L. (1986b) Milk from contented cows – operation of a dairy farm. In: Broom, D.M., Sambraus, H.H. and Albright, J.L. (eds) *The Encyclopedia of Farmed Animals.* Equinox, Oxford, UK, pp. 56–57.

Albright, J.L. (1987) Dairy animal welfare: current and needed research. *Journal of Dairy Science* 70, 2711–2731.

Albright, J.L. (1993) Feeding behaviour of dairy cattle. *Journal of Dairy Science* 76, 485–498.

Albright, J.L. (1994) Behavioral considerations – animal density, concrete/flooring. In: *Proceedings National Reproduction Workshop.* American Association Bovine Practitioners. Rome, Georgia, pp. 171–176.

Albright, J.L. (1995) Flooring in dairy cattle facilities. In: *Animal Behavior and the Design of Livestock and Poultry Systems.* NRAES-84, Northeast Regional Agricultural Engineers Service, Cornell University, Ithaca, New York, pp. 168–182.

Albright, J.L. and Arave, C.W. (1997) *The Behaviour of Cattle.* CAB International, Wallingford, UK, 306 pp.

Albright, J.L. and Beecher, S.L. (1981) Computerized dairy herd disposal study – behaviour and other traits. *Journal of Dairy Science* 64 (Suppl. 1), 91 (abstract).

Albright, J.L. and Fryman, L.R. (1964) Evolution of dairying in Illinois with special emphasis on milking systems. *Journal of Dairy Science* 47, 330–335.

Albright, J.L. and Grandin, T. (1993) Understanding dairy cattle behaviour to improve handling and production: videotape. *Journal of Dairy Science* 76 (Suppl. 1), 235 (abstract).

Albright, J.L., Dillon, W.M., Sigler, M.R. and Wisker, J.E. (1991) Dairy farm analysis and solution of stray voltage problems. *Agri-Practice* 12, 23–27.

Albright, J.L., Cennamo, A.R. and Wisniewski, E.W. (1992) Animal behaviour considerations. In: *Proceedings National Conference on Milking Center Design.* Northeast Regional Agricultural Engineers Service, Cornell University, Ithaca, New York, pp. 114–123.

Albright, J.L., Bickert, W.G., Blauwiekel, R., Morrill, J.L., Olson, K.E. and Stull, C.L. (1999) Guidelines for dairy cattle husbandry. In: Mench, J.A. (ed.) *Guide for the Care and Use of Agricultural Animals in Agricultural Research and Teaching,* 1st revised edn. Federation of Animal Science Societies, Savoy, Illinois, 120 pp.

Aneshansley, D.J., Gorewit, R.G., and Price, L.R. (1992) Cow sensitivity to electricity during milking. *Journal of Dairy Science* 75, 2733–2744.

Anon. (1971) Five types of rotary milkers. *New Zealand Journal of Agriculture* 123(3), 34–35.

Appleman, R.D. (1991) *Effects of Electrical Voltage/Current on Farm Animals: How to Detect and Remedy Problems.* Agriculture Handbook 696, United States Department of Agriculture, Agriculture Research Service. Superintendent of Documents, Government Printing Office, Washington, DC.

Arave, C.W., Albright J.L. and Sinclair, C.L. (1974) Behaviour, milk yield and leucocytes of dairy cows in reduced space and isolation. *Journal of Dairy Science* 59, 1497–1501.

Armstrong, D.V. (1992) Milking parlour efficiencies for various parlour designs. In: *Proceedings Milking Center Design National Conference*. Northeast Regional Agricultural Engineering Service, Cooperative Extension, Cornell University, Ithaca, New York, pp. 68–77.

Armstrong, D.V. (1994) Heat stress interaction with shade and cooling. *Journal of Dairy Science* 77, 2044–2050.

Armstrong, D.V. and Quick, A.J. (1986) Time and motion to measure milking parlour performance. *Journal of Dairy Science* 69, 1169–1177.

Armstrong, D.V. and Welchert, W.T. (1994) Dairy cattle housing to reduce stress in a hot climate. In: Buchlin, R. (ed.) *Dairy Systems for the 21st Century. Proceedings 3rd International Dairy Housing Conference*. American Society of Agricultural Engineers, St Joseph, Michigan, pp. 598–604.

Armstrong, D.V., Wiersma, F., Gingg, R.G. and Ammon, D.S. (1984) Effect of manger shade on feed intake and milk production in a hot arid climate. *Journal of Dairy Science* 67 (Suppl.1), 211 (abstract).

Armstrong, D.V., Wiersma, F., Fuhrmann, T.J. and Tappan, J.M. (1985) Effect of evaporative cooling under a corral shade on reproduction and milk production in a hot, arid climate. *Journal of Dairy Science* 68 (Suppl.1), 167 (abstract).

Armstrong, D.V., Gamroth, M.J., Welchert, W.T. and Wiersma, F. (1989) *Parallel Milking Parlour Performance and Design*. Technical Paper PR-89–101, American Society of Agricultural Engineers, St Joseph, Michigan.

Armstrong, D.V., Gamroth, M.J., Smith, J.F., Welchert, W.T. and Wiersma, F. (1990) *Parallel Milking Parlour Performance and Design Considerations*. Technical Paper 904042, American Society of Agricultural Engineers, St Joseph, Michigan.

Baker, F.H. (ed.) (1981) *Scientific Aspects of the Welfare of Food Animals*. Report 91, Council for Agriculture Science and Technology, Ames, Iowa.

Battaglia, R.A. (1998a) Dairy cattle management techniques. In: *Handbook of Livestock Management*, 2nd edn. Prentice Hall, Upper Saddle River, New Jersey, pp. 179–207.

Battaglia, R.A. (1998b) *Handbook of Livestock Management*, 2nd edn. Prentice Hall, Upper Saddle River, New Jersey, 589 pp.

Bickert, W.G. (1977) *Milking Parlour Types and Mechanization*. Technical Paper 77–4552. American Society of Agricultural Engineers, St Joseph, Michigan.

Bickert, W.G. and Smith, J.F. (1998) Free stall barn design and management for cow comfort. In: *Proceedings of the Midwest Dairy Management Conference*. Minneapolis, Minnesota, pp. 103–118.

Bolinger, D.J., Albright, J.L., Morrow-Tesch, J., Kenyon, S.J. and Cunningham, M.D. (1997) The effects of restraint using self-locking stanchions on dairy cows in relation to behaviour, feed intake, physiological parameters, health and milk yield. *Journal of Dairy Science* 80, 2411–2417.

Britt, J.H., Armstrong, J.D. and Scott, R.G. (1986) Estrous behaviour in ovarectimized Holstein cows treated repeatedly to induce estrous during lactation. *Journal of Dairy Science* 69 (Suppl. 1), 91 (abstract).

Buchlin, R.A., Turner, L.W., Beede, D.K., Bray, D.R., and Hemken, R.W. (1991) Methods to relieve heat stress for dairy cows in hot, humid climates. *Applications of Engineering in Agriculture* 7, 241–247.

Curtis, S.E. (1983) *Environmental Management in Animal Agriculture*. Iowa State University Press, Ames, Iowa, 409 pp.

de Passille, A.M., Rushen, J., Ladewig, J. and Petherick, C. (1996) Dairy calves discrimination of people based on previous handling. *Journal of Animal Science* 74, 969–974.

Fellows, T. (1971) Rotary sheds in Europe. *New Zealand Journal of Agriculture* 123(3), 49–53.

Fraser, A.F. (1980) *Farm Animal Behaviour*, 2nd edn. Baillière Tindall, London, 291 pp.

Gooding, B. (1971) Farmers combine. *New Zealand Journal of Agriculture* 123(3), 36–38.

Gorewit, R.C., Aneshansley, J.D., Ludington, R.A., Pellerin, R.A and Shao, X. (1989) AC voltage effects in water bowls: effects on lactating Holsteins. *Journal of Dairy Science* 72, 2184–2192.

Gorewit, R.C., Aneshansley, J.D. and Price, L.R. (1992) Effects of voltages on cows over a complete lactation. *Journal of Dairy Science* 75, 2719–2725.

Grandin, T. (1990) Calves you sell should be old enough to walk. *Hoard's Dairyman* 135, 776.

Grant, R.J., Colenbrander, V.F. and Albright, J.L. (1990) Effect of particle size in forage and rumen cannulation upon chewing activity and laterality in dairy cows. *Journal of Dairy Science* 73, 3158–3164.

Gustafson, G.M. (1993) Effects of daily exercise on the health of tied dairy cows. *Preventative Veterinary Medicine* 17, 209–223.

Hart, B.L. (1985) *The Behaviour of Domestic Animals*. W.H. Freeman, New York, USA, 390 pp.

Hemsworth, P.H. and Coleman, G.J. (1998) *Human–Livestock Interactions: The Stockperson and the Productivity and Welfare of Intensively Farmed Animals*. CAB International, Wallingford, UK, 152 pp.

Hemsworth, P.H., Barnett, J.L., Beveridge, L. and Mathews, L.R. (1995) The welfare of extensively managed dairy cattle: a review. *Applied Animal Behaviour Science* 42, 161–182.

Holmes, B.J. and Graves, R.E. (1994) Natural ventilation for cow comfort and increased profitability. In: *Dairy Systems for the 21st Century. Proceedings 3rd International Dairy Housing Conference*. American Society of Agricultural Engineers, St Joseph, Michigan, pp. 558–568.

House, H.K. (1999) Mattresses: the case of cow comfort. *Hoard's Dairyman* 144, 11.

Jarrett, J.A. (1995) Rough concrete was making cows lame. *Hoard's Dairyman* 140, 697.

Kiley-Worthington, M. (1976) The tail movements of ungulates, canids and felids with particular reference to their causation and function as displays. *Behaviour* 56, 69–115.

Kilgour, R. (1985) The definition, current knowledge and implementation of welfare for farm animals – a personal view. In: Mickley, L. and Fox, M. (eds) *Advances in Animal Welfare* Science. Humane Society of the United States, Washington, DC, pp. 31–46.

Lamb, R.C., Baker, R.O., Anderson, J.J. and Walters, J.L. (1979) Effects of forced exercise on two-year-old Holstein heifers. *Journal of Dairy Science* 62, 1791–1797.

Lefcourt, A.M. (ed.) (1991) *Effects of Electrical Voltage/Current on Farm Animals: How to Detect and Remedy Problems*. Agriculture Handbook Number 696. United States Department of Agriculture, Washington, DC.

Livestock Conservation Institute (1992) *Proper Handling Techniques for Non-ambulatory Animals*. Downer Task Force, Madison, Wisconsin, 19 pp.

Matthews, L.R., Phipps, A., Verkerk, G.A., Hart, D., Crockford, J.N., Carragher, J.F., and Harcourt, R.G. (1995) *The Effects of Tail Docking and Trimming on Milker Comfort and Dairy Cattle Health, Welfare and Production.* Contract Research Report to Ministry of Agriculture and Food. Wellington, Animal Behaviour and Welfare Research Centre, AgResearch Ruakura, Hamilton, New Zealand.

Metcalf, J.A., Roberts, S.J., and Sutton, J.D. (1992) Variations in blood flow to and from the bovine mammary gland measured using transit time ultrasound and dye dilution. *Research in Veterinary Science* 53, 59–63.

Midwest Plan Service (1985) *Dairy Housing and Equipment Handbook*, 4th edn. Iowa State University, Ames, Iowa.

Midwest Plan Service (1997) *Dairy Freestall Housing and Equipment Handbook*. 7th edn. Iowa State University, Ames, Iowa, 136 pp.

Munksgaard, L., de Passille, A.M., Rushen, J., Thoxberg, K. and Jensen, M.B. (1995) The ability of dairy cows to distinguish between people. In: Rutter, S.M., Rushen, J., Randle, H.D. and Eddison, J.C. (eds) *Proceedings 29th International Congress. International Society for Applied Ethology.* ISAE, Exeter, UK, and Universities Federation of Animal Welfare, Potters Bar, UK, p. 14.

O'Callaghan, M.A. (1916) *Farm Implement and Machinery Review* (Australia), October.

Pajor, E., Rushen, J. and de Passille, A.M. (1998) Development of aversion learning techniques to evaluate dairy cow handling practices. Presented at Journées de Recherche et Colloque en Productions Animales. Hôtel des Gouverneurs, Sainte-Foy, Quebec, Canada, 21–22 May (abstract).

Pajor, E., Rushen, J. and de Passille, A.M. (1999) Aversion learning techniques to evaluate dairy cow handling practices. 91st Annual Meeting American Society of Animal Science, Indianapolis, Indiana, USA, July 21–24. *Journal of Animal Science* 77, (abstract).

Plumb, C.S. (1893) *Does it Pay to Shelter Milk Cows in Winter?* Bulletin 47, Purdue University Agricultural Experiment Station. West Lafayette, Indiana.

Quaife, T. (1999) Rotary parlours are here to stay. *Dairy Herd Management* 36(3), 44–48.

Rankin, G.W. (1925) *The Life of William Dempster Hoard.* W.D. Hoard and Sons Press, Fort Atkinson, Wisconsin.

Reid, N.S.C. (1977) *To Endeavour to Correlate any Common Factors which Exist in Herdsmen in High Yielding Dairy Herds.* Nuffield Farming Scholarship Report, London, 21 pp.

Roman-Ponce, H.W., Thatcher, W., Buffington, D.E., Wilcox, C.J. and Van Horn, H.H. (1977) Physiological and production responses of dairy cattle to a shade structure in a subtropical environment. *Journal of Dairy Science* 60, 424–430.

Schultz, T.A., Collar, L.S. and Morrison, S.R. (1985) Corral misting effects on heat stressed lactating cows. *Journal of Dairy Science* 68 (Suppl.1), 239 (abstract).

Seabrook, M.F. (1972) A study to determine the influence of the herdman's personality on milk yield. *Journal of Agricultural Labour Science* 1, 45–59.

Seabrook, M.F. (1977) Cowmanship. *Farmers Weekly*, London, 24 pp.

Seabrook, M.F. (1980) The psychological relationship between dairy cows and dairy cowmen and its implications for animal welfare. *International Journal for the Study of Animal Problems* 1, 295–298.

Seabrook, M.F. (1991) The human factor – the benefits of humane and skilled stockmanship. In: Carruthers, S.P. (ed.) *Farm Animals: It Pays To Be Humane.* Centre for Agricultural Strategy, University of Reading, Reading, UK, pp. 62–70.

Seabrook, M.F. (1994) Psychological interaction between the milker and the cow. In: Bucklin, R. (ed.) *Dairy Systems for the 21st Century*. American Society of Agricultural Engineers, St Joseph, Michigan. pp. 49–58.

Smith, J.F., Armstrong, D.V., Gamroth, M.J. and Harner, J., III (1998) Factors affecting milking parlour efficiency and operator walking distance. *Applied Engineering in Agriculture* 14, 643–647.

Smith, J.F., Harner, J.P., Brook, M.L., Armstrong, D.V., Gamroth, M.J., Meyer, M.J., Boomer, G., Bethard, G. and Putnam, D. (1999) Many issues involved in dairy relocation and expansion. *Feedstuffs* 71(28), 11–21.

Stephens, D.B. (1980) Stress and its measurement in domestic animals: a review of behaviour and physiological studies under field and laboratory situations. *Advances in Veterinary Science and Comparative Medicine* 24, 179.

Underwood, W., McClary, D. and Kube, J. (1995) The bovine perfect sleeper or use of shredded rubber filled polyester mattresses to prevent injury to dairy cattle housed in tie stalls. *Bovine Practitioner* 29, 143–148.

Whittlestone, W.G., Kilgour, R., de Langen, H. and Duirs, G. (1970) Behavioural stress and the cell count of milk. *Journal of Milk and Food Technology* 33, 217–220.

Wilson, P. (1998) An animal's first attack should be its last. *Hoard's Dairyman* 143, 787.

Wood, M.T. (1977) Social grooming in two herds of monozygotic twin dairy cows. *Animal Behaviour* 25, 635–642.

# Cattle Transport

9

## Vivion Tarrant[1] and Temple Grandin[2]

[1]The National Food Centre, Teagasc, Dunsinea, Castleknock,
Dublin 15, Eire; [2]Department of Animal Sciences, Colorado
State University, Fort Collins, CO 80523, USA

## Introduction

Transportation by its nature is an unfamiliar and threatening event in the life
of a domestic animal. It involves a series of handling and confinement
situations which are unavoidably stressful and can lead to distress, injury or
even death of the animal unless properly planned and carried out. Transport
often coincides with a change of ownership whereby responsibility for the
animal's welfare may be compromised.

Cattle are transported by road, rail, sea and air for the purposes of
breeding, fattening and slaughter. In Europe, the biggest trade is in the road
transport of cattle to slaughter, and most of the research data available relate
to this situation. There is a tendency towards centralization of modern
slaughter facilities, with corresponding increases in transportation times. In
the USA, transport distances between feedlots and slaughter plants are
relatively short, but weaned calves and yearlings often travel 1000–3000 km
to the feedlots.

The transport process begins with assembly and includes loading,
confinement with and without motion, unloading and penning in a new
and unfamiliar environment. During transport, animals are exposed to
environmental stresses including heat, cold, humidity, noise and motion.
Additional stress may be caused by social regrouping.

Regulations governing the transport of domestic animals vary from
country to country. For example, the European Union requires that journey
times shall not exceed 8 h. However, this may be extended, if the transporting
vehicle meets additional requirements, to 14 h of travel, after which a rest
period of at least 1 h with water is specified for adult animals. They may then
be transported for a further 14 h. Two 9-h periods with 1 h rest for watering is

the maximum permitted for unweaned calves. In the USA, there are no rest-stop requirements. In Canada, a rest stop is required after 48 h of travel. In practice, unless resting facilities are adequate and the animals are unloaded with care, rest stops may be counter-productive and serve only to prolong the overall journey time.

Because transport is associated with a variety of physical and emotional stimuli, many of which are novel and some of which are aversive, it is recognized as a common cause of distress (Fraser, 1979). The response of cattle to transport varies, depending on the situation, and ranges from a moderate and readily identifiable stress response that may not cause concern about the animal's welfare to extreme responses that signify distress and cause major concern about welfare and economic losses.

Fear is a very strong stressor. Both previous experience and genetic factors affecting temperament interact in complex ways to determine how fearful an animal may become when it is handled or transported (Grandin, 1997). In this respect, cattle produced under intensive or extensive production systems react quite differently to transport. In air transport James (1997) observed that cattle were unperturbed by flight and remained calm.

## Measurement of Transport Stress

An objective measurement of transport stress may be attempted, using behavioural, physiological and pathological indicators. Data on behaviour during transport are scarce, but are useful because they provide information about how the animals adapt and cope and show where modifications are necessary to improve transport equipment. More data are available on the physiological responses, at least for the more common transport situations. For example, changes in heart rate, blood composition (electrolytes, hormones, metabolites and enzymes) and live-weight are used to assess the response of livestock to transport. The glucocorticoid content in blood is a good index for the reaction of animals to any environmental challenge (Dantzer and Mormede, 1979). Elevated plasma cortisol and glucose concentrations reflect activation of the pituitary–adrenal axis, whereas increased heart rate and raised plasma glucose and non-esterified fatty acids reflect activation of the sympathetic–adrenomedullary system.

These measures tend to respond similarly to physical and psychological stressors, both of which may occur during transport. However, some types of physical activity may not necessarily compromise welfare, whereas psychological events producing the same sympathetic response may be interpreted less favourably (Jacobson and Cook, 1998).

The yield and the chemical composition of the carcass and offal also provide evidence about the animal's condition before slaughter. The greatest cause for concern about road transport is the injury or bruising that may occur on long or possibly short journeys, together with the respiratory infections and morbidity associated with 'shipping fever' after long journeys.

## Experimental Subdivision of the Transportation Process

To identify stressful or hazardous steps in the road transport operation, Kenny and Tarrant (1987a,b) investigated several components of the transport process. These were, in order of increasing complexity, repenning in a new environment, loading/unloading, confinement on a stationary vehicle and confinement on a moving vehicle. Social regrouping was used as an additional variable. The experimental animals were slaughter-weight Friesian steers or bulls and they were subjected to 1-h transport treatments while under continuous direct observation of behaviour.

The frequency of social behaviours increased greatly in response to the least complex treatment, the repenning of cattle in an unfamiliar environment. The social behaviours observed were sexual, e.g. mounts and 'chin resting', as well as aggressive behaviours, e.g. butts, pushes, threats and mock fights.

As the complexity of the transport treatment increased, the frequency of interactions between the bulls was decreased, being inhibited by close confinement on the truck and further decreased by motion. This pattern of behaviour was observed in both non-mixed and socially regrouped animals. Eldridge (1988) also reported higher social activity in slaughter-weight cattle during repenning and declining activity during trucking.

The increase in physical activity was reflected in increased activity of the muscle enzyme creatine kinase (CK) detected in the plasma. CK leaks into the bloodstream from muscle tissue during unaccustomed or vigorous exercise or as a result of muscle damage caused by bruising.

In marked contrast with the social interactions, solitary behaviours such as exploration and urination increased in frequency as the complexity of the transport treatment increased. A higher frequency of urination may indicate fear. The plasma cortisol concentration also increased with complexity of the transport treatment, suggesting a growing stress response.

On the basis of these and other experiments it was concluded that young adult cattle showed increased disturbance or stress in the following order of treatment: repenning < stationary confinement < confinement in a moving truck. This ranking was based on an increasing plasma cortisol concentration, suppression of social interactions and increasing urination, and applied equally to non-mixed and regrouped animals. The data identified the more stressful operations in road transport. However, there was no evidence that any of the transport treatments were harmful or caused major distress to the cattle.

Loading and unloading using a tailgate ramp is a cause of stress and injury for smaller animals, such as pigs (van Putten, 1982), and appears to be the most stressful component of the transport chain for pigs (van Putten and Elshof, 1978; Augustini and Fischer, 1982). However, the loading ramp did not present a major obstacle for cattle and loading/unloading was accomplished without difficulty, the main physiological effect being an increase in heart rate, which was inevitable in view of the physical exertion involved in mounting the ramp (Kenny and Tarrant, 1987a). Adult cattle were able to negotiate a wide variety of ramp designs without difficulty (Eldridge *et al.*, 1986). Difficulties at loading in commercial transport are often caused by overloading, with the last few cattle being driven forcefully on board.

The above results were obtained using relatively docile cattle from intensive production systems. These results cannot be extrapolated to cattle raised in extensive systems with infrequent contact with stockpersons. For wild cattle, loading and unloading may be more stressful than riding in the vehicle. Extensively raised beef cattle will have higher heart rates and cortisol levels when they are restrained and handled compared with tame dairy cows (Lay *et al.*, 1992a,b). In wild beef cattle, handling stresses were almost as severe as hot-iron branding stress (Lay *et al.*, 1992a, b). Agnes *et al.* (1990) report that loading calves up a 30° angle ramp and simulated truck noise elicited cortisol levels similar to simulated transport.

Eldridge *et al.* (1988) concluded that, once cattle adapted to the journey, road transport was not a major physical or psychological stressor. These authors recorded heart rates of beef heifers by radiotelemetry during road transport at different loading densities. The overall mean heart rates while travelling were only 15% above those recorded while animals were grazing at pasture. Similar results were obtained for bulls and steers undergoing short-haul road transport (Tennessen *et al.*, 1984).

## Cattle Behaviour in Moving Vehicles

The behaviour of cattle on moving vehicles is of interest in view of the evidence that confinement on a moving vehicle is the most stressful step in the transportation process, at least for intensively reared cattle. Earlier reports on behaviour during rail transportation noted heightened activity immediately after loading, and characteristic standing and lying behaviour in moving and stationary vehicles (Bisschop, 1961). Experiments in the USA with railcars equipped with hay and water troughs indicate that the animals will eat and drink during transport (Irwin and Gentleman, 1978).

### Restlessness

This is indicated by the frequency of changes in position in the vehicle. Restlessness increased with social regrouping on the truck, but not with motion (Kenny and Tarrant, 1987a,b). Changes in position were frequently triggered by social interactions, such as chin-resting and mounting, and also, when the truck was moving, by driving events, particularly cornering.

## Standing orientation

The most common direction for cattle to face on a truck is either perpendicular or parallel to the direction of motion, with the diagonal orientations infrequently used (Kenny and Tarrant, 1987a,b; Eldridge *et al.*, 1988; Lambooy and Hulsegge, 1988; Tarrant *et al.*, 1988, 1992). This may indicate that cattle have a preferred orientation to improve security of balance on a moving vehicle. Bisschop (1961) found that cattle align themselves across the direction of travel during rail transport; however, Kilgour and Mullord (1973) found no clear preference by young beef cattle during road travel. On long journeys, the most common standing orientation was perpendicular to the direction of travel, and there was a strong bias against diagonal orientations (Tarrant *et al.*, 1992).

Cattle tend not to lie down in trucks while they are moving (Warriss *et al.*, 1995). In 1-h and 4-h journeys to slaughter, no animals lay down in 18 loads of Friesian steers or bulls transported at low or medium stocking density (Kenny and Tarrant, 1987a,b; Tarrant *et al.*, 1988). At high stocking density, especially approaching maximum density, cattle occasionally went down, apparently involuntarily. Towards the end of long (24-h) road journeys with Friesian steers, several cattle lay down during the final 4–8 h of the journey. This was observed at all stocking densities, but only at high stocking density were animals trapped down and unable to rise. The stocking densities used in the first author's research were 200, 300 and 600 kg m$^{-2}$. The highest stocking density is the maximum amount of cattle that can be put on a truck when it is easy to close the rear gate.

Honkavaara (1998) noted that even one restless animal was sufficient to cause continuous movements of the group; as a result, no animal could lie down during transport. However, when cattle were transported in stock crates designed to hold two animals per pen, an animal often lay down after 2 or 3 h of transport (Honkavaara, 1993). This may indicate a preference for lying down when circumstances permit.

## Maintenance of balance

Loss of balance on moving vehicles is a major consideration in cattle transport, in view of the hazard associated with large animals going down during transport and the risk of injury or suffocation. Observations show that minor losses of balance occur regularly and that cattle quickly respond by shifting their footing to regain their balance.

The relationship between loss of balance and driving events during 24-h journeys with Friesian steers is shown in Table 9.1. Eighty per cent of losses of balance were accounted for by braking, gear changes and cornering. Similar results were obtained for Freisian bulls and steers on shorter road journeys (Kenny and Tarrant, 1987a,b; Tarrant et al., 1988).

Table 9.1 shows that cornering is the driving event that caused most losses of balance in cattle transported at high stocking density, whereas braking was a greater hazard at low density.

These data show that losses of balance are under the direct control of the driver. Eldridge (1988) observed that the heart rate of beef heifers was lower, indicating reduced stress, when the vehicle was travelling smoothly on highways, compared with rougher country roads or suburban roads with frequent intersections.

Factors likely to influence security of balance during unsteady driving are the slipperiness of the floor surface and the availability of support from adjacent structures, including vehicle sides and partitions, and other animals. It may be advisable to withhold water during the last 6 h before loading (Wythes, 1985), thus resulting in a drier truck floor, giving cattle a better footing during the journey.

The major factors determining the well-being of cattle in road transport are vehicle design, stocking density, ventilation, the standard of driving and the quality of the roads. The importance of frequent inspection of the livestock and of careful driving cannot be overemphasized. Resting periods, with access to water, are necessary when the journey exceeds 24 h. Long journeys are practical under good conditions, whereas even short journeys are hazardous under poor conditions.

**Table 9.1.**   The association between losses of balance on a moving vehicle and driving events during 24-h journeys with Friesian steers. Data are percentages of total losses of balance. (From Tarrant et al., 1992.)

| Driving event | Stocking density | | |
| --- | --- | --- | --- |
| | Low | Medium | High |
| Braking | 55 | 58 | 19 |
| Gear changes | 21 | 17 | 19 |
| Starting/stopping | 9 | 15 | 0 |
| Cornering | 5 | 6 | 50 |
| Bumping | 2 | 2 | 0 |
| All other events | 1 | 1 | 0 |
| Uneventful | 6 | 2 | 12 |

## Falls

The major risk in cattle transport is that of cattle going down under foot. This risk is greatly increased at high stocking density (Tarrant *et al.*, 1988, 1992). The normal response to a loss of balance is a change or shift of footing in order to regain balance. Shifts were inhibited at high stocking density (Table 9.2) and there was a corresponding increase in struggles and falls at high stocking density. These unstable situations were caused either by driving events, typically cornering or braking, by standing on a fallen animal or by strenuous and usually unsuccessful attempts to change position in a full pen.

When cattle went down at high stocking density, they were trapped on the floor by the remaining cattle 'closing over' and occupying the available standing space. Several unsuccessful attempts by fallen animals to stand up were observed. A 'domino' effect was created when standing animals lost their footing by trampling on a fallen animal. The substantial increase found in carcass bruising at high stocking density was explained by these observations.

## Stocking density on trucks

Freedom of movement was severely restricted at high loading densities, with only 16 changes of position observed per group of cattle per hour of transport, compared with 109 per group per hour at low stocking density (Tarrant *et al.*, 1988). Similar results were obtained on long road journeys (Tarrant *et al.*, 1992). Exploratory, sexual, aggressive behaviours were inhibited at high stocking densities, with the exception of mounting and pushing, which increased in frequency with stocking density.

The preferred orientations adopted by animals during long-distance transportation were frustrated as the stocking density was increased. Thus, In addition to reducing mobility, an increase in the stocking density also prevented cattle from facing in the preferred direction. These effects may combine to increase the rate of loss of balance and falling, as discussed above.

**Table 9.2.** The effects of stocking density in the truck on losses of balance by Friesian steers during 24-h journeys by road (from Tarrant *et al.*, 1992).

| Losses of balance | Stocking density | | |
|---|---|---|---|
| | Low | Medium | High |
| Shifts | 153 | 142 | 26 |
| Struggles | 5 | 4 | 10 |
| Falls | 1 | 1 | 8 |

In 4-h and 24-h road journeys to the abattoir, the cortisol and glucose content in the plasma of Friesian steers increased with stocking density, indicating increasing stress (Tarrant *et al.*, 1988, 1992). The activity of the muscle enzyme CK in the bloodstream also increased with stocking density, reflecting muscle damage. Carcass bruising increased with stocking density (Table 9.3).

An unexpected finding was that dressed carcass weight was significantly reduced at high loading density (Eldridge and Winfield, 1988); this weight loss was only partly explained by the higher trimming of bruised tissue from the carcass at high density.

It is clear from the above that stocking density is important for animal welfare during transportation and becomes critical at high stocking densities. High stocking density on trucks was clearly associated with reduced welfare and carcass quality, when compared with medium and low stocking densities. Attempts to reduce transport costs by overloading of trucks are offset by reduced carcass weight, downgrading of carcasses owing to bruising and increased risk of serious injury or death during travel (Eldridge and Winfield, 1988).

At low stocking densities, e.g. half-loads, cattle will travel very well unless subjected to poor driving techniques, such as sudden braking or swerving, and emergency stops. Eldridge and Winfield (1988) transported steers of 400 kg live-weight on road journeys of 6 h at high (0.89 m$^2$ per animal), medium (1.16 m$^2$ per animal) and low (1.39 m$^2$ per animal) stocking densities. Bruise scores at the high and low stocking densities were four and two times greater, respectively, than at the medium space allowance (8.2, 4.6 and 1.9 bruise scores, respectively; $P < 0.01$). The results show that the medium stocking density was superior to the low density and indicate that an optimum density may be defined by such experiments. The medium stocking densities used by Eldridge and Winfield (1988) can be found in Grandin (1981a) and Federation of Animal Science Societies (1999). These medium stocking densities used by

**Table 9.3.** The effect of stocking density during 24-hour road journeys on plasma constituents and carcass bruising in Friesian steers. Values for plasma cortisol, glucose and CK are the difference between the pre- and post-transport values. (From Tarrant *et al.*, 1992.)

| | Stocking density | | | |
|---|---|---|---|---|
| Plasma constituent | Low | Medium | High | Level of statistical significance |
| Plasma cortisol (ng ml$^{-1}$) | 0.1 | 0.5 | 1.1 | $P < 0.05$ |
| Plasma glucose (mmol l$^{-1}$) | 0.81 | 0.93 | 1.12 | $P < 0.15$ |
| Plasma CK (units l$^{-1}$) | 132 | 234 | 367 | $P < 0.001$ |
| Carcass bruise score | 3.7 | 5.0 | 8.5 | $P < 0.01$ |

Eldridge and Winfield (1988) are very similar to a stocking density formula published by Randall (1993). Randall qualified the use of the equations, saying they should be used for trips of under 5 h. Knowles (1999) contains an excellent diagram which compares stocking densities recommended by the Farm Animal Welfare Council (FAWC) and the European Union with those of Randall. Knowles (1999) concludes that welfare can be poor if stocking density is either too high or too low.

## Carcass Bruising

High financial losses are incurred by the livestock industry as a result of carcass bruising (Dow, 1976; Hails, 1978; Grandin, 1980; Wythes and Shorthose, 1984; Eldridge and Winfield, 1988). Carcass bruises cause an estimated loss of edible meat and carcass devaluation of $11.73 animal$^{-1}$ in the USA (Smith *et al.*, 1994). Bruising is an impact injury that can occur at any stage in the transportation chain and may be attributed to poor design of handling facilities, ignorant and abusive stockmanship and poor road driving techniques during transportation (Grandin, 1983a). Cattle should be marketed in a manner that minimizes the number of times that they are handled or restrained immediately prior to slaughter, particularly when cattle are transported more than 325 km to slaughter (Hoffman *et al.*, 1998). Cattle that were handled roughly had greatly elevated bruising compared with cattle that were handled gently (Grandin, 1981b).

The skill of the driver and the quality of the road appear to be more important than the distance travelled. Economic incentives can greatly reduce bruising. Cattle sold by live-weight had twice as many bruises compared with cattle sold on a carcass basis (Grandin, 1981b). Producers selling on a carcass basis have bruises deducted from their payments. Stocking density is an important consideration, and high stocking density was associated with a twofold or greater increase in carcass bruising in both short-haul (Eldridge and Winfield, 1988; Tarrant *et al.*, 1988) and long-haul road transport (Tarrant *et al.*, 1992) (see also Table 9.3). Barnett *et al.* (1984) considered that cattle with elevated blood corticosteroid concentrations as a result of chronic stress may be more susceptible to bruise damage than other cattle.

Wythes (1985) and Shaw *et al.* (1976) found that horned cattle had twice as much bruising. Contrary to popular belief, cutting the tips of the horns does not reduce bruising (Ramsey *et al.*, 1976). Cows in late pregnancy suffered more bruising and produced tougher meat than those in early pregnancy or those which were not pregnant (Wythes, 1985). The Dutch Road Transport Act prescribed that adult cows and heifers should be separated by a gate between every two animals when the transport lasts longer than 10 h (Lambooy and Hulsegge, 1988). However, the transporters do not adhere to this rule and carry five to ten cattle per compartment. Experimentation showed that loose transport of eight heifers per pen is preferential to penning in pairs between gates, mainly because of lower risks of injury and lower frequency of lesions at contact points, e.g. hips and knees. Honkavaara (1995) found more carcass damage in cattle transported in trailers than in cattle transported in the front cars.

## Vehicle Design

Information on the design of loading/unloading ramps and stock crates for single- and double-deck trucks and trailers is available (Anon., 1977; Wythes, 1985; Grandin, 1991). Practical experience in the USA indicated that there were fewer bruises in trailers which had doors that opened up to the full width of the trailer for unloading. Stock carried by rigid vehicles tend to experience a rougher ride than stock transported by an articulated trailer. This is mainly because rigid-body trucks, which are smaller and easier to handle, are generally driven faster than large articulated vehicles (Anon., 1977; Fig. 9.1). Vibration stress can also be reduced by installing a pneumatic suspension (Singh, 1991). These systems must be kept in good repair, because a damaged pneumatic suspension may produce higher vibration than a vehicle with leaf springs (Singh, 1991). Practical experience has shown that a well-maintained pneumatic suspension system will reduce stress. Over-inflated tyres will also increase vibration in a livestock truck (Stevens and Camp, 1979). Drivers over-inflate tyres to prolong the life of the tyres, but this practice is probably

**Fig. 9.1.**   Australian double-deck articulated cattle truck.

detrimental to livestock. Cattle in the USA are hauled in aluminium trailers (Fig. 9.2). A lack of bedding on the aluminium floor can cause toe abscesses (Sick *et al.*, 1982).

European Union regulations permit longer journey times where the transporting vehicle meets certain requirements. These include sufficient bedding on the floor, appropriate feed for the journey time, equipment for connection to a water supply during stops, adjustable ventilation, direct access to the animals and movable partitions for creating separate compartments.

## Ventilation

Heat builds up rapidly in a stationary vehicle. Vehicles should be kept moving and rest stops should be kept to a minimum. Even during cool weather, a beef animal's temperature will rise when the vehicle is stationary (Stevens *et al.*, 1979). In aeroplanes and ships heat can rapidly build up to fatal levels when the vehicle is stationary (Stevens and Hahn, 1977; Grandin, 1983b; Connell, 1984). Ventilation recommendations for ships and aircraft are given in Muller (1985), Stevens (1985) and Animal Transportation Association (1992). Muirhead (1985) found that there are areas of no air movement in moving trucks. Natural ventilation in trucks through openings in the side walls results in non-uniform air circulation at animal head level in practice (Honkavaara,

**Fig. 9.2.** Typical aluminium-sided trailer which is used in the USA for transporting both cattle and pigs. Numerous holes provide abundant ventilation.

1998); therefore, controlled mechanical ventilation systems are recommended for vehicles on long and short journeys. In the USA, trailers have adequate ventilation because the side of the trailer has numerous small holes (see Fig. 9.2). During cold weather one-third to one-half of the holes are covered with either plastic plugs or plywood.

## Meat Quality

Good-quality beef has a final pH value close to 5.5. At pH values of 5.8 and above, both the tenderness and the keeping quality of the fresh chilled meat are adversely affected. High-pH meat is unsuitable for the premium trade in vacuum-packed fresh meats and, depending on the commercial use of the product, dark-cutting meat may be discounted by 10% or more (Tarrant, 1981).

High meat pH is caused by an abnormally low concentration of lactic acid in the meat, which in turn is a reflection of low muscle glycogen content at slaughter. Post-mortem production of lactic acid requires an adequate content of glycogen in the muscles at slaughter. Ante-mortem glycogen breakdown is triggered by increased adrenaline release in stressful situations or by strenuous muscle activity. Circumstances that trigger one or both of these glycogen breakdown mechanisms will deplete muscle glycogen, especially in the fast twitch fibres (Lacourt and Tarrant, 1985; Shackleford *et al.*, 1994), and will result in high-pH dark-cutting meat unless a recovery period from stress is allowed. In practice, in many abattoirs, restful conditions with access to feed cannot be provided. Furthermore, the rate of muscle post-stress glycogen repletion is slower in cattle than in other species (McVeigh and Tarrant, 1982), so it is better to avoid the problem than to attempt to remedy it.

The animal behaviour most closely associated with glycogen depletion and dark-cutting beef is mounting activity. This behaviour is stimulated by social regrouping, as in mixed penning of young bulls (McVeigh and Tarrant, 1983; Warriss *et al.*, 1984) and by heat (oestrus) in groups of females (Kenny and Tarrant, 1988). Modifications of the holding pens aimed at reducing mounting activity during penning before slaughter have been successful in preventing dark-cutting in beef (Kenny and Tarrant, 1987c). Social regrouping prior to transport causes a much higher incidence of dark cutters in bulls compared with steers (Price and Tennessen, 1981; Tennessen *et al.*, 1985). Short periods of mixing greatly increases dark cutters in bulls, but dark-cutting will increase in steers if they are mixed for more than 24 h (Grandin, 1979). Scanga *et al.* (1998) found that dark-cutting increased if there were sharp temperature fluctuations or temperature extremes 24–72 h prior to slaughter. Practical experience in large slaughter plants has also shown that feedlot cattle that spend the night in the plant lairage have more dark cutters. Other factors which can greatly influence the occurrence of dark-cutting beef in fed cattle is excessive use of growth promotants (Scanga *et al.*, 1998).

Short road journeys are not likely to cause dark-cutting (Eldridge and Winfield, 1988), except where trauma occurs, for example, when an animal goes down (Tarrant *et al.*, 1992). Warnock *et al.* (1978) also found much higher meat pH values in the carcasses of 'downer' cows, compared with cows that did not go down (6.3 vs. 5.7).

Long-distance road or rail transport of cattle caused a small elevation of meat pH and a corresponding increase in the incidence of dark cutters (Wythes *et al.*, 1981; Tarrant *et al.*, 1992; Honkavaara, 1995). This was reversed by resting and feeding for 2 days or longer before slaughter (Shorthose *et al.*, 1972; Wythes *et al.*, 1980).

Other effects of transport on meat quality include an increase in toughness (Schaefer *et al.*, 1990) and a decrease in palatability (Jeremiah *et al.*, 1992; Schaefer *et al.*, 1992). The sensory quality of veal was lower after long-distance transport of 20-week-old calves (Fernandez *et al.*, 1996).

## Effect of Transport on Yield

The loss of live-weight and carcass yield during transportation of cattle is of both welfare and economic concern. Animals lose live-weight as a consequence of excretion, evaporation and respiratory exchange (Dantzer, 1982).

In cattle hauled an average of 1023 km, almost half the shrinkage is loss of carcass weight (Self and Gay, 1972). Most of the live-weight losses during transportation may be attributed to the effect of withdrawal of feed and water. The gut contents can account for 12–25% of the animal's live-weight. Fasting of 396 kg steers for 12, 24, 48 and 96 h caused live-weight losses of 6, 8, 12 and 14% (Wythes, 1982). Similarly, in the USA, slaughter-weight cattle transported for 5 and 26 h, lost 2 and 6.3% of the body weight, respectively (Mayes *et al.*, 1979). In 24-h journeys by road, the live-weight losses in cattle were about 8% (Shorthose, 1965; Lambooy and Hulsegge, 1988; Tarrant *et al.*, 1992). Recovery of body weight to pre-transport values took 5 days (Warriss *et al.*, 1995).

In 24-h road journeys under cool ambient conditions (4–16°C), there was evidence of dehydration, as shown by increases in red blood cell count, haemoglobin, total protein and packed-cell volume (Tarrant *et al.*, 1992; Warriss *et al.*, 1995). Dehydration is, therefore, a factor in loss of live-weight and also carcass weight during transportation. Lambooy and Hulsegge (1988) found slightly increased haematocrit and haemoglobin values in pregnant heifers transported by road for 24 h. The heifers had access to water and feed after 18 h of transport, and water uptake per animal ranged from 1 to 6 litres. Shorthose (1965) calculated that the approximate rate of carcass weight loss in steers was 0.75% day$^{-1}$ for transport and holding times lasting from 3 to 8 days. Providing water *ad libitum* to fasted livestock reduces shrink (Hahn *et al.*, 1978). The effect of giving cattle access to water after a long journey in hot weather (25 and 36°C) was examined by Wythes (1982). Access to water for

3.5 h or longer before slaughter allowed muscle water content to increase, and this was reflected in heavier carcasses. Providing cattle with an oral electrolyte in their drinking water reduced both carcass shrink and dark-cutting (Schaefer *et al.*, 1997). In a major study of 4685 calves and yearlings, animals subjected to the increased stress of moving through a saleyard had greater shrink than animals purchased directly from the ranch of origin (Self and Gay, 1972).

Collectively, the physiological changes observed in cattle during transport and handling, which include changes in blood cells, blood metabolites and enzymes, electrolyte balance, dehydration and increased heart rate, suggest that treatments designed to attenuate stress should be considered as a means of protecting animal welfare and benefiting carcass quality and yield (Schaefer *et al.*, 1990, 1997). The application of oral electrolyte therapy, especially if similar in constituents to interstitial fluid, seems to attenuate these physiological changes and results in less carcass shrink and reduced dark-cutting.

## Bovine Respiratory Disease and 'Shipping Fever'

The most important disease associated with the transportation of cattle is shipping fever, which is attributed to the stress caused by transporting cattle or calves from one geographical region to another. In North America, where feedlot fattening of beef cattle is common, it is estimated that 1% of cattle die as a consequence of transport stress and its aftermath (Irwin *et al.*, 1979). Bovine respiratory disease (shipping fever) is responsible for 50% death losses in the feedlot and 75% of the sickness (Edwards, 1996). Feedlot cattle with bovine respiratory disease gain less weight (Morck *et al.*, 1993). Shipping fever has also been reported in most European countries and in Asia (Hails, 1978). Differences between marketing systems in the USA and Australia that predispose cattle to shipping fever are discussed by Irwin *et al.* (1979). To reduce losses, calves hauled long distances should be fed a 50% concentrate diet before shipping (Hutcheson and Cole, 1986).

Economic losses caused by deaths are minor compared with the cost of prophylactic treatment of affected cattle and poor growth in those that recover. The main symptoms of shipping fever are those of the bovine respiratory disease complex. This syndrome is characterized by fever, dyspnoea and fibrinous pneumonia, less often by gastroenteritis and only occasionally by internal haemorrhage. Fed cattle that had lung lesions from pneumonia (shipping fever) at slaughter gained less weight, carcasses were downgraded for less marbling and the meat was tougher (Gardner *et al.*, 1998). Other researchers have also found that the presence of lung lesions at slaughter reduces weight gain (Wittum *et al.*, 1995).

The pathogenesis of bovine respiratory disease involves a sequential cascade of events initiated by stress, which may have lowered the animal's resistance to infection. Very little research has been done on the detrimental effects on the immune system of heat, cold, crowding, mixing, noise and restraint (Kelley, 1980; Kelley *et al.*, 1981). Rumen function is impaired by transit stress. Transport imposes a greater stress on the rumen than feed and water deprivation (Galyean *et al.*, 1981). This impairment may be explained by a decrease in rumination during transport. Kent and Ewbank (1991) reported that during transport rumination greatly decreased in 3-month-old calves. In extensively reared beef cattle, the stress of transport had a greater detrimental effect on the animal's physiology than the stress of feed and water deprivation for the same length of time (Kelley *et al.*, 1981; Blecha *et al.*, 1984). Stress-induced changes in host resistance may explain the physiological basis of shipping fever in cattle. Tarrant *et al.* (1992) observed an increase in total white blood cell count and neutrophil numbers and a reduction in lymphocyte and eosinophil numbers in cattle after long journeys. The reduction in lymphocytes may result in a loss of resistance to infection in cattle after long journeys. For example, Murata (1995) found that serum collected after 48 h of transportation had an immunosuppressive effect on peripheral-blood neutrophils, decreasing their bactericidal activity. Transporting beef calves immediately after weaning can increase stress. Crookshank *et al.* (1979) found that calves transported immediately after weaning had higher cortisol levels compared with calves that were weaned and placed in feedlot pens for 2 weeks prior to transport. Both weaning and transport affect the humoral immune response of calves (MacKenzie *et al.*, 1997).

In a study of 45,000 6-month-old calves transported to feedlots, Ribble and colleagues (1995) found that differences between short and long hauls explained little, if any, of the variation among truck-loads of calves in the risk of fatal fibrinous pneumonia. They suggested that other elements of the transportation process may be more stressful and therefore responsible for shipping fever.

Research on 7845 calves has shown that sickness in 6-month-old calves that have been transported long distances can be greatly reduced by weaning and vaccinating the calves 5–6 weeks prior to long-distance transport (National Cattlemen's Association, 1994). Unvaccinated calves that are shipped the same day they are weaned will have more respiratory sicknesses (National Cattlemen's Association, 1994). Death losses due to respiratory disease were 0.16% in vaccinated and preweaned calves, 0.98% in calves which were still bawling after being removed from the cow and 2.02% in calves bought from order buyers and auctions (National Cattlemen's Association, 1994). The calves in this study experienced one of the worst winters on record at a Kansas feedlot. Preweaning and vaccination prior to shipment may provide greater advantages to animals exposed to extreme weather conditions.

Practical experience shows that cattle from pastures that were deficient in minerals had more death losses than cattle that received mineral supplements (Peltz, 1999). Supplementation of newly arrived calves with vitamin E, chromium or an antioxidant can reduce sickness and improve performance (Barajas and Ameida, 1999; Purnell, 1999; Stovall *et al.*, 1999). A large dose of 1600 IU day$^{-1}$ of vitamin E in the feed was most effective. Since newly arrived feeder cattle which are showing signs of sickness often have reduced feed intake, they should be fed diets with increased nutrient density and be supplemented with extra vitamins and minerals to help reduce sickness (Galyean *et al.*, 1999; Loerch and Fluharty 1999; Sowell *et al.*, 1999).

When wild, extensively raised calves are transported, dealers who transport thousands of calves on trips ranging from 1000 km to 2000 km have found that the animals are less likely to get sick if they are transported within 32 h without a rest stop (Grandin, 1997). This may possibly be due to the fact that some of the calves have not been vaccinated prior to transport. Another factor is that loading and unloading may be stressful to calves which are not accustomed to handling. Factors unrelated to transport or handling may also affect susceptibility to shipping fever. Calves which received an adequate passive immunity from the mother cow's colostrum are more resistant to bovine respiratory disease (Wittum and Perino, 1995). This implies that maternal traits and adequate milk production affect susceptibility to disease due to transport stresses later in life.

## Conclusions

Under good conditions of transportation and on short journeys by road, cattle show clear physiological and behavioural signs of stress. These are not necessarily signs of distress, but may represent a normal adaptation to an uncertain and threatening situation.

If transportation conditions deteriorate, the observed physiological and behavioural changes intensify and pathological changes can occur. It is not uncommon for transport conditions to deteriorate to such an extent that considerable suffering and economic losses occur. Deterioration may be accidental or may be the result of bad practice or ignorance. Evidence of malpractice includes lack of suitable equipment for animal handling and lack of adequate training of stockmen and truck drivers, together with inadequate supervision.

The detrimental effects of transport can be highly variable, depending on climate, driving skill, stocking density, fear stress and vaccinations. Since conditions are so variable, there is a need for surveys of hundreds of loads conducted under commercial conditions in each country. Rest-stop schedules that work well in Europe may increase stress in wild cattle in Australia, the USA and South America. Improved facilities are urgently needed to take advantage of the behavioural characteristics of animals, to improve their throughput and to reduce damage to the carcass.

In intensively raised cattle, the main stress in transportation is caused by confinement on the moving vehicle. Loading/unloading using conventional ramps is not a source of stress in relatively tame cattle. Loss of balance on a moving vehicle is associated with specific driving events, notably cornering and braking. Loss of balance may result in cattle going down in transit. When this happens at high stocking density, animals are trapped on the floor, with very serious consequences.

Low stocking density *per se* is not a cause of stress but leaves cattle more vulnerable to careless driving and emergency stops. Cattle normally prefer to stand during transport and may orientate themselves across the direction of travel.

The evidence of dehydration and fatigue after road transport for 24 h suggests that any extension of journey time or deterioration in transport conditions would be detrimental to the welfare of the animals. Meat quality was decreased due to higher meat pH after long distance transport.

It is concluded that transportation is inevitably associated with a stress response but that distress may be avoided by good management and the use of suitably designed and equipped transport vehicles and facilities.

# References

Agnes, F., Sartorelli, P., Abdi, B.H. and Locatelli, A. (1990) Effect of transport loading noise on biochemical variables in calves. *American Journal of Veterinary Research* 51, 1679–1681.

Animal Transportation Association (1992) *Handbook of Live Animal Transportation*. Fort Washington, Maryland.

Anon. (1977) *Livestock Building Project: Stockyard and Transport Crate Design*. Report of the National Material Handling Bureau, Department of Productivity, Australia, July, 46 pp.

Augustini, C. and Fischer, K. (1982) Physiological reaction of slaughter animals during transport. In: Moss, R. (ed.) *Transport of Animals Intended for Breeding, Production and Slaughter*. Martinus Nijhoff, The Hague, pp. 125–135.

Barajas, R. and Ameida L. (1999) Effect of vitamin E and chromium-methionine supplementation on growth performance response of calves recently arrived to a feedlot. *Journal of Animal Science* 77 (Suppl. 1) 269 (abstract).

Barnett, J.L., Eldridge, G.A., McCausland, I.P., Caple, I.W., Millar, H.W.C., Truscott, T.G. and Hollier, T. (1984) Stress and bruising in cattle. *Animal Production in Australia, Proceedings of the Australian Society of Animal Production* 15, 653.

Bisschop, J.H.R. (1961) Transportation of animals by rail (1): the behavior of cattle during transportation by rail. *Journal of the South African Veterinary Medical Association* 32, 235–261.

Blecha, F., Boyles, S.L. and Riley, J.G. (1984) Shipping suppresses lymphocyte blastogenic responses in Angus and Brahman cross feeder calves. *Journal of Animal Science* 59, 576–583.

Connell, J. (1984) *International Transport of Farm Animals Intended for Slaughter*. Report EUR 9556, Commission of the European Communities, Brussels, 67 pp.

Crookshank, H.R., Elissalde, M.H., White, R.G., Clanton, D.C. and Smalley, H.E. (1979) Effect of transportation and handling of calves upon blood serum composition. *Journal of Animal Science* 48, 430–436.

Dantzer, R. (1982) Research on farm animal transport in France: a survey. In: Moss, R. (ed.) *Transport of Animals Intended for Breeding, Production and Slaughter*. Martinus Nijhoff, The Hague, pp. 218–231.

Dantzer, R. and Mormede, P. (1979) Le Stress en Élevage Intensif. Masson, Paris, 117 pp.

Dow, J.K.D. (1976) A survey of the transport and marketing conditions of slaughter cattle, sheep and pigs in the United Kingdom. Animal Health Trust, London, 23 pp.

Edwards, A.J. (1996) Bovine Practice 30, 5.

Eldridge, G.A. (1988) Road transport factors that may influence stress in cattle. In: Chandler, C.S. and Thornton, R.F. (eds) Proceedings of the 34th International Congress of Meat Science and Technology. CSIRO, Brisbane, Queensland, pp. 148–149.

Eldridge, G.A. and Winfield, C.G. (1988) The behaviour and bruising of cattle during transport at different space allowances. Australian Journal of Experimental Agriculture 28, 695–698.

Eldridge, G.A., Barnett, J.L., Warner, R.D., Vowles, W.J. and Winfield, C.G. (1986) The handling and transport of slaughter cattle in relation to improving efficiency, safety, meat quality and animal welfare. In: Research Report Series No. 19. Department of Agriculture and Rural Affairs, Victoria, Australia, pp. 95–96.

Eldridge, G.A., Winfield, C.G. and Cahill, D.J. (1988) Responses of cattle to different space allowances, pen sizes and road conditions during transport. Australian Journal of Experimental Agriculture 28, 155–159.

Federation of Animal Science Societies (1999) Guide for the Care and Use of Agricultural Animals in Agricultural Research and Teaching. Federation of Animal Science Societies, Savoy, Illinois.

Fernandez, X., Monin, G., Culioli, J., Legrand, I. and Quilichini, Y. (1996) Effect of duration of feed withdrawal and transportation time on muscle characteristics and quality in Friesian–Holstein calves. Journal of Animal Science 74, 1576–1583.

Fraser, A.F. (1979) The nature of cruelty to animals. Applied Animal Ethology 5, 1–4.

Galyean, M.L., Lee, R.W. and Hubbart, M.W. (1981) Influence of fasting and transit on rumen function and blood metabolites in beef steers. Journal of Animal Science 53, 7–18.

Galyean, M.L., Perino L.J. and Duff G.C. (1999) Interaction of cattle health/immunity and nutrition. Journal of Animal Science 77, 1120–1134.

Gardner, B.A., Dolezal, H.G., Bryant, L.K., Owens, F.N., Nelson, J.L., Schutte, B.R. and Smith, R.A. (1998) Health of finishing steers: effects on performance carcass traits and meat tenderness. In: 1998 Animal Science Research Report p-965. Oklahoma Agricultural Experiment Station, Stillwater, Oklahoma, pp. 37–45.

Grandin, T. (1979) The effect of pre-slaughter handling and penning procedures on meat quality. Journal of Animal Science 49 (Suppl. 1), 147 (abstract).

Grandin, T. (1980) Bruises and carcass damage. International Journal of Studies in Animal Problems 1, 121–137.

Grandin, T. (1981a) Livestock Trucking Guide. Livestock Conservation Institute, Madison, Wisconsin.

Grandin, T. (1981b) Bruises on southwestern feedlot cattle. Journal of Animal Science 53 (Suppl. 1), 213 (abstract).

Grandin, T. (1983a) Welfare requirements of handling facilities. In: Baxter, S.H., Baxter, M.R. and MacCormack, J.A.C. (eds) Farm Animal Housing and Welfare. Martinus Nijhoff, The Hague, pp. 137–149.

Grandin, T. (1983b) A Survey of Handling Practices and Facilities Used in the Export of Australian Livestock. Department of Primary Industry, Canberra, Australia. Reprinted in Proceedings Animal Air Transportation Association 1984. Dallas, Texas.

Grandin, T. (1991) Design of loading facilities and holding pens. *Applied Animal Behaviour Science* 28, 187–201.

Grandin, T. (1997) Assessment of stress during handling and transport. *Journal of Animal Science* 75, 249–257.

Hahn, G.L., Clark, W.D., Stevens, D.G. and Shanklin, M.D. (1978) *Interaction of Temperature and Relative Humidity on Shrinkage of Fasting Sheep, Swine and Beef Cattle*. Paper No. 78.6010, American Society of Agricultural Engineers, St Joseph, Michigan.

Hails, M.R. (1978) Transport stress in animals: a review. *Animal Regulation Studies* 1, 289–343.

Hoffman, D.E., Spire, M.F., Schwenke, J.R. and Unruh, J.A. (1998) Effect of source of cattle and distance transported to a commercial slaughter facility on carcass bruises in mature beef cows. *Journal of the American Veterinary Medical Association* 212, 668–672.

Honkavaara, M. (1993) Effect of a controlled ventilation stockcrate on stress and meat quality. *Meat Focus International* 2(12), 545–547.

Honkavaara, M. (1995) The effect of long distance transportation on live animals. In: Hinton, M.H. and Rowlings, C. (eds) *Factors Affecting the Microbial Quality of Meat 1. Disease Status, Production Methods and Transportation of the Live Animal*. University of Bristol Press, Bristol, UK, pp. 111–115.

Honkavaara, M. (1998) Animal transport. In: *Proceedings XVIII Nordic Veterinary Congress*. Helsinki, pp. 88–89.

Hutcheson, D.P. and Cole, N.A. (1986) Management of transit-stress syndrome in cattle: nutrition and environmental effects. *Journal of Animal Science* 62, 555–560.

Irwin, M.R. and Gentleman, W.R. (1978) Transportation of cattle in a rail car containing feed and water. *Southwestern Veterinarian* 31, 205–208.

Irwin, M.R., McConnell, S., Coleman, J.D. and Wilcox, G.E. (1979) Bovine respiratory disease complex: a comparison of potential predisposing and etiologic factors in Australia and the United States. *Journal of the American Veterinary Medical Association* 175, 1095–1099.

Jacobson, L.H. and Cook, C.J. (1998) Partitioning psychological and physical sources of transport-related stress in young cattle. *Veterinary Journal* 155, 205–208.

James, M. (1997) Report on the transportation of cattle to the Isle of Man by air. *State Veterinary Journal* 7(4), 1–2.

Jeremiah, L.E., Schaefer, A.L. and Gibson, L.L. (1992) The effects of antemortem feed and water withdrawal, antemortem electrolyte supplementation, and postmortem electrical stimulation on the palatability and consumer acceptance of bull beef after aging (6 days at 1°C). *Meat Science* 32, 149.

Kenny, J.F. and Tarrant, P.V. (1987a) The physiological and behavioural responses of crossbred Friesian steers to short-haul transport by road. *Livestock Production Science* 17, 63–75.

Kenny, J.F. and Tarrant, P.V. (1987b) The reaction of young bulls to short-haul road transport. *Applied Animal Behaviour Science* 17, 209–227.

Kenny, J.F. and Tarrant, P.V. (1987c) The behaviour of young Friesian bulls during social regrouping at the abattoir. Influence of an overhead electrified wire grid. *Applied Animal Behaviour Science* 18, 233–246.

Kenny, J.F. and Tarrant, P.V. (1988) The effect of oestrus behaviour on muscle glycogen concentration and dark-cutting in beef heifers. *Meat Science* 22, 21–31.

Kelley, K.W. (1980) Stress and immune function: a bibliographic review. *Annales de Recherches Veterinaires* 11, 445–478.

Kelley, K.W., Osborne, C. Everman, J., Parish, S. and Hinrichs, D. (1981) Whole blood leucocytes vs. separated mononuclear cell blastogenes in calves, time dependent changes after shipping. *Canadian Journal of Comparative Medicine* 45, 249–258.

Kent, J.E. and Ewbank, R. (1991) The behavioural response of 3-month-old calves to 18 hours of road transport. *Applied Animal Behaviour Science* 26, 289 (abstract).

Kilgour, R. and Mullord (1973) Transport of calves by road. *New Zealand Veterinary Journal* 21, 7–10.

Knowles, T.G. (1999) A review of road tranport of cattle. *Veterinary Record* 144, 197–201.

Lacourt, A. and Tarrant, P.V. (1985) Glycogen depletion patterns in myofibres of cattle during stress. *Meat Science* 15, 85–100.

Lambooy, E. and Hulsegge, B. (1988) Long-distance transport of pregnant heifers by truck. *Applied Animal Behaviour Science* 20, 249–258.

Lay, D.C., Friend, T.H., Bowers, C.L., Grissom, K.K. and Jenkins, O.C. (1992a) Behavioral and physiological effects of freeze and hot iron branding on crossbred cattle. *Journal of Animal Science* 70, 330–336.

Lay, D.C., Friend, T.H., Bowers, C.L. Grissom, K.K. and Jenkins, O.C. (1992b) A comparative physiological and behavioral study of freeze and hot-iron branding using dairy cows. *Journal of Animal Science* 70, 1121–1125.

Loerch, S.C. and Fluharty, F.L. (1999) Physiological changes and digestive capabilities of newly received feetlot cattle. *Journal of Animal Science* 77, 1113–1119.

MacKenzie, A.M., Drennan, M., Rowan, T.G., Dixon, J.B. and Carter, S.D. (1997) Effect of transportation and weaning on the humeral immune response of calves. *Research in Veterinary Science* 3, 227–230.

McVeigh, J. and Tarrant, P.V. (1982) Glycogen content and repletion rates in beef muscle, effect of feeding and fasting. *Journal of Nutrition* 112, 1306–1314.

McVeigh, J.M. and Tarrant, P.V. (1983) Effect of propranolol on muscle glycogen metabolism during social regrouping of young bulls. *Journal of Animal Science* 56, 71–80.

Mayes, H.F., Asplund, J.M. and Anderson, M.E. (1979) Transport stress effects on shrinkage. In: *ASAE Paper No. 79*, p. 6512.

Morck, D.W., Merrill, J.K. and Thorlakson, B.E. (1993) Prophylactic efficacy of tilmicosin for bovine respiratory trait disease. *Journal of the American Veterinary Medical Association* 202, 273–277.

Muirhead, V.V. (1985) *An Investigation of the Internal and External Aerodynamics of Cattle Trucks.* NASA Contract 170400, Ames Research Center, Dryden Flight Center, Edwards, California.

Muller, W. (1985) *Ventilation Requirements During Transport of Animals.* Proceedings, Animal Air Transportation Association, Dallas, Texas.

Murata, H. (1995) Suppression of bovine neutrophil function by sera from calves transported from over a long distance by road. *Animal Science and Technology (Japan)* 66 (11), 976–978.

National Cattlemen's Association (1994) *Strategic Alliances Field Study in Coordination with Colorado State University and Texas A&M University.* National Cattlemen's Beef Association, Englewood, Colorado.

Peltz, H. (1999) True immunity in the feedyard relies on cow–calf nutrition. *Drover's Journal* 127(5), 26.

Price, M.A. and Tennessen, T. (1981) Pre slaughter management and dark cutting in the carcasses of young bulls. *Canadian Journal of Animal Science* 61, 205–208.

Purnell, R. (1999) Vitamin value: vitamin E may help incoming calves respond positively to treatment, gain faster and get sick less. *Beef* (Minneapolis, Minnesota, USA) July, 26.

Ramsey, W.R., Neischke, H.R.C. and Anderson, B. (1976) *Australian Veterinary Journal* 52, 285–286.

Randall, J.M. (1993) Environmental parameters necessary to define comfort for pigs, cattle and sheep in livestock transporters. *Animal Production* 57, 299–307.

Ribble, C.S., Meek, A.H., Shewen, P.E., Jim, G.K. and Guichon, P.T. (1995) Effect of transportation on fatal fibrinous pneumonia and shrinkage in calves arriving at a large feedlot. *Journal of the American Veterinary Medical Association* 207, 612–615.

Scanga, J.A., Belk, K.E., Tatum, J.D., Grandin, T. and Smith, G.C. (1998) Factors contributing to the incidence of dark cutting beef. *Journal of Animal Sciences* 76, 2040–2047.

Schaefer, A.L. and Jeremiah, L.E. (1992) Effect of diet on beef quality. In: *Proceedings of the 13th Western Nutrition Conference*, University of Saskatchewan, Saskatoon, pp. 123–128.

Schaefer, A.L., Jones, S.D.M., Tong, A.K.W. and Young, B.A. (1990) Effect of transport and electrolyte supplementation on ion concentrations, carcass yield and quality in bulls. *Canadian Journal of Animal Science* 70, 107.

Schaefer, A.L., Jones, S.D. and Stanley, R.W. (1997) The use of electrolyte solutions for reducing transport stress. *Journal of Animal Science* 75, 258–265.

Self, H.L. and Gay, N. (1972) Shrink during shipment of feeder cattle. *Journal of Animal Science* 35, 489–494.

Shackelford, S.D., Koohmaraie, M., Wheeler, T.L., Cundiff, L.V. and Dikeman, M.E. (1994) Effect of biological type of cattle on the incidence of dark, firm and dry condition in the longissimus muscle. *Journal of Animal Science* 72, 337–343.

Shaw, F.D., Baxter, R.I. and Ramsey, W.R. (1976) The contribution of horned cattle to carcass bruising. *Veterinary Record* 98, 255–257.

Shorthose, W.R. (1965) Weight losses in cattle prior to slaugher. *CSIRO, Food Preservation Quarterly* 25, 67–73.

Shorthose, W.R., Harris, P.V. and Bouton, P.E. (1972) The effects of some properties of beef or testing and feeding cattle after a long journey to slaughter. *Proceedings of the Australia Society of Animal Production* 9, 387–391.

Sick, F.L., Bleeker, C.M., Mouw, J.K. and Thompson, W.S. (1982) Toe abscesses in recently shipped cattle. *Veterinary Medicine and Small Animal Clinician*, 1385.

Singh, S.P. (1991) *Vibration Levels in Commercial Truck Shipments*. Paper No. 91–6016, American Society of Agricultural Engineers, St Joseph, Michigan.

Smith, G.C., Morgan, J.B. and Tatum, J.D. (1994) *National Non-fed Beef Quality Audit*. National Cattlemen's Association, Englewood, Colorado, pp. 7–213.

Sowell, B.F., Branine, M.E., Bowman, J.G.P., Hubbert, M.E., Sherwood, H.E. and Quimby, W. (1999) Feeding and watering behavior of healthy and morbid steers in a commercial feedlot. *Journal of Animal Science* 77, 1105–1112.

Stevens, D.G. (1985) Ventilation requirements for transport of livestock. In: *Proceedings Animal Air Transportation Association*. Dallas, Texas.

Stevens, D.G. and Camp, T.H. (1979) *Vibration in a Livestock Vehicle*. Paper No. 79-6511, American Society of Agricultural Engineers, St Joseph, Michigan.

Stevens, D.G. and Hahn, G.L. (1977) *Air Transport of Livestock: Environmental Needs.* Paper No. 77-4523. American Society of Agricultural Engineers, St Joseph, Michigan.

Stevens, D.G., Stermer, R.A. and Camp, T.H. (1979) *Body Temperature of Cattle During Transport.* Paper No. 79-6511, American Society of Agricultural Engineers, St Joseph, Michigan.

Stovall, T.C., Han, H. and Gill, D.R. (1999) Effect of Agrado on the health and performance of transport-stressed heifer calves. *Journal of Animal Science* 71 (Suppl. 1), 272 (abstract).

Tarrant, P.V. (1981) The occurrence, causes and economic consequences of dark-cutting in beef – a survey of current information. In: Hood, D.E. and Tarrant, P.V. (eds) *The Problem of Dark Cutting in Beef.* Martinus Nijhoff, The Hague, pp. 3–34.

Tarrant, P.V. (1988) Animal behaviour and environment in the dark cutting condition. In: Febiansson, S.U., Shorthose, W.R. and Warner, R.D. (eds) *Dark-cutting in Cattle and Sheep. Proceedings of an Australian Workshop.* AMLRDC, Sydney, Australia, pp. 8–18.

Tarrant, P.V., Kenny, F.J. and Harrington, D. (1988) The effect of stocking density during 4 hour transport to slaughter, on behavior, blood constituents and carcass bruising in Friesian steers. *Meat Science* 24, 209–222.

Tarrant, P.V., Kenny, F.J., Harrington, D. and Murphy, M. (1992) Long distance transportation of steers to slaughter: effect of stocking density on physiology, behavior and carcass quality. *Livestock Production Science* 30, 223–238.

Tennessen, T., Price, M.A. and Berg, R.T. (1981) Bulls vs. steers: effect of time of mixing on dark cutting. In: *58th Annual Feeders Day Report.* University of Alberta, Canada.

Tennessen, T., Price, M.A. and Berg, R.T. (1984) Comparative responses of bulls and steers to transportaiton. *Canadian Journal of Animal Science* 64, 333–338.

Tennessen, T., Price, M.A. and Berg, R.T. (1985) The social interactions of young bulls and steers after regrouping. *Applied Animal Behaviour Science* 14, 37–47.

van Putten, G. (1982) Handling of slaughter pigs prior to loading and during loading on a lorry. In: Moss, R. (ed.) *Transport of Animals Intended for Breeding, Production and Slaughter.* Martinus Nijhoff, The Hague, pp. 15–25.

van Putten, G. and Elshof, W.J. (1978) Observations on the effects of transport on the well-being and lean quality of slaughter pigs. *Animal Regulation Studies* 1, 247–271.

Warnock, J.P., Caple, I.W., Halpin, C.G. and McQueen, C.S. (1978) Metabolic changes associated with the 'downer' condition in dairy cows at abattoirs. *Australian Veterinary Journal* 54, 566–569.

Warriss, P.D., Kestin, S.C., Brown, S.N. and Wilkins, L.J. (1984) The time required for recovery from mixing stress in young bulls and the prevention of dark-cutting beef. *Meat Science* 10, 53–68.

Warriss, P.D., Brown, S.N., Knowles, T.G., Kestin, S.C., Edwards, J.E., Dolan, S.K. and Phillips, A.J. (1995) Effects of cattle transport by road for up to 15 hours. *Veterinary Record* 136, 319–323.

Wittum, T.E. and Perino, L.J. (1995) Passive immune status at 24 hours portpartum and long term health and performance of calves. *American Journal of Veterinary Research* 56, 1149–1154.

Wittum, T.E., Woollen, N.E. and Perino, L.J. (1995) Relationship between treatment for respiratory disease and the presence of pulmonary lesions at slaughter and rate of gain in feedlot cattle. *Journal of Animal Science* 73 (Suppl. 1), 238.

Wythes, J.R. (1982) The saleyard curfew issue. *Queensland Agricultural Journal* November–December, pp. 1–5.

Wythes, J.R. (1985) Cattle transportation strategies. In: *Proceedings of the First National Grazing Animal Welfare Symposium*. Australian Veterinary Association, Brisbane, April.

Wythes, J.R. and Shorthose, W.R. (1984) *Marketing Cattle: Its Effects on Liveweight, Carcasses and Meat Quality*. No. 46, Australian Meat Research Committee, Sydney, New South Wales.

Wythes, J.R., Shorthose, W.R., Schmidt, P.J. and Davis, C.B. (1980) Effects of various rehydration procedures after a long journey on liveweight, carcasses and muscle properties of cattle. *Australian Journal of Agricultural Research* 31, 849–855.

Wythes, J.R., Arthur, R.J., Thompson, P.J.M., Williams, G.E. and Bond, J.H. (1981) Effect of transporting cows varying distances on liveweight, carcass traits and muscle pH. *Australian Journal of Experimental Agriculture and Animal Husbandry* 21, 557–561.

# Behavioural Principles of Sheep Handling* | 10

## G.D. Hutson

*Department of Animal Production, Institute of Land and Food Resources, University of Melbourne, Parkville, Victoria 3052, Australia*

## Introduction

The relationship between humans and sheep is probably more than 6000 years old (Hulet *et al.*, 1975). No doubt, the original relationship was one-sided, with humans hunting herds of wild sheep for food and clothing. But gradually, during the process of domestication, hunting changed to herding, and herding changed to farming. This transition from hunter to farmer has brought about a change in behaviour and attitude towards sheep, so that humans are now responsible for the day-to-day care and well-being of sheep flocks. However, the transition has been incomplete, and many of the techniques of the hunter/herder are still used during sheep handling.

There are about 1.2 billion sheep in the world (Lynch *et al.*, 1992). Most of these sheep are handled at least twice a year for two essential treatments – shearing and crutching (shearing of the breech and hind-legs to prevent fouling by faeces) – and generally they are handled more often. Sheep movement is usually prompted by the use of fear-evoking stimuli, and the treatment is usually stressful and aversive (Hutson, 1982b; Hargreaves and Hutson, 1990a, 1997). The handling procedure involves mustering, often with dogs and motor bikes, movement through yards, races and sheds, and finally, administration of a treatment, often involving isolation, manipulation and restraint. Treatments can be apparently mild, such as classing, drafting (sorting), drenching, dipping, vaccination and jetting (spraying), or more prolonged and stressful, such as foot trimming, weaning and shearing.

---

* This chapter is dedicated to the memory of Ron Kilgour, whose pioneering work on sheep behaviour, welfare and handling stimulated much of the research which is reviewed here.

©CAB *International* 2000. *Livestock Handling and Transport,* 2nd edn
(ed. T. Grandin)

# Behavioural Characteristics Important for Handling

Kilgour (1976) described the sheep as a:

> defenceless, vigilant, tight-flocking, visual, wool-covered ruminant, evolved
> within a mountain grassland habitat, displaying a follower-type dam–offspring
> relationship, with strong imitation between young and old in establishing range
> systems of tracks and best forage areas, showing seasonal breeding and a separate
> adult male sub-group structure.

He claimed that most behaviour seen in sheep on farms could be traced to one
or other of these characteristics. I think this description of the sheep is best
encapsulated in three words – flocking, following and vision; and I would add
one more – intelligence. These four characteristics form the basis for all principles of sheep handling.

## Flocking

The ancestral sheep adapted to mountain grasslands have evolved by natural
and artificial selection into breeds occupying diverse habitats and varying climates. The social organization of sheep in the wild is probably exemplified by
the feral Soay sheep on St Kilda. Grubb and Jewel (1966) found that ewes
tended to form large groups with fairly well-defined home ranges. Males associated in smaller numbers (two to three) and at mating moved from group to
group. Young males left the groups of ewes at about 1 year old to form their
own male–male groups.

   In domestic flocks this normal social organization is disrupted by removing lambs before natural weaning and by keeping sheep in flocks of uniform
age and sex. Despite these disruptions, sheep still aggregate to form flocks.
Crofton (1958) studied aerial photographs of domestic sheep and found that
the distance between individual grazing sheep varied, but they were orientated
so that an angle of about 110° was subtended between the head of each sheep
and two others in front of it. This angle corresponded to the angle between the
optic axes of the eyes and implied that sheep grazed that way because that was
the way their eyes were pointing. No doubt, vision is important in maintaining
contact within the flock, but the 110° angle seems to be fanciful and has not
been confirmed by other studies. However, Crofton's conclusions that sheep
aggregate and that vision is important in social spacing remain valid.

   Sheep maintain social spacing and orientation, even when confined in
pens. Hutson (1984) found that standing sheep orientated themselves so that
they were parallel to and facing in the same direction as their nearest neigh-
bour. Lying sheep lay down parallel to and next to the sides of open-sided pens,
but select positions at random in covered pens.

   Various factors will affect flock behaviour and structure while grazing,
including breed, stocking rate, topography, vegetation, shelter and distance to

water (Kilgour *et al.*, 1975; Squires, 1975; Stolba *et al.*, 1990; Lynch *et al.*, 1992). The adaptive advantages of the behaviour are clear. It provides more efficient exploitation of seasonal food resources and protection from predators. Sheep respond to the sight of a predator by flocking and flight (Kilgour, 1977). In the wild, sheep have long flight distances, but in confined spaces this distance will vary according to the space available for escape. When confined in a 2 m wide laneway the flight distance of sheep to an approaching man was 5.7 m compared with 11.4 m in a 4 m wide laneway (Hutson, 1982a). Flight distance was not affected by flock size, density or speed of approach. Individual sheep had longer flight distances than flocks.

One consequence of this social structure is that sheep have the opportunity to form stable relationships. Hinch *et al.* (1990) and Rowell (1991) reported that long-term social bonds may form between lambs and their mothers. Sheep may also develop a group identity. When groups of strange sheep have been mixed, they have kept to their own groups for several weeks before full integration occurred (Arnold and Pahl, 1974; Winfield *et al.*, 1981). The concept of a socially stable flock has important consequences for handling. Kilgour (1977) noted that when separated from a group an individual will run toward other sheep, irrespective of the position of the handler or dog. A sheep isolated from the flock may also show signs of tonic immobility or escape, depending on the type of restraint (Syme, 1985). Syme (1981) found that 30% of Merino sheep responded either physically or vocally to less than 5 min of isolation from the mob. Kilgour (1977) suggested that four or five sheep were required in a group before the group showed signs of social cohesion. Penning *et al.* (1993) found that sheep in small groups spent less time grazing and that a minimum group size of three, preferably four, was required for studies of grazing behaviour.

## Following

The following response of sheep is present at birth, when a newborn lamb will follow its dam during her daily activities. This response may play an important part in maintaining ewe–lamb contact, especially if the ewe has twin lambs. However, when following behaviour was tested experimentally in a circular runway there was great variability in the response (Winfield and Kilgour, 1976). Some lambs followed the surrogate ewe and others did not follow at all. The response was strongest for lambs between 4 and 10 days of age, which suggests that the generalized following response may be replaced by a more specific response to the dam. Nevertheless, following conspecifics persists in the life of a sheep. Scott (1945) described 'allelomimetic behaviour' as very common in sheep. This term refers to any type of activity which involves mutual imitation. In sheep it includes walking and running together, following one another, grazing together, bedding down together and bouncing stiff-legged past an obstacle together. Behaviour within the flock also tends to be synchronized, with sheep feeding, drinking, resting and ruminating at the same time. This synchronization of behaviour by grazing herbivores might be the result of social facilitation (Rook and Penning, 1991).

Leadership also occurs in sheep flocks and may be related to independence (Arnold, 1977). Individual sheep have been identified consistently among the leaders or among the tail-enders in small sheep flocks (Squires and Daws, 1975). Hutson (1980b) identified about 10% of sheep in small flocks as leaders. It is unlikely that these sheep deliberately led the flock, but more likely that they moved independently of other sheep and were then followed by them. Leadership will also depend on the setting, the size and composition of the flock and the purpose of movement (Syme, 1985). Sheep can be trained successfully to lead other sheep. Bremner (1980) used operant conditioning techniques to train sheep to walk through yards, push open and unlatch gates, accept leash restraint and lead mobs of sheep.

**Vision**

Sheep have excellent eyesight, as noted by Geist (1971): 'a popular myth circulates in North America that sheep vision is equal to that of a man aided by 8-power binoculars'. A wide visual field is a common characteristic of ungulates and may be an adaptation by prey animals to enable early detection of predators (Walls, 1942). In sheep, the angle between the optic axis and the midline is about 48° (Whitteridge, 1978), which indicates that the sheep should have a wide, although not panoramic, monocular field and a binocular field of about 60°. Piggins and Phillips (1996) measured the visual field of Welsh mountain and Cambridge sheep with a retinoscope and recorded a monocular field of 185°, with binocular overlap of 61.7°. In practice, Hutson (1980a) found that the visual field of merino sheep ranged from 190° to 306°, with a mean of 270°. The main causes of obstruction to rearward vision were ears, horns and wool growth. The binocular field of sheep, where the field of vision from each eye overlapped, ranged from 4° to 77°, with a mean of 45°. The main obstruction to the binocular field was the snout.

Stereoscopic vision is the perception of three dimensions in space. For an animal to have good stereoscopic vision, it is essential that it has good binocular vision, but in addition the optic nerve fibres must decussate incompletely at the optic chiasm. Clarke and Whitteridge (1973) worked out the projections of the retina on the visual cortex of sheep. In area 18, they found an overlap of about 30° on each side of the midline and most cells with fields up to 25° out were driven binocularly. In addition, visual acuity, as estimated from peak retinal ganglion cell density, was twice as high as that of the cat. Clarke and Whitteridge suggested that highly developed stereoscopic vision probably forms the basis for the sure-footedness of ungulates. Tanaka et al. (1995) determined visual acuity scores in three sheep ranging from 0.085 to 0.19. These values indicate that sheep could resolve visual detail at about one-twelfth to one-fifth the standard threshold for humans.

Depth perception is the discrimination of a drop-off or depth downward as opposed to straight ahead. Depth can be detected using several cues: an animal may detect a difference in the density of light from similar surfaces at different depths; head movements or a change in position as the animal looks will produce motion parallax; and an animal may possess true stereoscopic vision. Walk and Gibson (1961) and Lemmon and Patterson (1964) tested sheep on the visual cliff test and found good perception of depth. One-day-old lambs placed on a sheet of glass without visual support showed an immediate protective response – freezing, stiffening the forelegs and backing off the glass.

Sheep also possess some form of colour vision, but it is not known to what extent they rely on colour for environmental discriminations. Morphological evidence suggests that the sheep is a dichromat. Jacobs *et al.* (1998) used electroretinographic techniques to study the spectral sensitivity of cones in the sheep retina. Two cone types were identified – an S cone, with a spectral peak of 445.3 nm, and an M/L cone, with a spectral peak of 552.2 nm. The authors concluded that the sheep has the requisite photopigment basis to support dichromatic colour vision. A similar conclusion has been drawn for other ungulates, including cows and goats (Jacobs *et al.*, 1998), and pigs (Neitz and Jacobs, 1989).

The behavioural evidence for colour vision does not concur totally with the morphological evidence and suggests that colour sensitivity may favour the longer-wavelength end of the spectrum. Alexander and Stevens (1979) found that ewes could distinguish coloured lambs from grey-shaded lambs if the lambs were red, orange, yellow and white. However, ewes with blue, green or black lambs performed poorly. Munkenbeck (1982) used an operant conditioning technique to demonstrate that sheep could discriminate wavelengths at 30 nm intervals in the range 520 to 640 nm. Tanaka *et al.* (1989a,b) also used an operant conditioning technique to demonstrate that sheep could distinguish between the three primary colours and the same shades of grey. However, 11 sessions (330 trials) were required to reach the criterion on the red discrimination and 20 sessions (600 trials) for blue, and one subject failed to discriminate between green and grey after 64 sessions (1920 trials). Thus the limited behavioural evidence supports colour vision in the yellow–orange–red end of the spectrum, but provides conflicting evidence for the blue–green end of the spectrum. However, it should be borne in mind that in behavioural tests of colour vision it is notoriously difficult to effectively eliminate non-colour cues (Neitz and Jacobs, 1989).

Kendrick and Baldwin (1987) and Kendrick (1991) have investigated visual recognition in the sheep, using single-cell electrophysiological recording techniques. They investigated the responses of single neurons in the temporal cortex of sheep to various visual images. A small population of cells responded specifically to images of dog and human faces. Other cells responded to the sight of a human shape rather than the face. These cells did not distinguish between humans, their sex, what they were wearing, whether the back view or front view was presented or if the head and shoulders were covered. Thus decisions about an appropriate behavioural response to a potential predator could be made quickly at the level of sensory analysis. However, there is no doubt that sheep could also learn these discriminations. Baldwin (1981) has demonstrated that sheep can perform complex visual-discrimination learning tasks in the laboratory using geometrical symbols and Kendrick *et al.* (1995) have shown that sheep can discriminate in a Y maze between the projected images of faces of different sheep and humans. Davis *et al.* (1998) used an operant conditioning technique to demonstrate that sheep can discriminate between individual humans.

## Other senses

Olfaction is less important to sheep in handling situations, although sheep have well-developed olfactory apparatus and are able to make keen olfactory discriminations. Baldwin and Meese (1977) demonstrated that sheep were able to distinguish between conspecifics, using a range of secretory and excretory products. Blissitt *et al.* (1990) found that rams could discriminate between fresh oestrus and non-oestrus urine odours. The odour of dog faeces appears to have an innate repellent effect. Arnould and Signoret (1993) reported that sheep refused to eat food contaminated with dog faecal odour and did not habituate to repeated exposure. Geist (1971) claimed that the sense of smell was well enough developed for sheep to be able to scent a man 350 yards away under favourable conditions. Despite this extravagant claim, the distance over which olfactory discriminations are effective probably limits the usefulness of this sense in handling situations. Alexander (1978) found that ewes could distinguish their own lambs from aliens at close quarters, but when the lamb was more than 0.25 m away the ewes could no longer make the discrimination. Franklin and Hutson (1982a) reported that interdigital gland secretion did not influence path choice of sheep moving through a Y maze.

   Wollack (1963) investigated auditory acuity in three sheep using a conditioned leg-flexion response. Auditory sensitivity increased at frequencies from 10 Hz to 10,000 Hz, with a slight decrease around 1000 Hz. There was a rapid decrease in sensitivity from 10,000 to 40,000 Hz. Ames and Arehart (1972) determined auditory threshold curves by electroencephalogram (EEG) changes and behavioural responses and found maximum sensitivity at 7000 Hz. Shillito (1972) showed that the dominant frequencies in a lamb's bleat and a ewe's rumble were between 1000 and 4000 Hz and not at the peak sensitivity of hearing. Maximum sensitivity of hearing may therefore be attuned to auditory detection of predators and danger, rather than the calls of other sheep.

   Sheep show little response towards sonic booms and jet aircraft noise. Espmark *et al.* (1974) found the strongest reaction to sonic booms in standing sheep, who flung up their heads and started running, forming a 'bunch' with other sheep and moving off together. However, the sheep quickly adapted, so that after three exposures (five booms per day) they barely responded. Some sheep still reacted with a short and fast run, but then immediately resumed their previous activity. The sheep showed little or no response to subsonic aircraft noise ranging from 75 to 109 dB. Ewbank and Mansbridge (1977) also found that grazing lowland sheep tended to run together in response to simulated sonic booms at their first few exposures, but quickly adapted. In contrast, hill sheep scattered, but they too adapted. Weisenberger *et al.* (1996) reported that heart rates of captive mountain sheep exposed to simulated jet aircraft noise increased but returned to resting levels in 1–3 min.

Vocal communication between sheep seems to be of relatively minor importance, since sheep do not give a vocal alarm call and they vocalize only in specific situations, such as isolation from the flock (Torres-Hernandez and Hohenboken, 1979), during courtship of oestrous ewes, when rams utter a low-pitched rumble (Banks, 1964), and in ewe–lamb recognition (Alexander and Shillito, 1977; Dwyer *et al.*, 1998). Sheep vocalizations have no attractive effect on movement along races (Franklin and Hutson, 1982b).

## Intelligence

Many farmers deride the intelligence of sheep with remarks such as 'sheep are stupid'. However, this apparent stupidity can nearly always be attributed to the overriding presence of the protective flocking instinct (Kilgour and Matthews, 1983; Hutson, 1994). There have been many studies which confirm the above-average learning ability of the sheep. They can be conditioned easily in classical conditioning experiments. Liddell and Anderson (1931) were probably the first to use sheep in conditioning experiments. They conditioned sheep to make reflex leg movements in response to the beat of a metronome after eight to nine pairings of the metronome with an electrical shock. Sheep can also be conditioned to perform operant responses, and will press panels with their muzzles to make shape discriminations (Baldwin, 1981) or push through a weighted door to obtain food (Jackson *et al.*, 1999), operate foot treadles to obtain sodium solutions (Abraham *et al.*, 1973) and push cards off buckets to make colour discriminations (Bazely and Ensor, 1989). They can reach high rates of responding on fixed ratio schedules for preferred foods (Hutson and van Mourik, 1981; Hutson and Wilson, 1983).

In general, sheep perform well on tasks involving discrimination between left and right turns in U and T mazes (Kratzer, 1971). Liddell (1925, 1954) found that individual sheep could learn to run through a simple maze in a few trials. The maze consisted of three parallel alleys, one of which was a cul-de-sac. The sheep was required to find its way up the central alley and then down one of the outer alleys to a food reward. However, when the problem was made more difficult by reversing the position of the blind alley at every trial, the sheep could not learn to run through the maze without error for four consecutive trials. This was despite continued testing of some of the sheep, three times a week for 3 years. However, Liddell reported that some of the sheep found running the maze to be a 'self-rewarding activity'. More recent studies by Hosoi *et al.* (1995) have shown that sheep exhibit a strong lateral preference in simple T mazes, pay little attention to either intra- or extramaze cues and probably do not use maze cues for decision-making.

The speed and duration of learning in sheep quite clearly depend upon the nature of the task. Discrimination learning for natural objects like food and the faces of socially familiar animals is much faster than for geometrical symbols, novel objects such as bottles or socially unfamiliar animals (Kendrick, 1998). Learning to associate symbols or novel objects with food can take anything from 10 to 40 trials and learning is often only retained for a few hours or days (Kendrick *et al.*, 1996). In contrast, once lambs have learnt that wheat is a palatable food, they can retain this information for up to 34 months (Green *et al.*, 1984). Kendrick (1998) has suggested that the sheep's brain is adapted to efficiently learn associations between natural objects and reward, but not novel associations between artificial objects and reward.

Sheep appear to have an excellent spatial learning ability. Sandler *et al.* (1968) found that crossbred ewes learned the solution to a simple detour problem in a single trial. Hutson (1980b) found that sheep in a group could learn a route through yards in a relatively small number (four or five) of trials. Rushen and Congden (1986a) suggested that the increased transit time taken by sheep to move along a race towards repeated aversive treatments reflected the limited learning abilities of the sheep. But this surprising conclusion is not justified by their experimental results, where one trial was sufficient to demonstrate an aversion to the most severe treatment. A more likely explanation is that repetition of the treatment itself was responsible for the increase and that cumulative experience of aversive treatments influenced transit times. Sheep will remember an aversive experience for at least 12 weeks (Rushen, 1986a) and up to 1 year (Hutson, 1985a).

When sheep have been tested on natural spatial-memory tasks involving food finding, they have performed extremely well. Edwards *et al.* (1996) reported that sheep had the ability to retain information on the spatial distribution of a food resource after just a single exposure. Maximum efficiency was achieved in three to four trials. Sheep could learn the location of a food patch with and without cues, but learned faster when a cue was present. Associations with cues appeared to act independently of memory of spatial locations (Edwards *et al.*, 1997). When the location of the cue and food patch was switched randomly, sheep used spatial memory first to find the new location (Edwards *et al.*, 1996). Sheep can use spatial memory under even more complex conditions, and learn the location of hidden food in featureless environments with only distant landmarks (Dumont and Petit, 1998).

Sheep are capable of single-trial learning, even at the cellular level. Kendrick (1990) reported that cells in the hypothalamus respond to the sight, but not the smell, of known palatable foods and not to non-food objects. Initially, the cells do not respond to the sight of an unknown food. But, if a sheep eats the food just once and likes it, the cells will respond the next time it is seen, even if the sheep has not seen it for a month or more! Similarly, Provenza and Burritt (1991) have demonstrated single-trial learning in lambs for conditioned aversions to palatable foods treated with the toxin lithium chloride. Naturally occurring plant compounds, such as oxalic acid, can also induce conditioned aversions after a single exposure (Kyriazakis *et al.*, 1998). Food aversions can even be conditioned in anaesthetized sheep, which suggests that non-cognitive feedback processes are involved (Provenza *et al.*, 1994).

In summary, sheep have excellent learning ability, can be easily conditioned, perform sensory discriminations, acquire aversions, can learn simple mazes and have good short- and long-term memories.

## Implications of Behavioural Characteristics of Sheep for Handling

These four characteristics of the sheep – its flocking behaviour, following behaviour, vision and intelligence, form the basis of all behavioural principles of sheep handling. I shall consider these principles in relation to the three key elements of an integrated sheep-handling system – the design of the handling environment, the handling technique and the reason for being handled – the handling treatment.

### Design

Hutson (1980c) recommended that the most crucial design criterion was to give sheep a clear, unobstructed view towards the exit or towards where they are meant to move. This often becomes more evident by taking a sheep's eye view of the facility. Most behavioural principles of sheep handling are probably related to this criterion. Thus sheep movement is generally better on the flat, rather than up- or downhill (Hitchcock and Hutson, 1979b), away from buildings and dead ends (Kilgour and Dalton, 1984), in wide, straight races (Hutson and Hitchcock, 1978) and in well-lit areas (Hitchcock and Hutson, 1979a). Sheep will stop and investigate any novel visual stimulus or change in appearance of a race. Therefore, shadows or discontinuities on the ground (Hutson, 1980c), changes in race construction material or changes in floor type, e.g. from slats to concrete (Kilgour, 1971), should be avoided. Handling facilities should be painted one solid colour to avoid contrasts (Grandin, 1980). Judicious use of covered and open panels can direct movement and vision. Ramps should have covered sides, and movement inside sheds should be across the direction of the grating so that sheep can obtain a better grip with their feet and cannot see through the floor or perceive heights (Hutson, 1981a).

Learning, flocking and following behaviour also affect design. Thus, sheep should always be moved through yards and sheds along the same route and in the same direction as they will learn where they are meant to go (Hutson, 1980b). Sheep flow is better in wide races where sheep can move as a group rather than in single file (Hutson, 1980a). The sight of stationary sheep will slow down sheep movement through an adjacent race (Hutson, 1981b), but sheep will be attracted by the sight of other sheep or alternative visual stimuli, including mirrors, films, photographs and models (Franklin and Hutson, 1982c).

Dogs should be used cautiously, if at all, in confined handling situations, because sheep turn and face dogs when they cannot escape from them (Holmes, 1980). A 5 min exposure to a barking dog is used as a standard stimulus to induce stress in laboratory studies of sheep and elicits an abrupt elevation in adrenocorticotrophic hormone (ACTH) and cortisol concentrations (Cook, 1996; Komesaroff *et al.*, 1998).

## Handling technique

Humans have two conflicting roles in sheep handling: one is to act as a forcing stimulus and the other is to administer the treatment. The findings of Whateley *et al.* (1974) suggest that sheep react to these roles. They found that the relative ease of handling of different breeds reflected that breed's tolerance of humans and dogs, but breeds that handled well in paddocks and yards resisted physical restraint.

### The human as a forcing stimulus

The traditional motivation used to move sheep is the repeated application of fear-inducing stimuli. Sheep handlers use dogs, the natural predator of sheep, or auditory and visual signals, such as shouting and waving, to frighten sheep to move. Baskin (1974) describes herders waving arms and clothing, throwing rocks and even hoisting their caps on sticks to appear unusually tall. The aim is to frighten the animals and stimulate the flight response. However, fear-inducing stimuli do not always have the desired effect of prompting movement. Webb (1966) studied a range of stimuli and devices for driving sheep, including coloured and flashing lights, white noise, sinusoidal sound, electric shock, a mechanical sweep and air blasts. Sheep ignored the lights and sound and did not react violently to any of the shocks. Lambs quickly lost their fear of the sweep. Some lambs jumped over the pipe delivering air blasts and the stimulus could not prevent other lambs following. McCutchan *et al.* (1992) evaluated a mild electric shock as a prompt for sheep movement in a single-file race and found that sheep responded in an unpredictable manner to the stimulus. Some sheep moved forwards, some reversed backwards and some did not respond.

Vandenheede and Bouissou (1993, 1996) have shown that rams were less fearful than ewes but that wethers were more fearful than rams in various test situations.

It appears that the effectiveness of forcing stimuli declines as the sheep approach the area where they are treated. More force is then applied, and both handler and sheep get more aroused in an escalating vicious circle effect. Occasionally the sheep must be physically moved to the treatment area. The most likely explanation for this effect is the dual role of the human handler in forcing and treating sheep. The human is trying to apply more fear to make sheep move towards a fearful stimulus. In addition, increased force will result in greater arousal and less predictable and more erratic responses from the sheep, including stopping, freezing, fleeing, baulking, sitting, turning, reverse movement and jumping (Holmes, 1984a; Syme, 1985; Vette, 1985).

Alternative techniques which utilize different motivations, such as the flocking/following response, or positive rather than negative reinforcement could be used. For example, Bremner *et al.* (1980) have successfully trained sheep to lead other sheep, and Hutson (1985a) has reported that food rewards can encourage voluntary movement and improve sheep handling efficiency. Kiley-Worthington and Savage (1978) used classical conditioning to an auditory alarm to prompt movement of dairy cows. Vette (1985) has described a novel attempt to improve voluntary sheep movement into a single-file race. A rotating circular carousel holding four to six decoy sheep stimulated sheep to move into the race. Another novel attempt, using an artificial wind, was less successful (Hutson and van Mourik, 1982).

### The human as a handler

Many years ago, I speculated that future improvements in sheep handling would rely on an animal perspective, which implied that temperament studies should concentrate on the handler (Hutson, 1985b). For example, Seabrook (1972) reported that in dairy herds more milk was obtained by dairymen classified as confident introverts than by non-confident extroverts. Clear differences were noted in the willingness of cows to enter the milking parlour and return from pasture. Thus, it is quite evident that some people have inappropriate personalities to be animal handlers and others will need prolonged training. Hemsworth, in Chapter 14 of this volume, has identified similar relationships between behaviour and attitude of handlers towards pigs and subsequent production. There is no doubt that similar principles apply to sheep handling, although handling itself is less frequent.

It is generally assumed that sheep-handling skills are acquired by experience, although there has always been debate about whether good stock-handlers are born or made (Kilgour, 1978). Ewbank (1968) has suggested that they have an understanding of animal psychology that is probably based on acute powers of observation. For example, good stock-handlers:

**1.** Make the minimum possible use of fearful stimuli, avoid using loud noises which animals will associate with handling procedures, avoid punishing animals and use positive reinforcements (Hutson, 1985b).
**2.** Act quickly and decisively, because if handling is fumbled animals become more difficult to restrain (Ewbank, 1968).
**3.** Are aware of the flight distance of animals and utilize the strategy of reverse movement, i.e. by moving towards confined animals, they can prompt movement in the opposite direction more effectively than by frightening them from behind (Hutson, 1982a).
**4.** Are aware of the importance of arousal in animal handling and have the ability to predict animal responses in any situation (Holmes, 1984a).

Much of this knowledge is common sense, but it is essential that these techniques are made explicit for the training of inexperienced handlers. An excellent manual is now available which details sheep handling skills for New Zealand conditions but has universal application (Holmes, 1984a). A videotape is also available (Holmes, 1984b).

### Handling treatment

Although facility design and handling technique are very important in sheep handling, the main problem with obtaining efficient throughput of sheep is the nature of the handling treatment inflicted on the sheep. Hutson and Butler (1978) found that race efficiency fell from 93% to 73% after sheep had experienced inversion for 30 s in a handling machine. This suggested that many routine handling treatments were aversive and functioned as negative reinforcers of free movement through the handling system (Hutson, 1982b).

Recent research in Australia has focused on the stressfulness of various sheep-handling procedures. This research has been prompted by welfare concerns and the potential introduction of new technologies, such as robot shearing (Trevelyan, 1992). Although attempts to replace the shearer with a robot have now been abandoned, we have much more knowledge about the relative stressfulness and aversiveness of different handling treatments. Various physiological and behavioural techniques have been used.

Physiological measures of stressfulness include plasma cortisol, β-endorphin, haematocrit and heart rate. Kilgour and de Langen (1970) were the first to measure plasma cortisol concentrations in sheep for different handling procedures. They found a great deal of individual variation, but some treatments stressed sheep more than others. Dog chasing, especially when bitten, and prolonged shearing produced the highest cortisol levels. Fulkerson and Jamieson (1982) compared patterns of cortisol release following various stressors and reported the most severe stress was associated with shearing, less stress was imposed by yarding and handling and there was no effect attributable to feeding or fasting. Fell and Shutt (1988) used salivary cortisol to assess acute stressors and ranked treatments in decreasing order of stressfulness as shearing, stop–start transport, steady transport, sham shearing, isolation, cold, jetting and yarding.

β-Endorphin has also been used to monitor the stress response of sheep to potentially painful handling or surgical procedures. Jephcott *et al.* (1986) found significant rises in β-endorphin after electroimmobilization in comparison with a control handling procedure, and Shutt *et al.* (1987) reported a threefold increase in β-endorphin concentrations 15 min after tail-docking in lambs, and a maximal eight- to tenfold increase in response to castration and/or mulesing with tail-docking. Shutt and Fell (1988) found significant increases in plasma β-endorphin in wethers at 5 min and 15 min following mulesing, and suggested that an endorphin-induced analgesic response lasted for about 1 h. However, the extent to which β-endorphin modulates pain perception is still unknown and controversial. In another study, Fell and Shutt (1989) found elevated β-endorphin concentrations 24 h after mulesing. Anil *et al.* (1990) reported a twofold increase in plasma β-endorphin concentrations in response to electrical stunning and a further increase after animals regained consciousness. Fordham *et al.* (1989) reported that transport did not increase plasma β-endorphin concentrations in lambs above concentrations obtained after mustering with a dog.

Heart rate has also been used as an indicator of response to handling. Webster and Lynch (1966) found a steady increase in heart rate for the week following shearing. Syme and Elphick (1982) reported that sheep unresponsive to social isolation had lower heart rates than responsive sheep when standing alone in a race. Also, heart rates have been reported to increase in response to visual isolation, transportation, introduction into a new flock, human approach and human approach with a dog (Baldock and Sibly, 1990).

Hargreaves and Hutson (1990a) evaluated the stress response of sheep to routine handling procedures, using plasma cortisol and haematocrit. Both parameters were significantly elevated after shearing compared with untreated sheep, but declined to basal levels within 90 min after treatment. They concluded that sheep perceived shearing, crutching and drafting as more stressful than drenching or dipping. Drenching and dipping were the only treatments in which sheep stayed together as a group. In a further analysis of the shearing procedure, Hargreaves and Hutson (1990b,c) used several physiological measures, including haematocrit, plasma cortisol, plasma glucose and heart rate, to assess the stress response to components of the procedure. In a series of treatments, sheep were separated from other sheep, isolated, exposed to a human, blood-sampled, up-ended, exposed to shearing noise and partially shorn. Haematocrit, plasma cortisol and glucose increased significantly after shearing but not with isolation. Shearing was the only treatment which elevated heart rate significantly above pre-treatment values. The response to noise alone was less pronounced than to actual wool removal. It was concluded that wool removal was more stressful than any of the other manipulations involved in conventional shearing.

Isolation may not be as stressful as initial investigations (Kilgour and de Langen, 1970) suggested. Hargreaves and Hutson (1990b) did not detect a cortisol or haematocrit response to 4 min of isolation and suggested that individual handling and familiarity with the pre-treatment routine may have attenuated this response. In contrast, Parrott *et al.* (1994) reported an increase in plasma cortisol in response to 60 min of isolation, although the magnitude of the response was less than that to standing in water or simulated transport. Parrott *et al.* (1988a) also reported an increase in plasma cortisol in response to 120 min of isolation, which was reduced slightly by mirrored panels (Parrott *et al.*, 1988b). Coppinger *et al.* (1991) and Minton *et al.* (1995) reported that isolation coupled with restraint (binding of the legs with adhesive tape) for 6 h produced a robust cortisol response in lambs. However, it is known that restraint alone will produce an acute cortisol response (Niezgoda *et al.*, 1993). Cockram *et al.* (1994) found that isolation for 24 h produced a significant increase in plasma cortisol after 1.5, 3 and 9 h, but not 6 and 24 h. The response diminished on subsequent exposures and was not significant at the seventh and fourteenth repetition. The authors also noted that the response to the first period of isolation may have been affected by movement to the isolation pen and by exposure to a novel environment.

Clearly, it is hazardous comparing different experiments because of different protocols, but several points emerge from these studies of isolation. Transient isolation associated with handling treatments may not be as stressful as isolation for long periods, and sheep may habituate to long periods of isolation. Habituation may occur if sheep learn that there is an eventual escape and are able to predict the frequency and duration of the stressor (Cockram *et al.*, 1994). The response to isolation may also be modified by the presence of the handler. Boivin *et al.* (1997) noted that lambs vocalized and moved less when in the presence of a shepherd than when isolated. These observations support the hypothesis of Price and Thos (1980), who suggested that humans can serve as an effective substitute for a conspecific and reduce the distress of sheep in isolation.

Behaviour has been used to assess relative aversiveness of handling procedures in choice tests, in aversion learning tests and in approach/avoidance conflicts in an arena test. Rushen and Congdon (1986b) showed that two forced choices were sufficient for nine out of 12 sheep to discriminate between partial shearing and electroimmobilization, although the forcing stimulus itself may have influenced the outcome of such tests. Three sheep were indifferent. Grandin *et al.* (1986) reported that sheep preferred restraint on a squeeze-tilt table to electroimmobilization. Rushen (1986b) has also compared more conventional treatments, using the forced-choice method, and ranked the treatments in decreasing order of preference as: human presence, physical restraint, isolation, capture in isolation and inversion in isolation.

The willingness of sheep to move along a race towards a treatment area has been used to assess the aversiveness of handling treatments in aversion learning tests. Hutson and Butler (1978) found that a single experience of inversion for 30 s in a handling machine was sufficient to make sheep hesitate about moving along the race again. Hutson (1985a) reported that the time spent pushing sheep along a race and into a sheep-handling machine increased with successive trials when sheep were restrained by clamping, and increased at a greater rate when they were clamped and inverted. Rushen (1986a) reported that the longer transit times of sheep to move to a treatment site indicated that electroimmobilization was more aversive than physical restraint, with or without electrodes attached, and that the degree of aversion decreased with experience of electroimmobilization. Rushen and Congdon (1986a) found that simulated shearing together with electroimmobilization was more aversive than either immobilization or simulated shearing alone. Aversiveness to electroimmobilization was extinguished after five non-treatment trials (J.R. Stollery, 1990, personal communication). Stafford *et al.* (1996) used an aversion test with rams to show that part-shearing was more aversive than free movement and that electroejaculation was intermediate between the two.

Fell and Shutt (1989) used an arena test to assess the aversion of sheep to the human handler after the mules operation. An advantage of this method is that it is not influenced by the behaviour of the experimenter, as in forced-choice tests or aversion learning tests. Sheep were released into the mid-point of a $14 \times 4$ m arena three or four at a time. The handler who held the sheep during treatment stood quietly at one end of the arena with the remainder of the flock behind him. On entering the arena, control sheep turned and moved towards the handler, whereas mulesed sheep turned and moved in the opposite direction. Mulesed sheep continued to show a pronounced aversion to the handler up to 36 days after the operation. The aversion was no longer apparent after 114 days. It is not known whether the aversion is generalized to other humans.

## Reducing the Aversiveness and Stressfulness of Handling

A basic dilemma plagues sheep handling. Some of the treatments are stressful and produce pronounced aversions in sheep. For example, wool removal is the fundamental basis of sheep production and yet it is the prime contributor to the stress response to shearing. How can we improve the ease of handling and the welfare of sheep during handling if there is no prospect of changing the nature of the treatment itself? There are several possibilities:

**1.** Reduce the severity of the treatment, e.g. the function of the plunge or shower dip can now be performed with a pour-on chemical. Replacing the traditional method of wool severance by comb and reciprocating cutter will be more difficult. Chemical defleecing research, once a bright hope (Hudson, 1980), is now in decline, although an Australian company is experimenting with epidermal growth factor (Jarratt Biological Woold Harvesting, Sydney, 1999, personal communication). Alterations to the method of capture and restraint of sheep during shearing may be possible, and changes to the hand-piece which modify heat production, vibration and wool pull might modify the stress response to shearing.

**2.** Clearly, it is difficult to minimize the severity of surgical treatments. Mulesing is controversial (Townend, 1985), but non-surgical methods, such as topical application of a quaternary ammonium compound, appear to minimize the acute stress response and prevent the development of a lasting aversion to the handler (Chapman *et al.*, 1994). There is considerable debate over whether it is better to use the ring or the knife to castrate and tail-dock lambs (Barnett, 1988; Shutt and Fell, 1988; Shutt *et al.*, 1988; Mellor and Murray, 1989; Lester *et al.*, 1991, 1996), with the consensus appearing to favour the ring. Even so, the ring still causes considerable distress, so more benign methods should be actively sought (Lester *et al.*, 1996). Spatial and temporal separation of these painful treatments from other handling procedures is advised. Anaesthetics should be used. Kent *et al.* (1998) have demonstrated that injection of local anaesthetic into the neck of the scrotum or tail effectively reduced the acute pain produced by rubber-ring castration and tail-docking. Fell and Shutt (1989) recommended that mulesing of weaners should be carried out by contractors rather than regular handlers, and that handling should be minimized for several weeks following treatment. This recommendation should also be applied to the milder surgical treatments of castration and tail-docking. These procedures should not be performed in the usual handling area.

**3.** Train sheep, using food rewards. Hutson (1985a) reported that sheep trained to run through a handling system for a food reward required less labour for subsequent handling than unrewarded sheep. Grandin (1989) demonstrated that food rewards could also be used to entice previously restrained or electroimmobilized sheep to voluntarily enter and accept restraint in a squeeze-tilt table. Thus, with time, previously acquired aversions to even severe handling treatments will diminish, and the response of sheep to a handling procedure will approach that of naïve sheep (Hutson, 1985a). Siegel and Moberg (1980) reported that neonatal stress in lambs did not influence later performance or adrenocortical response in an active avoidance task. Similarly, Cook (1996) reported a reduction in the acute cortisol response of sheep that learned to avoid an electric shock and no change in their basal cortisol concentrations.

**4.** Habituation to the handling procedure. Simple repetition of a stimulus and an animal's habituation to it may be more important than whether it is perceived as being pleasant or unpleasant. Siegel and Moberg (1980) showed that the adrenocortical response of 8-month-old lambs decreased over ten sessions of avoidance conditioning to an electric shock. However, when Hargreaves and Hutson (1990e) exposed sheep to a sham shearing procedure on four occasions at 2-week intervals, the stress response was only slightly reduced by repetition. Peak cortisol response to the procedure was not affected, but concentrations declined more rapidly to baseline levels after four exposures. Adrenal responsiveness in response to an ACTH challenge was not affected. Fordham *et al.* (1989) reported no change in cortisol concentrations after twice-daily jugular venepuncture of lambs for 5 days. Thus, repeated exposure to stressors seems unlikely to modify handling responses.

**5.** Hargreaves and Hutson (1990d) investigated gentling as a method of reducing the aversion of sheep to subsequent handling. Gentling is often used with laboratory animals when repeated tactile, visual and auditory contact with a human makes them easier to handle. Adult sheep were gentled by being visually isolated, talked to and patted for 20 s day$^{-1}$ for 35 consecutive days. Gentling reduced the flight distance and the heart-rate response to humans, but did not reduce the aversion to a handling procedure (sham shearing). Mateo *et al.* (1991) confirmed these results by gentling sheep for 5 min day$^{-1}$ for 21 days. Gentling improved the approachability of sheep to humans but did not attenuate their response to restraint or shearing.

**6.** Condition sheep gradually to handling. The first handling experience is critical. Markowitz *et al.* (1998) reported that 40 min of positive human contact at 1–3 days reduced the subsequent timidity of lambs towards people. Miller (1960) found that rats trained to run down an alley for food could be induced to continue running, even when severely shocked in the goal box, so long as they had been exposed to a series of shocks of gradually increasing intensity. However, animals that received an intense shock at the outset showed a complete suppression of running. Although generalizations from rats to sheep should be treated cautiously, Miller's study has important implications for training sheep to handling procedures and indicates that training should be gradual, with exposure to innocuous treatments first before more stressful treatments such as shearing are carried out.

**7.** Simplify the learning procedure, so that all handling is done in one location, preferably the shed, using a uniform method of restraint. More research on the role of conditioning in the acquisition and extinction of behavioural and physiological responses to handling is required. In particular, studies of cues associated with the treatment, the place it is performed and the identity of the handler need to be done.

**8.** Finally, the conflicting roles of humans in sheep handling need to be resolved. I suggest that only dogs, machines and electrical and mechanical devices should be used as fear-inducing forcing stimuli. This would complete the transition from hunter/herder to farmer and allow humans to concentrate on their responsibility for the care and well-being of sheep.

# References

Abraham, S., Baker, R., Denton, D.A., Kraintz, F., Kraintz, L. and Purser, L. (1973) Components in the regulation of salt balance: salt appetite studied by operant behaviour. *Australian Journal of Experimental Biology and Medical Science* 51, 65–81.

Alexander, G. (1978) Odour, and the recognition of lambs by Merino ewes. *Applied Animal Ethology* 4, 153–158.

Alexander, G. and Shillito, E.E. (1977) The importance of odour, appearance and voice in maternal recognition of the young in Merino sheep (*Ovis aries*). *Applied Animal Ethology* 3, 127–135.

Alexander, G. and Stevens, D. (1979) Discrimination of colours and grey shades by Merino ewes: tests using coloured lambs. *Applied Animal Ethology* 5, 215–231.

Ames, D.R. and Arehart, L.A. (1972) Physiological response of lambs to auditory stimuli. *Journal of Animal Science* 34, 994–998.

Anil, M.H., Fordham, D.P. and Rodway, R.G. (1990) Plasma β-endorphin increase following electrical stunning in sheep. *British Veterinary Journal* 146, 476–477.

Arnold, G.W. (1977) An analysis of spatial leadership in a small field in a small flock of sheep. *Applied Animal Ethology* 3, 263–270.

Arnold, G.W. and Pahl, P.J. (1974) Some aspects of social behaviour in domestic sheep. *Animal Behaviour* 22, 592–600.

Arnould, C. and Signoret, J.-P. (1993). Sheep food repellents: efficacy of various products, habituation, and social facilitation. *Journal of Chemical Ecology* 19, 225–236.

Baldock, N.M. and Sibly, R.M. (1990) Effects of handling and transportation on the heart rate and behaviour of sheep. *Applied Animal Behaviour Science* 28, 15–39.

Baldwin, B.A. (1981) Shape discrimination in sheep and calves. *Animal Behaviour* 29, 830–834.

Baldwin, B.A. and Meese, G.B. (1977) The ability of sheep to distinguish between conspecifics by means of olfaction. *Physiology and Behaviour* 18, 803–808.

Banks, E.M. (1964) Some aspects of sexual behavior in domestic sheep, *Ovis aries*. *Behaviour* 23, 249–279.

Barnett, J.L. (1988) Surgery may not be preferable to rubber rings for tail docking and castration of lambs. *Australian Veterinary Journal* 65, 131

Baskin, L.M. (1974) Management of ungulate herds in relation to domestication. In: Geist, V. and Walther, F.R. (eds) *Behaviour of Ungulates and its Relation to Management*, vol. 2. International Union for Conservation of Nature, Morges, pp. 530–541.

Bazely, D.R. and Ensor, C.V. (1989) Discrimination learning in sheep with cues varying in brightness and hue. *Applied Animal Behaviour Science* 23, 293–299.

Blissitt, M.J., Bland, K.P. and Cottrell, D.F. (1990) Discrimination between the odours of fresh oestrus and non-oestrus ewe urine by rams. *Applied Animal Behaviour Science* 25, 51–59.

Boivin, X., Nowak, R., Desprès, G., Tournadre, H. and Le Neindre, P. (1997) Discrimination between shepherds by lambs reared under artificial conditions. *Journal of Animal Science* 75, 2892–2898.

Bremner, K.J. (1980) Follow my 'leader' – techniques for training sheep. *New Zealand Journal of Agriculture* September, 25–29.

Bremner, K.J., Braggins, J.B. and Kilgour, R. (1980) Training sheep as 'leaders' in abattoirs and farm sheep yards. *Proceedings of the New Zealand Society of Animal Production* 40, 111–116.

Chapman, R.E., Fell, L.R. and Shutt, D.A. (1994) A comparison of stress in surgically and non-surgically mulesed sheep. *Australian Veterinary Journal* 71, 243–247.

Clarke, P.G.H. and Whitteridge, D. (1973) The basis of stereoscopic vision in the sheep. *Journal of Physiology* 229, 22–23P.

Cockram, M.S., Ranson, M., Imlah, P., Goddard, P.J., Burrells, C. and Harkiss, G.D. (1994) The behavioural, endocrine and immune responses of sheep to isolation. *Animal Production* 58, 389–399.

Cook, C.J. (1996) Basal and stress response cortisol levels and stress avoidance learning in sheep (*Ovis ovis*). *New Zealand Veterinary Journal* 44, 162–163.

Coppinger, T.R., Minton, J.E., Reddy, P.G. and Blecha, F. (1991) Repeated restraint and isolation stress in lambs increases pituitary–adrenal secretions and reduces cell-mediated immunity. *Journal of Animal Science* 69, 2808–2814.

Crofton, H.D. (1958) Nematode parasite populations in sheep on lowland farms. VI. Sheep behaviour and nematode infections. *Parasitology* 48, 251–260.

Davis, H., Norris, C. and Taylor, A. (1998) Wether ewe know me or not: the discrimination of individual humans by sheep. *Behavioural Processes* 43, 27–32.

Dumont, B. and Petit, M. (1998) Spatial memory of sheep at pasture. *Applied Animal Behaviour Science* 60, 43–53.

Dwyer, C.M., McLean, K.A., Deans, L.A., Chirnside, J., Calvert, S.K. and Lawrence, A.B. (1998) Vocalisations between mother and young in sheep: effects of breed and maternal experience. *Applied Animal Behaviour Science* 58, 105–119.

Edwards, G.R., Newman, J.A., Parsons, A.J. and Krebs, J.R. (1996) The use of spatial memory by grazing animals to locate food patches in spatially heterogeneous environments: an example with sheep. *Applied Animal Behaviour Science* 50, 147–160.

Edwards, G.R., Newman, J.A., Parsons, A.J. and Krebs, J.R. (1997) Use of cues by grazing animals to locate food patches: an example with sheep. *Applied Animal Behaviour Science* 51, 59–68.

Espmark, Y., Falt, L. and Falt, B. (1974) Behavioural responses in cattle and sheep exposed to sonic booms and low-altitude subsonic flight noise. *Veterinary Record* 94, 106–113.

Ewbank, R. (1968) The behavior of animals in restraint. In: Fox, M.W. (ed.) *Abnormal Behaviour in Animals*. Saunders, London, pp. 159–178.

Ewbank, R. and Mansbridge, R.J. (1977) The effects of sonic booms on farm livestock. *Applied Animal Ethology* 3, 292.

Fell, L.R. and Shutt, D.A. (1988) Salivary cortisol and behavioural indicators of stress in sheep. *Proceedings of the Australian Society of Animal Production* 17, 186–189.

Fell, L.R. and Shutt, D.A. (1989) Behavioural and hormonal responses to acute surgical stress in sheep. *Applied Animal Behaviour Science* 22, 283–294.

Fordham, D.P., Lincoln, G.A., Ssewannyana, E. and Rodway, R.G. (1989) Plasma β-endorphin and cortisol concentrations in lambs after handling, transport and slaughter. *Animal Production* 49, 103–107.

Franklin, J.R. and Hutson, G.D. (1982a) Experiments on attracting sheep to move along a laneway. I. Olfactory stimuli. *Applied Animal Ethology* 8, 439–446.

Franklin, J.R. and Hutson, G.D. (1982b) Experiments on attracting sheep to move along a laneway. II. Auditory stimuli. *Applied Animal Ethology* 8, 447–456.

Franklin, J.R. and Hutson, G.D. (1982c) Experiments on attracting sheep to move along a laneway. III. Visual stimuli. *Applied Animal Ethology* 8, 457–478.

Fulkerson, W.J. and Jamieson, P.A. (1982) Pattern of cortisol release in sheep following administration of synthetic ACTH or imposition of various stressor agents. *Australian Journal of Biological Science* 35, 215–222.

Geist, V. (1971) *Mountain Sheep: a Study in Behaviour and Evolution*. University of Chicago Press, Chicago.

Grandin, T. (1980) Livestock behavior as related to handling facilities design. *International Journal for the Study of Animal Problems* 1, 33–52.

Grandin, T. (1989) Voluntary acceptance of restraint by sheep. *Applied Animal Behaviour Science* 23, 257–261.

Grandin, T., Curtis, S.E., Widowski, T.M. and Thurmon, J.C. (1986) Electro-immobilization versus mechanical restraint in an avoid–avoid choice test for ewes. *Journal of Animal Science* 62, 1469–1480.

Green, G.C., Elwin, R.L., Mottershead, B.E., Keogh, R.G. and Lynch, J.J. (1984) Long term effects of early experience to supplementary feeding in sheep. *Proceedings of the Australian Society of Animal Production* 15, 373–375.

Grubb, P. and Jewel, P.A. (1966) Social grouping and home range in feral Soay sheep. *Symposia of the Zoological Society of London* 18, 179–210.

Hargreaves, A.L. and Hutson, G.D. (1990a) The stress response in sheep during routine handling procedures. *Applied Animal Behaviour Science* 26, 83–90.

Hargreaves, A.L. and Hutson, G.D. (1990b) Changes in heart rate, plasma cortisol and haematocrit of sheep during a shearing procedure. *Applied Animal Behaviour Science* 26, 91–101.

Hargreaves, A.L. and Hutson, G.D. (1990c) An evaluation of the contribution of isolation, up-ending and wool removal to the stress response to shearing. *Applied Animal Behaviour Science* 26, 103–113.

Hargreaves, A.L. and Hutson, G.D. (1990d) The effect of gentling on heart rate, flight distance, and aversion of sheep to a handling procedure. *Applied Animal Behaviour Science* 26, 243–252.

Hargreaves, A.L. and Hutson, G.D. (1990e) Some effects of repeated handling on stress responses in sheep. *Applied Animal Behaviour Science* 26, 253–265.

Hargreaves, A.L. and Hutson, G.D. (1997) Handling systems for sheep. *Livestock Production Science* 49, 121–138.

Hinch, G.N., Lynch, J.J., Elwin, R.L. and Green, G.C. (1990) Long-term associations between Merino ewes and their offspring. *Applied Animal Behaviour Science* 27, 93–103.

Hitchcock, D.K. and Hutson, G.D. (1979a) The effect of variation in light intensity on sheep movement through narrow and wide races. *Australian Journal of Experimental Agriculture and Animal Husbandry* 19, 170–175.

Hitchcock, D.K. and Hutson, G.D. (1979b) The movement of sheep on inclines. *Australian Journal of Experimental Agriculture and Animal Husbandry* 19, 176–182.

Holmes, R.J. (1980) Dogs in works: friend or foe? *4 Quarter* 1(3), 16–17.

Holmes, R.J. (1984a) *Sheep and Cattle Handling Skills: A Manual for New Zealand Conditions.* Accident Compensation Corporation, Wellington, New Zealand.

Holmes, R.J. (1984b) *Stockhandling Skills* (22 min) video. Accident Compensation Corporation, Wellington, New Zealand.

Hosoi, E., Swift, D.M., Rittenhouse, L.R. and Richards, R.W. (1995) Comparative foraging strategies of sheep and goats in a T-maze apparatus. *Applied Animal Behaviour Science* 44, 37–45.

Hudson, P.R.W. (1980) Technology brings home the wool harvest. *New Scientist* 1218, 768–771.

Hulet, C.V., Alexander, G. and Hafez, E.S.E. (1975) The behaviour of sheep. In: Hafez, E.S.E. (ed.) *The Behaviour of Domestic Animals.* Baillière Tindall, London, pp. 246–294.

Hutson, G.D. (1980a) Visual field, restricted vision and sheep movement in laneways. *Applied Animal Ethology* 6, 175–187.

Hutson, G.D. (1980b) The effect of previous experience on sheep movement through yards. *Applied Animal Ethology* 6, 233–240.

Hutson, G.D. (1980c) Sheep behaviour and the design of sheep yards and shearing sheds. In: Wodzicka-Tomaszewska, M., Edey, T.N. and Lynch, J.J. (eds) *Behaviour in Relation to Reproduction, Management and Welfare of Farm Animals.* University of New England Publishing Unit, Armidale, NSW, pp. 137–141.

Hutson, G.D. (1981a) Sheep movement on slatted floors. *Australian Journal of Experimental Agriculture and Animal Husbandry* 21, 474–479.

Hutson, G.D. (1981b) An evaluation of some traditional and contemporary views on sheep behaviour. *Wool Technology and Sheep Breeding* 29(1), 3–6.

Hutson, G.D. (1982a) 'Flight distance' in Merino sheep. *Animal Production* 35, 231–235.

Hutson, G.D. (1982b) Sheep handling facilities. *Proceedings of the Australian Society of Animal Production* 14, 121–123.

Hutson, G.D. (1984) Spacing behaviour of sheep in pens. *Applied Animal Behaviour Science* 12, 111–119.

Hutson, G.D. (1985a) The influence of barley food rewards on sheep movement through a handling system. *Applied Animal Behaviour Science* 14, 263–273.

Hutson, G.D. (1985b) Sheep and cattle handling facilities. In: Moore, B.L. and Chenoweth, P.J. (eds) *Grazing Animal Welfare.* Australian Veterinary Association, Indooroopilly, Queensland, pp. 124–136.

Hutson, G.D. (1994) So who's being woolly minded now? *New Scientist* 1951, 52–53.

Hutson, G.D. and Butler, M.L. (1978) A self-feeding sheep race that works. *Journal of Agriculture (Victoria)* 76, 335–336.

Hutson, G.D. and Hitchcock, D.K. (1978) The movement of sheep around corners. *Applied Animal Ethology* 4, 349–355.

Hutson, G.D. and van Mourik, S.C. (1981) Food preferences of sheep. *Australian Journal of Experimental Agriculture and Animal Husbandry* 21, 575–582.

Hutson, G.D. and van Mourik, S.C. (1982) Effect of artificial wind on sheep movement along indoor races. *Australian Journal of Experimental Agriculture and Animal Husbandry* 22, 163–167.

Hutson, G.D. and Wilson, P.N (1983) A note on the preference of sheep for whole or crushed grains and seeds. *Animal Production* 38, 145–146.

Jackson, R.E., Waran, N.K. and Cockram, M.S. (1999) Methods for measuring feeding motivation in sheep. *Animal Welfare* 8, 53–63.

Jacobs, G.H., Deegan, J.F. and Neitz, J. (1998) Photopigment basis for dichromatic color vision in cows, goats, and sheep. *Visual Neuroscience* 15, 581–584.

Jephcott, E.H., McMillen, I.C., Rushen, J., Hargreaves, A. and Thorburn, G.D. (1986) Effect of electro-immobilisation on ovine plasma concentrations of ß-endorphin/ß-lipotrophin, cortisol and prolactin. *Research in Veterinary Science* 41, 371–377.

Kendrick, K.M. (1990) Through a sheep's eye. *New Scientist* 1716, 40–43.

Kendrick, K.M. (1991) How the sheep's brain controls the visual recognition of animals and humans. *Journal of Animal Science* 69, 5008–5016.

Kendrick, K.M. (1998) Intelligent perception. *Applied Animal Behaviour Science* 57, 213–231.

Kendrick, K.M and Baldwin, B.A. (1987) Cells in temporal cortex of conscious sheep can respond preferentially to the sight of faces. *Science* 296, 448–450.

Kendrick, K.M., Atkins, K., Hinton, M.R., Broad, K.D., Fabre-Nys, C. and Keverne, B. (1995) Facial and vocal discrimination in sheep. *Animal Behaviour* 49, 1665–1676.

Kendrick, K.M., Atkins, K., Hinton, M.R., Heavens, P. and Keverne, B. (1996) Are faces special for sheep? Evidence from facial and object discrimination learning tests showing effects of inversion and social familiarity. *Behavioural Processes* 38, 19–35.

Kent, J.E., Molony, V. and Graham, M.J. (1998) Comparison of methods for the reduction of acute pain produced by rubber ring castration or tail docking of week-old lambs. *Veterinary Journal* 155, 39–51.

Kiley-Worthington, M. and Savage, P. (1978) Learning in dairy cattle using a device for economical management of behaviour. *Applied Animal Ethology* 4, 119–124.

Kilgour, R. (1971) Animal handling in works – pertinent behaviour studies. In: *Proceedings 13th Meat Industry Research Conference*. Meat Industry Research Institute, Hamilton, New Zealand, pp. 9–12.

Kilgour, R. (1976) Sheep behaviour: its importance in farming systems, handling, transport and pre-slaughter treatment. In: Truscott, G.M.C.and Wroth, F.H. (eds) *Sheep Assembly and Transport Workshop*. Western Australian Department of Agriculture, Perth, pp. 64–84.

Kilgour, R. (1977) Animal behaviour and its implications: 1. Design sheepyards to suit the whims of sheep. *New Zealand Farmer* 98(6), 29–31.

Kilgour, R. (1978) Minimising stress on animals during handling. In: *Proceedings 1st World Congress on Ethology Applied to Zootechnics*. International Veterinary Association for Animal Production, E-IV-5, Madrid, Spain, pp. 303–322.

Kilgour, R. and Dalton, D.C. (1984) *Livestock Behaviour*. University of New South Wales Press, Sydney.

Kilgour, R. and de Langen, H. (1970) Stress in sheep resulting from management practices. *Proceedings of the New Zealand Society of Animal Production* 30, 65–76.

Kilgour, R. and Matthews, L.R. (1983) Sheep are not stupid. *New Zealand Journal of Agriculture* 147(4), 48–50.

Kilgour, R., Pearson, A.J. and de Langen, H. (1975) Sheep dispersal patterns on hill country: techniques for study and analysis. *Proceedings of the New Zealand Society of Animal Production* 35, 191–197.

Komesaroff, P.A., Esler, M., Clarke, I.J., Fullerton, M.J. and Funder, J.W. (1998) Effects of estrogen and estrus cycle on glucocorticoid and catecholamine responses to stress in sheep. *American Journal of Physiology* 275, E671-E678.

Kratzer, D.D. (1971) Learning in farm animals. *Journal of Animal Science* 32, 1268–1273.

Kyriazakis, I., Anderson, D.H. and Duncan, A.J. (1998) Conditioned flavour aversions in sheep: the relationship between the dose rate of a secondary plant compound and the acquisition and persistence of aversions. *British Journal of Nutrition* 79, 55–62.

Lemmon, W.B. and Patterson, G.H. (1964) Depth perception in sheep: effects of interrupting the mother–neonate bond. *Science* 145, 835–836.

Lester, S.J., Mellor, D.J., Ward, R.N. and Holmes, R.J. (1991) Cortisol responses of young lambs to castration and tailing using different methods. *New Zealand Veterinary Journal* 39, 134–138.

Lester, S.J., Mellor, D.J., Holmes, R.J., Ward, R.N. and Stafford, K.J. (1996) Behavioural and cortisol responses of lambs to castration and tailing using different methods. *New Zealand Veterinary Journal* 44, 45–54.

Liddell, H.S. (1925) The behavior of sheep and goats in learning a simple maze. *American Journal of Psychology* 36, 544–552.

Liddell, H.S. (1954) Conditioning and emotions. *Scientific American* 190, 48–57.

Liddell, H.S. and Anderson, O.D. (1931) A comparative study of the conditioned motor reflex in the rabbit, sheep, goat and pig. *American Journal of Physiology* 97, 339–340.

Lynch, J.J., Hinch, G.N. and Adams, D.B. (1992) *The Behaviour of Sheep*. CAB International, Wallingford, UK.

McCutchan, J.C., Freeman, R.B. and Hutson, G.D. (1992) Failure of electrical prompting to improve sheep movement in single file races. *Proceedings of the Australian Society of Animal Production* 19, 455.

Markowitz, T.M., Dally, M.R., Gursky, K. and Price, E.O. (1998) Early handling increases lamb affinity for humans. *Animal Behaviour* 55, 573–587.

Mateo, J.M., Estep, D.Q. and McCann, J.S. (1991) Effects of differential handling on the behaviour of domestic ewes (*Ovis aries*). *Applied Animal Behaviour Science* 32, 45–54.

Mellor, D.J. and Murray, L. (1989) Changes in the cortisol responses of lambs to tail docking, castration and ACTH injection during the first seven days after birth. *Research in Veterinary Science* 46, 392–395.

Miller, N.E. (1960) Learning resistance to pain and fear: effects of overlearning, exposure, and rewarded exposure in context. *Journal of Experimental Psychology* 60, 137–145.

Minton, J.E., Apple, J.K., Parsons, K.M. and Blecha, F. (1995) Stress-associated concentrations of plasma cortisol cannot account for reduced lymphocyte function and changes in serum enzymes in lambs exposed to restraint and isolation stress. *Journal of Animal Science* 73, 812–817.

Munkenbeck, N.W. (1982) Color vision in sheep. *Journal of Animal Science* 55 (Suppl. 1), 129 (abstract).

Neitz, J. and Jacobs, G.H. (1989) Spectral sensitivity of cones in an ungulate. *Visual Neuroscience* 2, 97–100.

Niezgoda, J., Bobek, S., Wronska-Fortuna, D. and Wierzchos, E. (1993) Response of sympatho-adrenal axis and adrenal cortex to short-term restraint stress in sheep. *Journal of Veterinary Medicine A* 40, 631–638.

Parrott, R.F., Thornton, S.N. and Robinson, J.E. (1988a) Endocrine responses to acute stress in castrated rams: no increase in oxytocin but evidence for an inverse relationship between cortisol and vasopressin. *Acta Endocrinologica* 117, 381–386.

Parrott, R.F., Houpt, K.A., Misson, B.H. (1988b) Modification of the responses of sheep to isolation stress by the use of mirror panels. *Applied Animal Behaviour Science* 19, 331–338.

Parrott, R.F., Misson, B.H. and de la Riva, C.F. (1994) Differential stressor effects on the concentrations of cortisol, prolactin and catecholamines in the blood of sheep. *Research in Veterinary Science* 56, 234–239.

Penning, P.D., Parsons, A.J., Newman, J.A., Orr, R.J. and Harvey, A. (1993) The effects of group size on the grazing time in sheep. *Applied Animal Behaviour Science* 37, 101–109.

Piggins, D. and Phillips, C.J.C. (1996) The eye of the domesticated sheep with implications for vision. *Animal Science* 62, 301–308.

Price, E.O. and Thos, J. (1980) Behavioral responses to short-term social isolation in sheep and goats. *Applied Animal Ethology* 6, 331–339.

Provenza, F.D. and Burritt, E.A. (1991) Socially induced diet preference ameliorates conditioned food aversion in lambs. *Applied Animal Behaviour Science* 31, 229–236.

Provenza, F.D., Lynch, J.J. and Nolan, J.V. (1994) Food aversion conditioned in anesthetized sheep. *Physiology and Behaviour* 55, 429–432.

Rook, A.J. and Penning, P.D. (1991) Synchronisation of eating, ruminating and idling activity by grazing sheep. *Applied Animal Behaviour Science* 32, 157–166.

Rowell, T.E. (1991) Till death us do part: long-lasting bonds between ewes and their daughters. *Animal Behaviour* 42, 681–682.

Rushen, J. (1986a) Aversion of sheep to electro-immobilization and physical restraint. *Applied Animal Behaviour Science* 15, 315–324.

Rushen, J. (1986b) Aversion of sheep for handling treatments: paired-choice studies. *Applied Animal Behaviour Science* 16, 363–370.

Rushen, J., and Congden, P. (1986a) Relative aversion of sheep to simulated shearing with and without electro-immobilisation. *Australian Journal of Experimental Agriculture* 26, 535–537.

Rushen, J., and Congdon, P. (1986b) Sheep may be more averse to electro-immobilisation than to shearing. *Australian Veterinary Journal* 63, 373–374.

Sandler, B.E., van Gelder, G.A., Buck, W.B. and Karas, G.G. (1968) Effect of dieldrin exposure on detour behavior in sheep. *Psychological Reports* 23, 451–455.

Scott, J.P. (1945) Social behaviour, organization and leadership in a small flock of domestic sheep. *Comparative Psychology Monographs* 18, 1–29.

Seabrook, M.F. (1972) A study to determine the influence of the herdsman's personality on milk yield. *Journal of Agricultural Labour Science* 1, 45–59.

Shillito, E.E. (1972) Vocalization in sheep. *Journal of Physiology* 226, 45P–46P.

Shutt, D.A. and Fell, L.R. (1988) The role of endorphins in the response to stress in sheep and cattle. *Proceedings of the Australian Society of Animal Production* 17, 338–341.

Shutt, D.A., Fell, L.R., Connell, R., Bell, A.K., Wallace, C.A. and Smith, A.I. (1987) Stress-induced changes in plasma concentrations of immunoreactive ß-endorphin and cortisol in response to routine surgical procedures in lambs. *Australian Journal of Biological Science* 40, 97–103.

Siegel, B.J. and Moberg, G.P. (1980) The influence of neonatal stress on the physiological and behavioral response of lambs during active-avoidance conditioning. *Hormones and Behaviour* 14, 136–145.

Squires, V.R. (1975) Ecology and behaviour of domestic sheep (*Ovis aries*): a review. *Mammal Review* 5, 35–57.

Squires, V.R. and Daws, G.T. (1975) Leadership and dominance relationships in Merino and Border Leicester sheep. *Applied Animal Ethology* 1, 263–274.

Stafford, K.J., Spoorenberg, J., West, D.M., Vermunt, J.J., Petrie, N. and Lawoko, C.R.O. (1996) The effect of electro-ejaculation on aversive behaviour and plasma cortisol concentration in rams. *New Zealand Veterinary Journal* 44, 95–98.

Stolba, A., Hinch, G.N., Lynch, J.J., Adams, D.B., Munro, R.K. and Davies, H.I. (1990) Social organization of Merino sheep of different ages, sex and family structure. *Applied Animal Behaviour Science* 27: 337–349.

Syme, L.A. (1981) Social disruption and forced movement orders in sheep. *Animal Behaviour* 29, 283–288.

Syme, L.A. (1985) *Intensive Sheep Management, with Particular Reference to the Live Sheep Trade*. Australian Agricultural Health and Quarantine Service, Department of Primary Industry, Canberra.

Syme, L.A. and Elphick, G.R. (1982) Heart-rate and the behaviour of sheep in yards. *Applied Animal Ethology* 9, 31–35.

Tanaka, T., Asakawa, K., Kawahara, Y., Tanida, H. and Yoshimoto, T. (1989a) Color discrimination in sheep. *Japanese Journal of Livestock Management* 24, 89–94.

Tanaka, T., Sekino, M., Tanida, H. and Yoshimoto, T. (1989b) Ability to discriminate between similar colors in sheep. *Japanese Journal of Zootechnical Science* 60, 880–884.

Tanaka, T., Hashimoto, A., Tanida, H. and Yoshimoto, T. (1995) Studies on the visual acuity of sheep using shape discrimination learning. *Journal of Ethology* 13, 69–75.

Torres-Hernandez, G. and Hohenboken, W. (1979) An attempt to assess traits of emotionality in crossbred ewes. *Applied Animal Ethology* 5, 71–83.

Townend, C. (1985) *Pulling the Wool.* Hale and Ironmonger, Sydney.

Trevelyan, J.P. (1992) *Robots for Shearing Sheep.* Oxford University Press, Oxford, 398 pp.

Vandenheede, M. and Bouissou, M.F. (1993) Sex differences in fear reactions in sheep. *Applied Animal Behaviour Science* 37, 39–55.

Vandenheede, M. and Bouissou, M.F. (1996) Effects of castration on fear reactions of male sheep. *Applied Animal Behaviour Science* 47, 211–224.

Vette, M. (1985) *Sheep Behaviour: an Analysis of Yard-to-restrainer Operations in Twenty-six New Zealand Export Meat Plants.* Technical Report, Meat Industry Research Institute of New Zealand, Hamilton, New Zealand.

Walk, R.D. and Gibson, E.J. (1961) A comparative and analytical study of visual depth perception. *Psychological Monographs* 75, 1–44.

Walls, G.L. (1942) *The Vertebrate Eye and its Adaptive Radiation.* Cranbrook Institute of Science, Bloomfield Hills, Michigan.

Webb, T.F. (1966) *Feasibility Tests of Selected Stimuli and Devices to Drive Livestock.* Report 52-11, Agricultural Research Service, Department of Agriculture, Beltsville, Maryland.

Webster, M.E.D. and Lynch, J.J. (1966) Some physiological and behavioural consequences of shearing. *Proceedings of the Australian Society of Animal Production* 6, 234–239.

Weisenberger, M.E., Krausman, P.R., Wallace, M.C., De Young, D.W. and Maughan, O.E. (1996) Effects of simulated jet aircraft noise on heart rate and behavior of desert ungulates. *Journal of Wildlife Management* 60, 52–61.

Whateley, J., Kilgour, R. and Dalton, D.C. (1974) Behaviour of hill country sheep breeds during farming routines. *Proceedings of the New Zealand Society of Animal Production* 34, 28–36.

Whitteridge, D. (1978) The development of the visual system in the sheep. *Archives Italiennes de Biologie* 116, 406–408.

Winfield, C.G. and Kilgour, R. (1976) A study of following behaviour in young lambs. *Applied Animal Ethology* 2, 235–243.

Winfield, C.G., Syme, G.J. and Pearson, A.J. (1981) Effect of familiarity with each other and breed on the spatial behaviour of sheep in an open field. *Applied Animal Ethology* 7, 67–75.

Wollack, C.H. (1963) The auditory acuity of the sheep (*Ovis aries*). *Journal of Auditory Research* 3, 121–132.

# Design of Sheep Yards and Shearing Sheds*

**11**

## Adrian Barber[1] and Robert B. Freeman[2]

[1]*Department of Agriculture, PO Box 81, Keith, South Australia, Australia:* [2]*Agricultural Engineering Section, University of Melbourne, Melbourne, Victoria, Australia*

## Introduction

There have been significant advances in the design of sheep yards and shearing sheds. Well-designed facilities enable thousands of sheep to be handled efficiently, humanely, and with a minimum of labour.

## Sheep Yards

Main pens and force pens should be built of material that is clearly visible and strong. Fence height is 90 cm for fences inside the yard area and 1 m for fences at the outside edge of the yards. Sheep density in the main pens is 1.5–2 sheep m$^{-2}$ and rises to 2.5–3 sheep m$^{-2}$ in the force pens.

Where a curved force pen (a bugle) (Fig. 11.1) is used to lead sheep into a drafting (sorting) or handling race, the curved approach should be constructed of solid covering material or sheeting on the inner side for the last 6 or 8 m. Many bugle curves work well with open panels on the outer side of the draft race and force pen. However, in some cases, it may be necessary to blind off the outer draft wall and the outside fence for the last 1–3 m from the race entrance.

When the force pen leads to a handling race, it should hold a few more than one or two times the number of sheep that can fit in the race. This leaves some left over after the race is filled to act as decoys when bringing the next lot of sheep into the force pen. Generally, a force pen should not hold more than 100 sheep. Otherwise, there are too many to control and start up the race.

CAB *International 2000. Livestock Handling and Transport,* 2nd edn
(ed. T. Grandin)

Triangular force pens are usually used in rectangular yards and can be built in single or double forms (Fig. 11.2). A single force has one fence as an extension of the drafting or handling race side, with the second fence set at a 30–40° angle. Double triangular forces have two-wing fences running back at

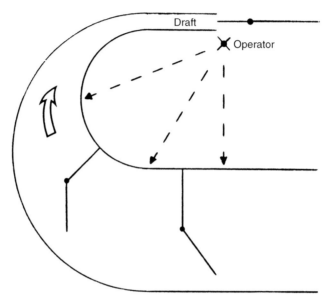

**Fig. 11.1.** Bugle force pen, showing operator position. The curved design provides the operator with easy access to the sheep.

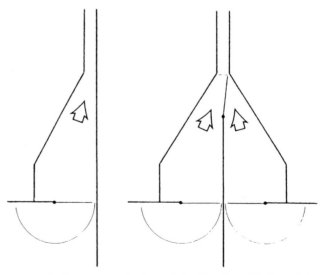

**Fig. 11.2.** Triangular force pens, single and double. One side is straight and the other is on a 30° angle.

similar angles and a central fence with a flip-flop gate at the race entrance to allow sheep entry from either side.

## Drafting Race

An efficient drafting (sorting) system allows the operator to identify and draft the sheep he or she wishes to separate with a minimum of errors. To do this accurately requires an even flow of sheep. For small flocks, a two-way sort is satisfactory, but in large-scale sheep enterprises, particularly where crossbred sheep for prime lamb production are run, a three-way sort, using two gates, may be necessary. Four- and five-way sorts can be built for special purposes, such as on stud properties.

The sorting race is approximately 3 m long, with the exit point showing a clear escape route for the sheep. Where a bugle force pen leads to the draft, the straight outer fence is usually 4–4.5 m long to allow for the in-flow curve.

Several points to be considered when building a drafting race are:

1. Closed-in sides;
2. Adjustable sides;
3. Tapered sides;
4. A durable floor (concrete or battens);
5. A stop gate at the outlet of the race;
6. Rubber dampers on the leading edge of the draft gate(s) to reduce noise;
7. A large diameter vertical roller at the race entrance to prevent sheep jamming (e.g. a 200-litre drum set with its surface almost flush with the angled fence);
8. A remote-control gate.

### Siting the drafting race

The direction of the drafting race should minimize the effects of sunlight and shadows. Sheep appear to run better northwards into the sun (their shadows are behind them) and the stock worker has a better chance to see ear tags or earmarks when the sun is shining on the heads of the approaching sheep. This will not work in the very early morning if the sun blinds the sheep.

The race should be directed away from, or parallel to, the shearing shed wall so that the sheep do not get an impression of approaching a dead end. The race should be on flat or slightly inclined upwards ground, but not inclined downwards. Sheep in the drafted pens need to be clearly visible to other sheep still coming through the race, and thus act as decoys to encourage sheep flow. In some yard layouts, the drafting race will face a shearing shed wall. This can be satisfactory, if the race is set well back and there is plenty of pen area between the race and the shed.

## Side panels

In most sheep yards, the drafting race has closed sheeting on both sides to prevent outward vision by the sheep. This is to direct the sheep's attention towards the exit – sheep will actually increase speed to get through the race.

However, in some yards, a race with open panels on the side opposite the stock worker has been found to work well. The sheep are encouraged to enter the race by seeing other sheep moving away from the drafting gate on the outer side.

Timber planks, steel sheeting, or weatherproof plywood are suitable materials for the closed sides of the race. The sheeting must be placed on the inner faces, so as to give a completely smooth surface. Bolts should be countersunk, or round-headed coach bolts should be used with the nuts to the outside.

## The draft gate

The draft gate is usually 1.3 m long, but may be in the range of 1–1.5 m. Where twin gates are used to give three-way drafting, the handles should be offset from the gate top to prevent the stock worker from jamming his or her hands between the gates. The handle should also have a looped knuckle guard, which is similar to the guard loops on handlebars of agricultural motor bikes, projecting forward to protect the stock worker's hands from charging sheep.

There is some debate on whether the draft gate should be made of open panels or closed sheeting. Reasons for using open rail gates include:

**1.** The oncoming sheep can see the previous sheep moving away from the draft and are more inclined to follow;
**2.** Open gates are lighter, and therefore, quicker and easier to use;
**3.** Open gates are less affected by winds blowing across the drafting race.

Reasons for using closed sheeting gates include:

**1.** Such gates act as a continuation of the drafting race, thus directing the sheep into the exit pen;
**2.** Closed gates prevent horns or legs from getting caught.

## Remote control sort gates

There are situations where it is an advantage to be able to draft sheep from behind, for example, picking daggy (dirty) sheep. This requires some type of remote control mechanism to operate the drafting gate (Fig. 11.3).

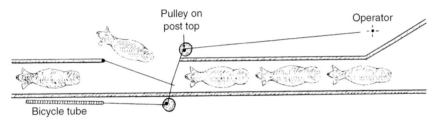

**Fig. 11.3.**   Remote control draft (sort) gate, with spring return.


## Sheep Handling Race

A multipurpose handling race for drenching, vaccinating, branding, classing, jetting (spraying), and other agricultural activities is needed in the sheep yards. The handling race usually leads off the same force pen as the drafting race. Alternatively, it can be constructed as a forward extension of the drafting race.

Several different types of handling races can be built.

**1.**   A single race of 52–64 cm wide where the stock worker is outside the race.
**2.**   A single race of 70–80 cm wide where the stock worker is inside the race.
**3.**   An adjustable-sided race of which the width can be varied between 45 and 80 cm.
**4.**   Double races side by side are satisfactory, but they must be worked correctly. A new batch of sheep is allowed to fill the first race while treated sheep are running out of the adjacent one. It can be difficult to fill one race if sheep are held in the second. The sheep will run about halfway along the race and then stop beside the other sheep they see through the rails.
**5.**   Triple races are successful where the sheep fill the two outer races and the stock worker is in the centre race.

A suitable handling race is 9–15 m long with sides 85 cm high. One or two stop gates are included along its length to prevent lambs from crowding. The race should have a concrete or batten floor that extends 70–100 cm on either side to form a walkway at the same level as the race. Either steel or timber can be used to build the handling race. Steel rails are generally preferred, as they allow greater access and vision by the operator through the race sides. One section of the race wall can be removable panels or a gate section. This provides for sheep to be diverted to other sheep handling facilities, for example, a cradle-crutching unit, an automatic jetting (spraying) race, or weighing scale. Sheep flow is often improved if the take-off point to these facilities is near the start of the race rather than the end. If possible, a drafting gate should be positioned at the exit of the handling race to allow separation of groups of sheep after an agricultural activity, such as classing.

# Bugle Yards

Bugle yards, or bugle races (force pens), work by leading the sheep around a curve while the sheep pathway narrows down to a draft or handling race (Figs 11.4 and 11.5).

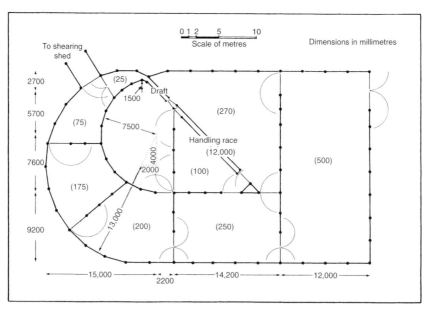

**Fig. 11.4.**    Diagram of a bugle sheep yard.

**Fig. 11.5.**    Bugle sheep yard with solid sides.

Bugle yards attempt to create an improved sheep flow compared with traditional square yards. This is a debatable point that has not been proved, which indicates that sheep yard shape is not the only factor required for efficient sheep flow. Other factors, such as sunlight and shade effects, ground slope, building materials, and position of the shearing shed, can also have an effect.

Bugle yards can be built with two- or three-way sorts. The main working areas of the yards are shown in the plans. Additional holding yards or mini-paddocks can be added to the sides of the plans to increase sheep holding and working capacities to suit individual properties.

Curved-type yards can turn to either the left or right. That is, for a given plan, there will be a matching mirror image.

The dimensions of a bugle force pen leading into a race are of critical importance in obtaining a satisfactory flow of sheep. If the curves narrow down too sharply, sheep can double up and become jammed. On the other hand, if the curved entry is too wide, sheep have the opportunity to turn back away from the race.

A right-angled triangle, 4 m (4000 mm) along the base and 2 m (2000 mm) vertically, is used to establish the curved fences. A radius of 7.5 m from the toe of the triangle gives the inner curve, and a radius of 13 m from the top of the triangle gives the outer curve. A 1.5 m radius curve leads into the sort race (see Fig. 11.4).

Three-way drafting is provided. Note that the centre batch of sheep does not have to travel the length of the handling race. The sheep are diverted by a panel gate behind the main sort gates into a storage pen beside the handling race. A common drafting problem in sheep yards that have the sort and handling races in line is that the centre batch of sheep being fed through the handling race may balk about halfway along. Sheep then bunch up back to the sort gates. The use of a fence, diagonal to the handling race, and an extra gate, to create an additional storage pen, overcomes this difficulty.

Curved-type yards can also be drafted in two ways (right or left) at the exit of the handling race.

The side and shape of the receiving pen and position of the gates may be altered to suite different sites.

# Circular Yard

The circular plan has a sheep pathway 2.5–3 m wide around the circumference of the yards, with a combined draft/handling race across the diameter (Fig. 11.6). The race allows three-way drafting at the main gates and two-way drafting at the exit of the handling race. The double bugle force pens leading to the race allows for multi-mob handling for shearing or crutching. The gates can be set so that sheep can bypass the race entrance. Being symmetrical in

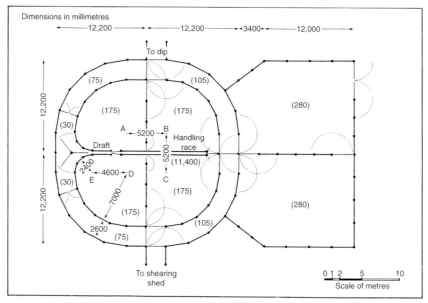

**Fig. 11.6.** Circular sheep yard.

shape, the yards can be reversed, with the race heading in the opposite direction, for fitting on particular sites.

The circular design gives a high degree of flexibility of sheep movements around the yards. The plan is ideal for stud properties. The circular raceways can be readily divided into small pens for displaying rams at field days and sales.

## Shearing Sheds

Older shearing sheds in Australia were designed with a centre board layout (Fig. 11.7). Sheep are moved from the holding pens into the small catching pens. The shearers catch the sheep one at a time and pull each to the shearing board (platform the shearers work on). After each sheep is shorn, it is dispatched by sliding it down a chute to a level below the holding pens. Some sheds have races for discharging shorn sheep.

After being shorn several times, sheep become reluctant to move from the holding pens into the catching pens. The front-fill catching pen was developed to help solve this problem (Fig. 11.8). Sheep are moved from the holding pen into a filling pen and are then moved into the catching pen through a gate near the front of the pen. Some other modern designs are the parallel-flow and diagonal-flow methods. To reduce walking distances for wool handlers, the shearing board in a modern shed is curved (Figs 11.8 and 11.9).

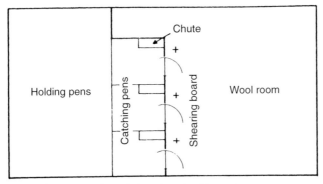

**Fig. 11.7.** Australian centre board shearing shed layout.

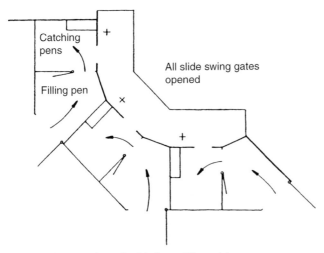

**Fig. 11.8.** Curved shearing board with front-fill catching pens.

The woolly sheep must walk up a ramp to get into the shed, because the floor is above ground level. The ramp should be 1.5–3 m wide, with a maximum slope of 20°. The sides of the ramp should be made of closed panels to restrict vision. Sheep will shy away from a visual cliff, such as that represented by a ramp with unprotected sides. To provide sufficient height for the discharge chutes, a minimum of 1.2 m of head room is required under the main floor.

When wood floor gratings are used, the battens should run at right angles to the direction the sheep are required to move. When sheep walk across the battens, it restricts their vision and helps to prevent them from seeing sunlight coming up through the floor.

A shearing shed must be large enough to shelter the daily throughput of the shed from rain. Woolly sheep can be held in an area of 2.5 sheep m$^{-2}$. Shorn sheep can be held at 3.5 sheep m$^{-2}$.

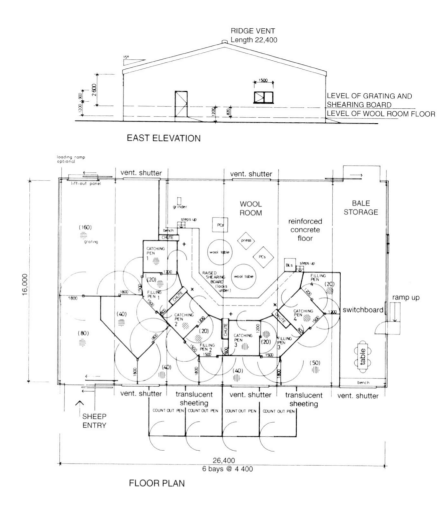

**Fig. 11.9.** Floor plan of three-stand (three shearers) curved board shearing shed. (Dimensions in millimetres.)

Careful attention must be paid to the design of the discharge chutes. A chute should start on a 45° angle and then change to a 20° angle at the halfway point. This design will help prevent injuries.

Shearing sheds should have good lighting to facilitate sheep movement. Translucent plastic panels installed in the roof will provide evenly diffused lighting, and will help prevent the distorting influence of strong contrasts of light and dark shadows.

## Acknowledgements

More detailed information on sheep yards and shearing sheds can be obtained in *Design of Shearing Sheds and Sheep Yards* (1986) by A.A. Barber and R.B. Freeman, published by Inkata Press, 18 Salmon Street (PO Box 146), Port Melbourne, Victoria 3207, Australia, on which this chapter is based.

Diagrams of sheep yards and shearing sheds are reproduced with permission of Inkata Press.

## Further Reading

Anon. (1987) Elevated race system edges towards success. *New Zealand Journal of Agriculture* 152, 20–24.
*Describes elevated race system for shearing sheds*
Belschner, H.G. (1957) *Sheep Management and Diseases*. Halstead Press, Sydney, Australia.
*Sheep yard designs for drafting (sorting) many different ways*
Burnell, R.E. (1967) Cradles make foot treatment easier. *New Zealand Journal of Agriculture* 114(4), 50–53.
*Describes sheep restraint devices*
Grandin, T. (1984) Sheep handling and facilities. In: Baker, F.H. and Mason, M.E. (eds) *Sheep and Goat Handbook*. Westview Press, Boulder, Colorado.
*Sheep handling and yard design*
Grandin, T. (1990) Handling can improve productivity, *National Wool Grower* (American Sheep Industry Foundation, Englewood, Colorado) 79(12), 12–14.
Jefferies, B. (1981) Sheep cradles can be used for many jobs. Fact Sheet, Agdex 430/724, Department of Agriculture, South Australia, Australia.
*Information on different types of sheep restraint devices*
McLaren, L. (1977) Sheep laneways. Fact Sheet, Agdex 430/10, Department of Agriculture and Fisheries, South Australia, Australia.
Midwest Plan Service. (1982) *Sheep Housing and Equipment Handbook*. Midwest Plan Service, Iowa State University, Ames, Iowa.
*Plan for easy to build tilting sheep restraint device*
Miller, K.R. and Meade, W.J. (1984) A simple restrainer for sheep for abdominal radiography. *New Zealand Veterinary Journal* 32, 146–148.
*Canvas restraint device which is easy to build*
Pearse, E.H. (1965) *Sheep and Property Management*. The Pastoral Review Pty Ltd. Sydney, Australia.
*Yard systems for drafting (sorting) in many different ways*
Ralph, I.G. (1975) Forward planning in sheep handling systems. *Pastoral Review* 85(10), 679–681.
*Design of tilting sheep restraint devices*

# Facilities for Handling Intensively Managed Sheep

**12**

## Huw Ll. Williams

Formerly of Department of Farm Animal and Equine Medicine
and Surgery, Royal Veterinary College, University of London,
Potters Bar, Hertfordshire EN6 1NB, UK

## Introduction

The massive annual migration of sheep between geographical areas is a
unique feature of the British sheep industry. It involves a vast number of
breeding stock from the hills and uplands consisting mainly of ewes which
have had three lambings in the adverse environment of the hills, and also
unbred crossbred ewe lambs and two-tooth ewes from the uplands. The latter
provide replacement animals for the lowland crossbred flocks; these half-bred
ewes are now the mainstay of lowland sheep farming. The vast majority of the
sheep involved in this annual movement are sold through specially arranged
regional sales and are then transported, often for long distances, to lowland
sheep units. There is also seasonal movement of breeding stock between
lowland farms, usually fairly local and for a short term. In addition to the
major movement of animals from hill areas, ewe lambs temporarily move
down to lowland farms over winter and return to the hills in the spring as
yearlings (Williams, 1978, 1999). Hill and many upland areas do not have the
resources to feed lambs for the fatstock market and consequently wether lambs
are sold to lowland units, usually through special lamb sales, for a period of
fattening. Movement of lambs also occurs between lowland farms before being
presented at fatstock markets or sold directly to abattoirs.

### Management changes in lowland sheep units

The interdependence that has developed between regions in terms of the
supply of breeding animals has allowed the majority of lowland sheep units in
the UK to concentrate solely on the management of animals in production and

thus eliminate the need to allocate a proportion of the grassland and other crops to the needs of young replacement stock.

Although the average size of the British flock is only 217 ewes, there has been a trend towards bigger flocks and more animals in the care of one person during the 1990s. Almost half the breeding ewes in the UK are now in flocks of over 500 ewes and it is not unusual to find flocks of 700–1000 ewes cared for by one shepherd. This development has led to more dependence on contractors to carry out the main flock routine tasks. Whereas in the 1980s contract work was limited to shearing, contractors may now be employed to shear, dag, dip, foot-trim, determine pregnancy and litter size and assist with or fully supervise the lambing. This places more emphasis on the need for proper handling facilities to enable the tasks to be completed efficiently and without compromising the welfare of the animals.

The economic pressures of the last two decades have led to more intensive forms of production in lowland units involving higher stocking rates, higher output of lambs per ewe and, in the case of late autumn/early winter lambing, early weaning of lambs coupled with fattening on *ad libitum* concentrates. High stocking densities, of up to 35 ewes ha$^{-1}$ at certain periods, involve controlled grazing and frequent movement between grazing areas (Brown and Meadowcroft, 1989). More intensive use of grassland has led to housing throughout the late autumn and winter period, simply to avoid damage to the pasture. When sheep are housed for this length of time, winter shearing is usually undertaken. Less intensively managed units house later, usually over the last 6 weeks of pregnancy, to provide more surveillance at lambing and more intensive care of lambs, particularly from large litters, over the neonatal period. The marked trend towards breeding from ewe lambs in their first autumn also demands more surveillance. It is not uncommon to provide 24-h surveillance over the peak lambing period; housing provides better working conditions and better opportunities for skilled staff to demonstrate their expertise (Williams, 1988).

## Handling

### The welfare aspects

Prettejohn (1988) has drawn attention to the continuing concern of the general public for animal welfare, particularly within the more intensive systems of management. The concern embraces general handling techniques and methods of carrying out various procedures during the course of the year. The first statutory provisions for the welfare of livestock were introduced in the Agriculture (Miscellaneous Provisions) Act 1968 and the first code for sheep was issued in 1977 and has recently been revised (MAFF, 1999). The codes encourage the highest standard of husbandry and emphasize the need for good handling and for a high degree of skill in carrying out the variety of tasks which have to be undertaken. These skills and stockmanship are normally acquired through a combination of technical training and that gained from working alongside experienced shepherds. These are vital requirements for those involved in the care of sheep in intensively managed units.

The early detection of abnormality, whether it be health or behaviour, is of paramount importance. Coupled with this is the need to segregate such animals quickly and to establish whether the whole flock requires close inspection or treatment. For example, this is vitally important in the control of foot-rot and other conditions resulting in lameness. Lameness has been identified as a major welfare problem in the UK sheep industry (Broom, 1988) and is given emphasis in the new code (MAFF, 1999). Vigilance and prompt action are particularly required where stocking density is high. As intensification increases, so the frequency of handling associated with the variety of tasks, including prophylaxis, increases and thus there is a particular need to provide good handling facilities, whether they be central and permanent or demountable and portable.

## Basic characteristics and behaviour of sheep

Good handling is based on a sound understanding of the sheep's behaviour. This allows the experienced shepherd or handler to carry out whatever manoeuvre is required with minimum effort and with minimum of disturbance and stress to the sheep. The shepherd should be able to predict and act in anticipation of what the sheep are likely to do in a given situation. Most of the manoeuvres connected with gathering, confining and handling will arouse a moderated level of fear and alertness. Provided they are undertaken skilfully, these responses should not advance into a state of stress; the sheep's physiological and behavioural reserves should prevent this from happening (Lynch *et al.*,1992).

The basic characteristics which are useful in the handling and management of sheep have been identified (Hutson, 1980; Kilgour and Dalton, 1984; Lynch *et al.*, 1992) as a strong flight reaction, the dominant role of vision in social organization and marked flocking/follower behaviour.

During the initial stages of gathering, particularly if they are scattered over a wide area, the flocking tendency can be exploited through the use of a trained dog. Flocking is usually followed by a state of alert, which is quite evident by the sheep's head and ear movements. Whilst in this state, the flock can be encouraged to move slowly forward along the main pathway of the drive by both human and dog moving towards them slowly but not near enough to initiate a flight reaction. The absence of side escape routes ensures their steady forward movement. It gives an impression of an escape route to those in the front of the flock and the remainder react according to their strong follower instinct. In a long drive, it is important not to take them near the point of exhaustion – an aspect which can be easily overlooked when some kind of vehicle or an unruly dog is used.

In and around the handling unit, a combination of basic characteristics is exploited, particularly the tendency to follow the leader and to move away from buildings and things that frighten them and towards what appears to be an open prospect and freedom. Brown and Meadowcroft (1989) suggest that these corridors to freedom should be slightly uphill, at a gradient of approximately 1:60.

In some situations, there is no doubt that trained animals can be used to lead the remainder of the flock through the sequence of pens (Hutson, 1980). This is quite useful in those lowland flocks for which replacement breeding stock is bought in from extensive upland breeding flocks. The principle has been used for centuries in those communities where the village flocks are taken out to pasture only during the day. Although training can be usefully used to ensure the steady flow of sheep through the unit, sheep have the capacity to recall aversive experiences (Rushen, 1986) and their behaviour will clearly indicate their apprehension during their approach to the area where this is normally undertaken. It is suggested that the procedures which initiate this kind of response, such as dipping and shearing, are carried out well to the side of the main pathway through the pens normally used for routine tasks involving minimal handling and restraint.

Visual contact is very important, particularly when individuals and small subgroups have to be separated from the main flock. The distress following separation can easily be alleviated by using pipe fences and gates in the area used for shedding and regrouping.


## The handling unit

Since there are a variety of tasks to be carried out on the flock during the course of a year, it is in the interest of efficient use of labour and the welfare of sheep to have an appropriate handling unit. It is necessary to provide adequate and safe holding and handling facilities (MAFF, 1990). The type of unit will vary according to the specific circumstances on the farm and the system of management. On many lowland farms, sheep farming is one of several enterprises operated under the same management and, in such a situation, competition for labour can easily arise during clashes of peak labour demand. A well-planned handling unit can greatly contribute to minimizing the labour requirement of a particular task and to providing good working conditions.

The basic requirements of a handling unit for a large flock are illustrated in Fig. 12.1. The type of pen arrangement and facility provided in each of these areas will vary considerably according to the specific requirements of the sheep unit and to the preferences of those operating the unit. A roof over the working areas provides protection during inclement weather. In the case of small flocks, the forcing (crowding) pen may be replaced by a small pretreatment pen

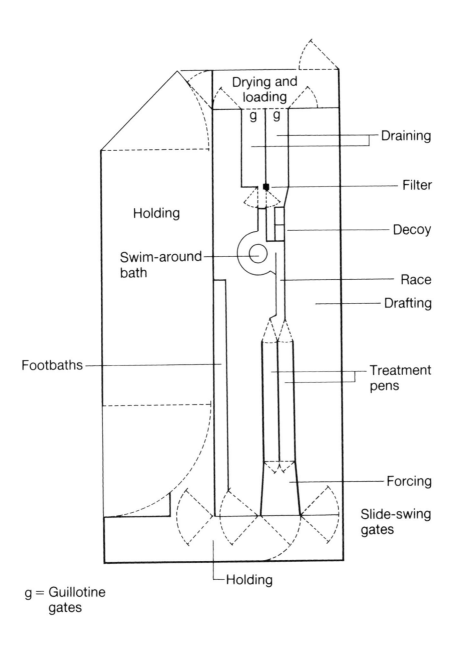

Drying and loading

g   g

Draining

Filter

Decoy

Race

Drafting

Holding

Swim-around bath

Footbaths

Treatment pens

Forcing

Slide-swing gates

Holding

g = Guillotine gates

**Fig. 12.1.** A plan for a handling unit for a large flock (National Agriculture Centre).

holding 10–20 sheep, so that they are conveniently placed for handling and treatment in the working area. The working areas should include, first, a pen for the handling, casting and treatment of individual sheep; as an alternative to the use of a race, some group treatments can also be carried out in this pen. Secondly, there should be a race which keeps the sheep in a single file.

The race can serve a multipurpose function, as many of the routine tasks can be undertaken in this facility with the operator standing on the outside. These include examination of mouth, head and back line, drenching, vaccination and application of pour-on parasiticides. When fitted with a footbath, it is also the means of controlling foot ailments. A race usually leads to a drafting (sorting) gate, which allows subdivision of the flock two or three ways into subsections of the holding area. Most units would provide the means of returning some or all of the flock to the holding and working areas. The race should have solid sides, so that the sheep are encouraged to go forwards towards the drafting gate (Brockway, 1975). The drafting gate should give an unobstructed view of one of the holding pens and possible sight of other sheep (Fig. 12.2). A race is usually 4.5–6 m long; if longer, it should be fitted with a central non-return gate. Where the race is connected directly to the initial loading area, the approach to the entrance should be funnelled and the entrance controlled by a guillotine gate operated by cord from the front of the race. It should also be possible to place equipment for specific tasks at the end of the race, for example weigh crate, side-tipping casting crate or artificial insemination (AI) crate. The Farm Buildings Information Centre (1983) provides comprehensive data concerning siting, layout and specification of handling pens.

**Fig. 12.2.** Portable sheep handling unit with a two-way drafting gate.

*Portable sheep handling units*

The basic principles relating to permanent handling facilities in terms of sheep psychology, siting, layout and working conditions also apply to the general features of a portable unit. It is, of course, considerably easier to adapt, modify and extend the latter compared with the former, particularly in relation to the needs of the various age-groups requiring handling and treatment during the grazing period (March–September). The majority of UK flocks have fewer than 300 ewes, and a portable handling unit, capable of being used in the fields and/or on a concrete floor (see Fig. 12.2), is ideally suited to dealing with most tasks without having to subject the flock to the stress of driving to a distant central unit. It also reduces the time the ewes are away from grazing. There are, however, certain disadvantages to such a unit, including the time taken for erection, the likelihood of damaging certain components, poor conditions underfoot, particularly in a wet season, and the lack of protection from the weather for the operators.

## Flock Management and Handling

There is a variety of reasons for handling the whole flock or subflock during the course of the year. These are: prevention of disease; weighing and body-condition scoring; drafting (sorting) for sale and regrouping; and shearing. The degree of physical restraint required to carry out these tasks is listed in Table 12.1. The tasks are carried out in an orderly sequence, governed largely by the timing of key events – mating, lambing, shearing and weaning. Of these key events, the timing of lambing, taking into consideration a wide spectrum of factors, such as type of soil, crops and seasonal availability, labour supply and marketing objectives, has a dominant role with regard not only to the timing of the tasks but also to the type of tasks which have to be undertaken. For example, ewes lambing in the autumn/early winter will go out to grass in the spring without their lambs and consequently their management during the grazing season, particularly in terms of stocking rates, parasite control and

**Table 12.1.** Classification of procedures according to degree of physical restraint.

| No restraint | Minimal restraint | Full restraint |
|---|---|---|
| Spraying/showering | Dosing | Foot trimming |
| Foot-bathing | Vaccinating | Docking |
| Weighing | Dipping | Castration |
| Drafting | Condition scoring | Shearing |
| Regrouping | Scanning | Dagging |
| Loading | | Artificial insemination |
| | | Tattooing, tagging, notching |

gathering to select lambs for marketing, is very different from spring lambing flocks.

## Restraint

An animal requiring close examination should be restrained by placing the open hand under the jaw and with the head held above the back line. It may then be manoeuvred sideways or backwards towards a corner of the pen to minimize sideways movement. During this manoeuvre, placing a hand on the opposite flank may help in its control. This is easily accomplished when sheep are already confined in a treatment pen. In a more open situation, such as a corner of a field or at a gateway, a crook may be used as an effective first-stage restraint, hooking the sheep just above the hock and raising the leg off the ground before moving forward quickly to grasp a hind leg above the hock. No part of the fleece should be used for restraint and only adult animals should be restrained by the horns. The horns of young animals are fragile and can be dislocated easily when used for restraint.

## Casting

Examination and treatment of the feet and of the underline, udder, scrotum and prepuce/penis involves casting. The resting position following casting is also used for the initial stages of shearing and for trimming wool in the udder area in preparation for lambing.

There are several methods of casting. The choice of method is usually determined by the size of the sheep and the physique of the operator. Irrespective of the method of casting, it is imperative, particularly in the interest of avoiding injury, that the operator takes up a proper position relative to the animal and for the animal to stand with the four feet forming a rectangle. The operator's knees should be close together against the central part of the ribcage, with the left hand under the jaw to keep the head up and to prevent forward movement. The right hand is used to grasp a fold of skin low down in the right flank region (Fig. 12.3a). This hand is used to lift the right hind leg off the ground and simultaneously a slight nudge of the knees prevents the sheep bracing on its left hind leg, the sheep is pulled momentarily on to the operator's knees before being gently turned on to its rump in one synchronous and gentle movement. The sheep then sits with its withers resting between the handler's knees (Fig. 12.3b). Slight pressure on the sternum helps to get the animal into a more relaxed position.

In the case of heavier sheep, one relies on unbalancing the animal prior to bringing it to rest on its rump and again with the withers between the operator's knees. In this method, the initial move is done by the left hand bringing the head round abruptly towards the ribcage and at the same time

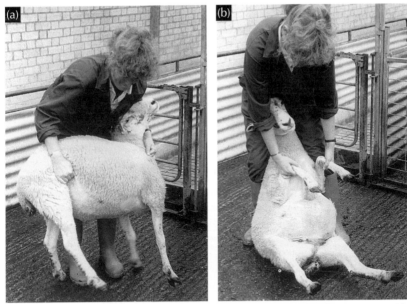

**Fig. 12.3.** (a) The initial stage of the lift method of casting. (b) The resting position following casting.

pressure is applied to the rump to get the animal to drop on to its rump before lifting the forequarters to get it into a sitting position. Alternatively, some operators find it easier to unbalance the animal by leaning over to grasp either the nearside or offside hind leg. Casting a high proportion of a large flock of a large breed can impose a great deal of physical stress on the operator, particularly when the ram stud is involved. There are various devices, such as a cradle to take the full weight of a cast animal or a mechanical device for casting with either a sideways movement or an up-and-over movement. In both cases, the sheep are held in the upside-down position until the particular technique is completed. The position allows two operators to work simultaneously (Fig. 12.4).

### Handling for control of skin parasites

All intensively managed sheep are treated two or three times during the year for protection or treatment against skin parasites. Treatments carried out in a spray race or shower unit or by jetting (spraying) with a handgun whilst in the race do not involve handling and are much less stressful than the enforced swim associated with plunge dipping. Plunge dipping, using an appropriate dip, is the only effective method of controlling sheep-scab, caused by a mite (*Acarus*). In 1992, sheep-scab ceased to be a notifiable disease. Under the Sheep Scab Order 1997, it is a criminal offence if owners fail to treat or move

**Fig. 12.4.**   Foot trimming in a sheep-turning crate.

visibly affected animals. The available evidence indicates that the disease continues to be a major problem and that it is widespread in all areas of England and Wales (Lewis, 1997). Design information on dip vats can be found in Shorrock (1981).

It is generally accepted that dipping is one of the most stressful tasks undertaken during the sheep year (Henderson, 1990). Sheep which are unwell or are in poor condition should not be dipped. It is inadvisable to dip sheep when they are heated, tired, wet, thirsty or full-fed (MAFF, 1980). Sheep should be brought into a yard and rested well before the commencement of dipping. They should not be overcrowded during this period. The sheep are then taken forward to a collecting pen or race leading to the bath. Where animals are lifted, as for casting, and then lowered into the bath, they must be firmly held (never by the fleece) and gently immersed, tail first. Those entering the bath by sliding or by an automatic tipping arrangement should be closely supervised in case some enter on their back. This close supervision should be maintained throughout their period of immersion, using a brush or crutch to control their movement, and particularly their total immersion during the dipping (Fig. 12.5). Exhausted animals should be given assistance to climb out of the bath. No overcrowding should occur in the draining pen. The sheep should then be released into a sheltered paddock or open covered yard, depending on weather conditions, and rested before returning to pasture.

## The handling of ewes and lambs at pasture

Daily surveillance of the flock during the grazing season invariably results on some days in the need to examine an individual or a few animals but does not warrant driving the whole flock to the central handling facility. In this kind of

**Fig. 12.5.** Dipping in progress. Note method of control (National Agriculture Centre).

situation, sheepdogs may be used to confine the flock in a corner of the field or, alternatively, the flock may be taken out of the field and into an alley or corridor, which may be temporarily penned off to allow animals to be caught and restrained and treated on the spot fairly quickly and without undue and prolonged stress. There is no doubt that management of sheep under extensive systems would be very difficult without properly trained dogs, particularly in terms of gathering and driving. In the more confined situations found in the majority of intensively managed lowland farms, sheepdogs are used to initiate flocking. They are used to confine animals in a specific area and to carry out the difficult manoeuvre of segregating individuals and subgroups from the remainder of the flock whilst at pasture. Sibly and Baldock (1987), in their investigation of the effects of a variety of handling procedures on heart rate in sheep, have clearly demonstrated that a human with a dog has a very large effect. Nothing can be more stressful and damaging to a group of sheep and exasperating to the handler than a dog which is not under proper control and poorly trained. With the advent of all-terrain vehicles, gathering and driving on many units is accomplished without the use of dogs. In some situations, older animals will readily respond to the sight and sound of a food bag and will follow the shepherd out of the field.

## Handling Associated with Lambing

Housing during the greater part of pregnancy has become an established practice in many intensively managed flocks to improve surveillance and to provide a high level of neonatal care for lambs from breeds and crossbreeds of high fecundity. There are flocks which house only during the last 6 weeks of pregnancy and there are those where lambing will occur in a paddock adjacent to a farm building into which the family group is taken for 48 h, depending on weather conditions. In all these alternative ways of managing pregnant and lambed ewes, it is imperative to plan the movement of ewes and lambs around lambing time – a period of frenetic activity which, more often than not, is coped with by minimal staff.

Figure 12.6 shows the flow of movement between the types of pens which are usually provided to meet the high standard of management required at this crucial time in the sheep production cycle. It provides optimal conditions in

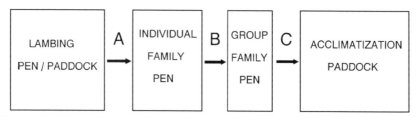

**Fig. 12.6.** Types of pens and movement between pens during lambing period.

terms of surveillance, parturient behaviour, hygiene, maternal behaviour and neonatal care of the lamb. Where ewes are permanently housed during late pregnancy, group size should be restricted to 30–40 ewes. This reduces the risk of injury and crushing at feeding time and when individual ewes have to be caught and attended to within the pen.

The routine tasks and type of handling undertaken in the pens and during the movement between the pens shown in Fig. 12.6 are as follows.

## Lambing pen/paddock

*Restraint*

The initial examination of individual ewes requiring assistance at lambing can be undertaken with the ewe standing. Most malpresentations will involve casting the ewe on to her side and with one knee pressing on the forequarters to restrict undue movement. Wet fostering of lambs will take place in this pen and also treatment of the navel cord.

*Movement A*

The family group is moved by dangling the lambs by the forelegs in front of the ewe, using the index finger between the knee joints to prevent damage to the skin. Ewes displaying normal maternal behaviour will follow to the site of the family pen. Inexperienced animals lambing for the first time may need to be restrained and manoeuvred to the pen where the offspring have already been placed.

## Individual family pen

This is where recognition and further bonding takes place. Provided the dam's udder and teats are normal, surveillance should concentrate on sucking and suckling behaviour. Ewes that reject their lambs or are used for late fostering may be transferred to a fostering stall which allows the lamb free movement on both sides. The ewe is usually restrained by some kind of yoke arrangement which prevents her turning round to butt the lambs. Water and foodstuffs are provided in the front of the stall.

*Feeding colostrum by stomach tube*

Failure to suck or inadequate intake of colostrum can initiate hypothermia in lambs after approximately 5 h. Lack of colostrum will result in low levels of passive immunity and thus a high risk of infections. Feeding colostrum from its own dam or from a frozen supply by stomach tube is now a common management practice, sometimes practised routinely to reduce the time spent on surveillance. A description of techniques for feeding colostrum to lambs is given in Eales and Small (1995) and Williams (1999).

### Docking and castrating

In the UK these procedures must be carried out in strict accordance with the law and by a competent trained operator (MAFF, 1990). The elastrator rubber-ring method is a widely used technique in the UK. It is usually carried out during the second day to avoid upsetting sucking behaviour during the first day. In the UK, it is an offence to use this technique without anaesthetic later than 7 days of age.

### Identification

Spraying, tattooing, plastic/metal tags or ear-notching tasks require only minimal restraint.

### Examination and treatment prior to turning out

Particular attention is given to udder and feet (trim and treat if necessary). It may be the policy to drench against a range of internal parasites before turning out to grass.

### Movement B

The ewe and her lamb(s) are driven to the group family pen, taking care that there are no escape routes for young ewes with poor maternal instincts.

## Group family pen

The main purpose of this pen is to test the mothering ability of the ewes and the general health status of the lambs. Some of the latter items listed for the preceding pen may be carried out in this pen. The short period spent in this pen is largely dictated by weather conditions, the health status of the family, the maternal behaviour of the ewes and the demand for space. For ewes lambing during the autumn and early winter and where the lambs are to be early weaned, this pen would cater for their needs throughout the suckling period.

### Movement C

This may involve the movement of only one or a few families at a time and sometimes transfer to a distant paddock for acclimatization. Where transport is required, it should allow the dams and offspring to be penned separately, thus preventing injury to the young lambs during the journey. Some arrangements allow the ewes to be penned separately, with the lambs in a compartment to the front of the ewe and clearly visible.

## Acclimatization paddock

After the first few days, surveillance would be restricted to feeding time, when there would be an opportunity of catching ewes requiring treatment while they are feeding in line at the troughs. A crook can be most useful to restrain ewes and to catch lambs requiring treatment.

### The handling of the early-weaned flock

The ewes and lambs remain together in pens of 20–30 ewe and lamb pairs for 5–6 weeks. During the second week, part of the pen, preferably centrally placed, would be partitioned off, using fence panels constructed from pipe to form a sanctuary or creep and feeding area for the lambs. A temporary source of heat, such as an infrared lamp, would encourage lambs to lie away from the ewes. Access to the creep area would be via an adjustable entry which would allow easy access to lambs only. A facility to close the entry point(s) would enable separation, handling and treatment of the lambs to be undertaken with minimal upset. Where group treatment of ewes has to be undertaken, they can be driven to the central handling facility without their lambs. At the time of weaning, which is usually abrupt, the ewes are taken out of the pen and transferred elsewhere to be dried off prior to going out to pasture in the spring. The lambs remain in their pen and the creep fences are removed. For subsequent group treatment and weighing, the whole pen may be driven to the central handling area. Minor veterinary treatments would only require confinement in a corner of their pen by means of portable fence panels.

## New Developments in Sheep Reproduction

There are clear indications that new developments in sheep breeding techniques are going to play a significant role in the achievement of genetic improvement in sheep (Meat and Livestock Commission, 1991, 1998). These techniques, which include AI, semen collection and embryo transfer, may involve severe restraint of both rams and ewes. Pregnancy diagnosis is accomplished by real time ultrasonic scanning. The sector scanner has a marked advantage, because it can be done with the ewe standing and preparation of the abdominal skin is usually not required. A raised walk-through crate, usually without a yoke restrainer is used (Fig. 12.7).

### Artificial insemination

There are two contrasting methods of AI – the intravaginal method and the laparoscopic method (Williams, 1995). Both involve the use of a restraining crate. In the case of intravaginal AI, sometimes referred to as cervical AI, the hindquarters of the ewe should be raised so that it is easier for the operator to

**Fig. 12.7.** Raised walk-through crate for ultrasound testing.

locate the cervix. Where a large number of ewes have to be inseminated, a custom-built pivoting and tipping crate, fitted with a yoke type of restrainer, is usually used (Fig. 12.8).

## Selection, Handling and Loading Animals for Sale

Prior to the transport of animals to a particular market as lambs, fatstock, breeding stock or culled stock, the flock or subflock is gathered to a handling point and appropriate actions, such as weighing, assessment of finish, examination for soundness, shedding and regrouping, are undertaken. The selected animals are then loaded for transport, in some cases a short on-farm journey, prior to transfer to a larger form of transport, sometimes shared with other owners.

**Fig. 12.8.**    Pivoting restraint crate for artificial insemination.

In the case of animals being presented for the fatstock market, the UK Meat and Livestock Commission (MLC) (1986, 1991) have given considerable publicity to the need for improved preslaughter handling. The need for careful handling during the on-farm phase, in the sequence of handling operations, is just as great as during the later phases, not only in the interest of animal welfare but also to produce high meat quality. The MLC estimates that serious bruising, abscesses and other forms of carcass damage cost the industry at least £3m annually.

Poor handling may lead to serious injury, stress-induced depression of carcass quality, skin blemishes and bruising, as a result of dragging or lifting by the fleece, slipping and falling over and minor injuries due to poor loading facilities. Stockmen and persons connected with the transport of animals should be familiar with the Welfare of Animals (Transport) Order 1997. The general provisions and requirements of the Order applicable to all species include protection during transport, space allowances, fitness to travel, feed and water and rest periods. The responsibilities of owners and transporters are clearly stated in the Order, as well as the documentation required. Knowles (1998) has reviewed the scientific literature relating to the transport of slaughter sheep.

Before loading is due to commence, all the pen and passageway arrangements and facilities for loading should be carefully checked. Particular attention should be given to the floor surface and slippery areas should be sanded or strawed down. Solid sides to the passageway help in that the only obvious way forward is towards the loading ramp (Fig. 12.9). The animals should have a clear run with no obstacles, puddles of water or alternate light and shadow areas to impede continuous forward movement.

**Fig. 12.9.** Loading sheep on to a truck.

It is advisable to load animals in small groups. Those with previous experience of transportation should be used as lead animals. They should not be packed too tightly, thus risking crushing or suffocation. On the other hand, animals held too loosely may fall and be injured as a result of sudden movement. Table 12.2 gives the suggested space allowances for various categories of sheep (MAFF, 1998). Adjustment of space allowance should be made, taking into account weather and driving conditions, the design and type of vehicle and whether the animals are horned or not. No subpen should be more than 3.1 m long.

## Recent Sheep Transport Research (compiled by Temple Grandin)

Knowles (1998) has reviewed the literature on sheep transport studies. He concluded that sheep are more tolerant of being transported by road than other species, such as pigs. Death losses in British sheep are low compared with those in Australia and South Africa. Mortality for sheep sold directly in the UK was 0.007% and 0.031% for animals marketed through auctions (Knowles *et al.*, 1998). Marketing direct from the farm reduces death losses, bruises and stress.

Knowles (1998) states that short rest stops for 1 h will be detrimental. Most research indicates that a rest stop must last for at least 8 h, to provide enough time for the sheep to eat and drink. Sheep will eat before they drink. If the rest stop is too short, they will not have time to drink. A complete recovery from 14 h of transport stress takes almost 5 days (Knowles, 1998).

**Table 12.2.** Space allowances for sheep during transport.

| Category | Weight (kg) | Area animal$^{-1}$ (m$^2$) |
| --- | --- | --- |
| Shorn | < 55 | 0.2–0.3 |
| | > 55 | > 0.3 |
| Unshorn | < 55 | 0.3–0.4 |
| | > 55 | > 0.4 |
| Pregnant ewes | < 55 | 0.4–0.5 |
| | > 55 | > 0.5 |

Studies on sheep transport recommend that enough space should be provided to allow all the sheep to lie down. There is disagreement between studies on the space requirements. Cockram *et al.* (1996) reports that 0.77 m$^2$ 100 kg$^{-1}$ of live-weight is required for all the sheep to lie down and Buchenaur (1996) reported that 1.14 m$^2$ 100 kg$^{-1}$ of live-weight was required. Differences between the two studies may possibly be due to the type of sheep and the wool length. Unshorn sheep with thick fleeces will need 25% more space.

Cockram *et al.* (1996) found that there were no differences in bruising when 35 kg sheep were transported at 0.22 m$^2$ animal$^{-1}$ and 0.4 m$^2$ animal$^{-1}$. This study refutes the common belief that sheep must be packed in a truck to prevent bruising. When stocking densities are being determined one should differentiate between a short trip and a long trip. Sheep do not need to lie down during a short journey. Studies of sheep behaviour indicate that sheep do not lie down immediately. They tend to lie down in increasing numbers during the first 5–10 h of journey (Knowles, 1998). Therefore it is reasonable to suggest that more space would be required to maintain better welfare on longer trips (Knowles, 1998). Knowles *et al.* (1998) reported that for 39.5 kg sheep very high stocking densities of 0.448 m$^2$100kg$^{-1}$ of live-weight caused an increase of creatine kinase during a 24-h journey. The effects of other stress measures were small. The stocking density in Knowles *et al.* (1998) was tighter than the stocking densities recommended by Grandin (1981) and the New Zealand Animal Welfare Code (Ministry of Agriculture and Fisheries, 1994). Both of these codes recommend tighter stocking densities than the British Ministry of Agriculture (MAFF, 1998). One can conclude that 0.448 m$^2$ 100 kg$^{-1}$ of live-weight is too tight, regardless of the trip length, but the recommendations by Grandin (1981) and the New Zealand welfare codes are probably adequate for trips of a few hours.

One of the difficulties in interpreting the results of studies on sheep stress is the difficulty in separating the stress caused by transport from stress caused by gathering on the farm, auctions or differing degrees of physiological stress. Broom *et al.* (1996) suggests that, if sheep were habituated to loading procedures, the stress from loading could be separated from the stress of transport. Grandin (1997) states that the highly variable results in transport studies may be due to differing levels of fear stress. Sheep accustomed to loading and handling may be less stressed by transport. This is most likely to be true for short trips, where fatigue and physical stress would be less of a factor.

## Conclusions

Intensively reared sheep in the UK require a greater variety of handling procedures compared with extensively reared sheep in Australia. Lameness is a major welfare concern and, as intensification increases, the frequency of handling procedures required to keep animals healthy will also increase. Ewes must be closely supervised during pregnancy and lambing. Specialized restraint and handling facilities are required for pregnancy testing and artificial insemination. Handling procedures and veterinary procedures must comply with strict UK animal welfare codes. Recent legislation concerning the health and welfare of farm animals covers all forms of local and international transport.

## Acknowledgements

My thanks to the Farm and Rural Buildings Centre, Stoneleigh, UK, for permission to reproduce Figs 12.1 and 12.5. My grateful thanks to Mrs Wendy LeStrange, Mr Rhidian Jones and Dr Sandra Ward for their help with illustrations involving animals and equipment, and to Mr David Gunn for the preparation of illustrations. My sincere thanks to Mrs Anne Palmer for her patience and skill in preparing the manuscript.

## References

Brockway, B. (1975) *Planning a Sheep Handling Unit*. Farm Buildings Information Centre, National Agriculture Centre, Kenilworth, UK.
Broom, D.M. (1988) The assessment of pain and welfare in sheep. *Proceedings of the Sheep Veterinary Society* 13, 41–45.
Broom, D.M., Goode, J.A., Hall, S.J.L., Lloyd, D.M. and Parrott, R.F. (1996) Hormonal and physiological effects of a 15-hour road journey in sheep: comparison with the responses of loading, handling and penning in the absence of transport. *British Veterinary Journal* 152, 593–604.
Brown D. and Meadowcroft, S. (1989) *The Modern Shepherd*. Farming Press, Ipswich, UK.
Buchenaur, D. (1996) *Proceedings of an International Conference Considering the Welfare of Sheep During Transport*. St Catherine's College, Cambridge, UK.
Cockram, M.S., Kent, J.E., Goodard, P.J., Waren, N.K., McGile, I.M., Jackson, R.E., Muwanga, G.M. and Pryytherch, S. (1996) *Animal Science* 62, 46.
Eales, F.A. and Small, J. (1995) *Practical Lambing and Lamb Care*. Blackwell Science, Oxford, UK.

Farm Buildings Information Centre (1983) National Agriculture Centre, Kenilworth, England.

Grandin, T. (1981) *Livestock Trucking Guide*. Livestock Conservation Institute, Bowling Green, Kentucky.

Grandin, T. (1997) The assessment of stress during handling and transport. *Journal of Animal Science* 75, 249–257.

Henderson, D.C. (1990) *The Veterinary Book for Sheep Farmers*. Farming Press, Ipswich, UK.

Hutson, G.D. (1980) Sheep behaviour and the design of sheep yards and shearing sheds. In: Wodzicka-Tomaszewska, M., Edey, T.N. and Lynch, J.J. (eds) *Behaviour in Relation to Reproduction, Management and Welfare of Farm Animals*. University of New England Publishing Unit, Armidale, New South Wales, pp. 137–141.

Kilgour, R. and Dalton, C. (1984) *Livestock Behaviour*. Granada, London.

Knowles, T.G. (1998) A review of the road transport of slaughter sheep. *Veterinary Record* 143, 212–219.

Knowles, T.G. (1998) A review of road transport of sheep. *Veterinary Record* 143, 212–219.

Lewis, C. (1997) Sheep scab – an update on history, control and present situation. *Proceedings of the Sheep Veterinary Society* 21, 99–101.

Lynch, J.J., Hinch, G.N. and Adams, D.B. (1992) *The Behaviour of Sheep*. CAB International, Wallingford, UK.

Meat and Livestock Commission (1986) *Sheep Yearbook*. MLC, Milton Keynes, UK.

Meat and Livestock Commission (1991) *Sheep Yearbook*. MLC, Milton Keynes, UK.

Meat and Livestock Commission (1998) *Sheep Yearbook*. MLC, Milton Keynes, UK.

Ministry of Agriculture and Fisheries (1994) *Code of Recommendations and Minimum Standards for the Welfare of Animals Transported within New Zealand*. Wellington, New Zealand.

Ministry of Agriculture, Fisheries and Food (MAFF) (1980) *Sheep Dipping and Spraying*. Leaflet 719, MAFF Publications, London.

Ministry of Agriculture, Fisheries and Food (MAFF) (1990) *Sheep: Codes of Recommendations for the Welfare of Livestock*. MAFF Publications, London.

Ministry of Agriculture, Fisheries and Food (MAFF) (1998) *Guidance on the Welfare of Animals (Transport) Order 1997*. MAFF Publications, London.

Ministry of Agriculture, Fisheries and Food (MAFF) (1999) *Sheep: Codes of Recommendations for the Welfare of Livestock*. MAFF Publications, London.

Prettejohn, M.W.H. (1988) Sheep welfare – the Ministry view. *Proceedings of the Sheep Veterinary Society*, 13, 52–56.

Rushen, J. (1986) Aversion of sheep for handling treatment: paired choice experiments. *Applied Animal Behaviour Science* 16, 363–370.

Shorrock, D.J. (1981) *Handling and Dipping Sheep*. Devon Sheep Community Group, Ministry of Agriculture, Fisheries, and Food, HMSO, Bristol, UK.

Sibly, R.M. and Baldock, N.M. (1987) Effects of farm handling procedures on heart rate in sheep. *Proceedings of the Sheep Veterinary Society* 12, 103–106.

Williams, H.L. (1978) Sheep. In: Scott, W.N. (ed.) *The Care and Management of Farm Animals*. Baillière Tindall, London, pp. 81–124.

Williams, H.L. (1988) Sheep. In: *Management and Welfare of Farm Animals*. Universities Federation for Animal Welfare, Baillière Tindall, London, pp. 62–111.

Williams, H.L. (1995) Sheep breeding and infertility. In: Meredith, M.J. (ed.) *Animal Breeding and Infertility*. Blackwell Scientific, Oxford, UK, pp. 354–434.

Williams, H.L. (1999) Sheep. In: Ewbank, R., Kim-Madslien, F. and Hart, C.B. (eds) *Management and Welfare of Farm Animals*. Universities Federation for Animal Welfare, UFAW, Wheathampstead, UK, pp. 83–117.

# Dogs For Herding and Guarding Livestock

**13**

## Lorna Coppinger and Raymond Coppinger

*School of Cognitive Science, Hampshire College, Amherst, MA 01002, USA*

## Introduction

Two completely different types of sheepdog assist livestock producers all over the world. Herding dogs are specialists at moving stock from place to place. Guarding dogs protect domestic stock from wild predators.

'The sheep dog is such a willing and uncomplaining worker and without him the farmer could not even begin to look after his sheep,' wrote Lt.-Col. K.J. Price in his foreword to a popular book on the subject (Longton and Hart, 1976). Longton and Hart noted that 'there are well over one thousand million sheep in the world, and one-third of this vast total is kept in countries where the Border collie is the chief work dog'.

Guarding dogs, well known among sheep and goat producers, particularly in the high pastures of countries all the way from Portugal to Tibet, also help to manage the world's ruminants. Although these large, placid protectors were all but unknown in the USA until the mid-1970s, they have since then been widely adopted throughout the USA and Canada. As a result, many flocks which had experienced a 10% or even greater loss to coyotes have enjoyed a marked reduction in predation.

Domesticated dogs and sheep appear together in archaeological excavations dating from 3685 BP (before present) (Olsen, 1985). They become part of written history in the Old Testament ('with the dogs of my flock'; Job 30: 1) and in the writings of Cato the Elder and Marcus Terentius Varro in the two centuries before Christ. These treatises on Roman farm management, translated by 'a Virginia farmer' (Anon., 1913), are so full of good information that, if another book had never been written on flock dogs, today's farmers could learn just about all they need from Cato and Varro.

©CAB *International* 2000. *Livestock Handling and Transport*, 2nd edn
(ed. T. Grandin)

Dogs . . . are of the greatest importance to us who feed the woolly flock, for the dog is the guardian of such cattle as lack the means to defend themselves, chiefly sheep and goats. For the wolf is wont to lie in wait for them and we oppose our dogs to him as defenders.

(Anon., 1913, p. 247).

Modern stock producers still need that information on how to choose a pup, what an adult should look like, what to feed, the value of a dog, breeding, raising pups and number of dogs per flock. Much of this ancient 'manual' suggests that the first sheepdogs were primarily guardians, rather than herders, although the difference is blurred.

The breeding and management of working farm dogs are not supported by an extensive technical farm-dog-specific literature. But they are supported by a rich, knowledgeable and generally professional trade literature based on individual experience. In recent years, resources available via computer on the World Wide Web have greatly increased the accessibility of expert information. Herding dogs have received the most attention, with books and articles on their selection and training widely available. This popularity is driven in part by the success of organized competitions among the owners of herding dogs, especially Border collies. Guarding dogs have been the subject of studies by biologists in the USA since the mid-1970s, and subsequently their use has greatly increased all over North America. Although general principles of breeding, management and health care are common to all domestic dogs, working stock dogs (and their handlers) have specific needs which benefit from technical assistance. The scientific literature about farm dogs has yet to address genetic improvement at the level so prevalent for other domestic livestock or to study their behaviour at the level enjoyed by canine pets.

Willis (1992) emphasized the lack of support for research in this field: 'Bearing in mind the enormous funding given to research into sheep breeding it is surprising that so little has been expended on understanding herding ability in dogs for without dogs most British/Australasian sheep farmers could not function.' Taking the need for research on dogs even further, Hahn and Wright (1998) and Hahn and Schanz (1996) emphasized the lack of studies into behaviour genetics of dogs, which limits the ability of professionals to provide useful techniques for breeders, trainers, veterinarians and end-users.

This lack might be even more serious for guarding dogs. The 10% loss to predators mentioned earlier is not uncommon in the USA; but businesses cannot sustain a 10% shrinkage for very many years. Research on guarding dogs needs to go beyond studies of breed differences and effect of dogs on predation. Two projects that would help the industry immediately are: (i) behavioural analyses focused on improving the success rate of guarding dogs; and (ii) closely monitored field trials of dogs learning to work in locations critical for the reintroduction and survival of endangered predatory wildlife. Subsequent transfer of information back to users, and especially to potential users, also needs improvement.

# Differentiating the Sheepdogs

It appears, from the old literature, that early 'shepherd dogs' were used both for guarding and herding. Today, distinctions between the two types are made more clearly. Herding dogs are not 24-h flock guardians, nor are guardians good at herding livestock. The reason becomes obvious when one considers morphological and behavioural differences between the two types (Table 13.1; Fig. 13.1).

Cato and Varro divided dogs into two kinds, hunting dogs ('used against wild beasts and game') and herd dogs ('used by the shepherd'). They described 'herd' dogs in terms of their guarding abilities. Other observers, as reported by Baur (1982), may have seen what they thought was a 'driving' dog when the dog was actually just 'following' the herd:

> 'These dogs take the entire care of the sheep, drive them out to pasture in the morning, keep them from straying during the day, and bring them home at night. These dogs have inherited a talent for keeping sheep, but the shepherds do not depend wholly on that'.
>
> (Anon., 1873, reported in Baur, 1982).

Baur commented that shepherds reinforced the dogs' talents with 'the old Spanish custom of using a foster-mother ewe to train [suckle] a puppy'.

Thomas (1983) distinguished between guarding and herding dogs, noting that the lack of wolves in England at the beginning of 'the early modern period'

**Table 13.1.** Some differences between herding dogs and guarding dogs.

|  | Herding dogs | Guarding dogs |
|---|---|---|
| Examples | Australian kelpie<br>Australian shepherd<br>Border collie<br>New Zealand huntaway | Anatolian shepherd<br>Great Pyrenees<br>Maremma<br>Šarplaninac |
| Morphology | 10–20 kg<br>Ears often pricked or 'tulip'; some breeds' ears hang down<br>Colour is usually dark with white or brown markings; some are white/grey with darker spots | 30–55 kg<br>Ears hang down<br><br>Colour is usually white or grey, although some breeds are brown with darker markings |
| Behaviour | 'Chase-and-bite'<br>Active around stock<br>Trained to respond to human commands<br>Have motor patterns specific to job: eye, clapping, heeling, heading or voicing, which are heritable | 'Never' chase or bite<br>Passive, even lethargic, around stock<br>Seldom trained to respond to commands<br>Have no job-specific motor patterns: form strong social bonds with stock through early, continuous association |

**Fig. 13.1.** Border collie (herding dog) and maremma (guarding dog).

resulted in dogs that drove sheep, while in France or Italy, where wolves still survive, sheep follow a shepherd, and a 'mastiff or wolfhound, rather than a sheepdog, [goes] in front as their protector'. An early dog expert, Dr Johannes Caius, wrote about 'The Shepherd's Dogge' (1576), describing a technique still used today by shepherds directing their herding dogs to move the stock:

> 'The dogge, either at the hearing of his master's voice, or at the wagging and whisteling of his fist, or at his shrill horse hissing, bringest the wandering weathers and straying sheepe into the self same place where his masters will and wishes is to have them.'

Kupper (1945) described dogs in the 19th-century American south-west as movers of sheep, although she did wonder about 'the wonderful sheep dogs' that would die of starvation rather than leave the flock.

In France, herding dogs are known as a fairly recent development, supplanting the guardian when large carnivores (bear, wolf) disappeared from western Europe (de Planhol, 1969; Laurans, 1975; Lory, 1989; Schmitt, 1989). Laurans advanced an explanation for herding dogs becoming more useful, for as population increased, '. . . in countries with many small parcels of land . . . the shepherd needed herding dogs in order to keep his gardens safe from damage by sheep'.

The most comprehensive modern book on farm dogs is that of Hubbard (1947), whose historical overview of working dogs differentiates between herding and guarding dogs and includes a long section on herding dogs and the competitive trials which are so popular in the British Isles and the USA.

Hubbard described 58 working breeds, arranged in three categories: pastoral dogs (47), draught dogs (5) and utility dogs (6).

> For the most part the pastoral dogs are Sheepdogs proper, that is, those breeds used for herding and controlling sheep only. In this group we find the well-known Collie, German Shepherd Dog and Old English Sheepdog; breeds invariably used only with sheep. Apart from these there are also Cattle Dogs, drovers' dogs or cow-herds' dogs; breeds like the Welsh Corgi, the various Bouviers of Flanders, the Australian Heeler, the Hungarian Pumi, and the Portuguese Cattle Dog. Furthermore, there are other breeds which are used in the protection, droving and controlling of other animals, such as some Russian Laiki (which herd reindeer), other Russian herding dogs (which control the dromedaries of Central Asia), and the many varieties of native races of South Sea Islands dogs (which round up the indigenous pigs of the islands). There are, of course, some Sheepdogs which work as well with cattle as with sheep, and a few, indeed, that work with goats and pigs as well as with sheep.

Today, producers in the USA and Australia report that their guarding dogs have been bonded also with llamas, ostriches, emus, turkeys and other unusual or rare species.

Discovering how modern herding dogs could be developed to such precision, and how their morphology and behaviour could be so distinctly differentiated from the guardians, is based on ethology, or the study of their behaviour. An early analysis, cut short by the Second World War, was carried out by Dawson (1965) in 1935. His goal was to study the 'inheritance of intelligence and temperament in farm animals', with dogs the experimental animal and sheep-herding one manifestation of intelligence, 'since it was of economic importance in agriculture'. Various cross-breedings were made between Hungarian pulis, German shepherds, Border collies, chows and a pair of Turkish guarding dogs. Dawson reported wide variation in reactions of dogs in all tests, but could not detect differences due to sex or between the larger breed groups. He noted 'marked indications that some of the behaviour traits were inherited'.

## Development of different behaviour patterns

More recently, looking specifically at differences between the two types of sheepdog, Coppinger *et al.* (1987a) wrote:

> As with juvenile wolves or coyotes, adult livestock conducting dogs displayed the first-half segment of a functional predatory system of motor patterns and did not express play or social bonding toward sheep; whereas, like wolf or coyote pups, adult livestock protecting dogs displayed sequences of mixed social, submissive, play and investigatory motor patterns and rarely expressed during ontogeny (even when fully adult) predatory behaviours. The most parsimonious explanation of our findings is that behavioural differences in the two types of livestock dogs are a case of selected differential retardation (neoteny) of ancestral motor pattern development.

In other words, herding dogs are selected to show hunting behaviours, such as eye, stalk, grip or heel. Guarding dogs are selected to show more of the wild ancestor's puppy-like or juvenile behaviour, preferring to stay in the 'litter' of livestock to which they are bonded and to react to novelty by barking an alarm. Guarding dogs are not attack dogs; they are defence dogs, although they have been aptly described as 'unbribable...extremely loyal, distrustful of strangers, and capable of attacking both wolves and bears' (Ružić, 1988).

Further investigations into the differences in behaviour between guarding and herding dogs were undertaken by Coppinger and Schneider (1995). They juxtaposed the behaviour of herding dogs, guarding dogs and sledge dogs to hypothesize that it is the timing of events during canine development which intensifies differences in innate motor patterns, and which in turn different-iates learning abilities of breeds. Guarding dogs, as noted above, seem to have been selected to mature at an early ontogenetic stage, before predatory sequences emerge. Otherwise they would not be trustworthy and could not be left alone with the stock. Herding dogs, represented in both studies by Border collies (Fig. 13.2), were selected to show the predatory sequences of eye, stalk and chase, but to mature before the dangerous crush–bite–kill patterns of the true predator.

Detailed profiles of both types, based on modern understanding of dog behaviour, appear in Coppinger and Coppinger (1998).

Different breeds of herding dogs exhibit varying degrees of predatory behaviour. Gathering dogs, such as the Border collie circle the livestock and

**Fig. 13.2.**   Border collie circling and staring at sheep. The dog is on the edge of the collective flight zone of the sheep (see Chapter 5). Some sheep are facing the dog, but others have started to turn away because the dog is entering their flight zone.

are less aggressive than heeling dogs, such as the Queensland blue heeler. The aggressive breeds exhibit a greater degree of predatory behaviour and their natural tendency is to chase and bite livestock instead of circling them. Aggressive breeds can be stressful to livestock and their use should be limited to cattle mustering in rough country.

### Neurotransmitter differences

These variations in motor patterns appear to be genetically predisposed or inherited within each breed. Willis (1992) reported on the heritability of several traits, briefly mentioning the herding and guarding abilities of sheepdogs. (NB: 'heritability' does not equal 'inherited'; 'heritability' includes genes plus environment as factors affecting variation.) He noted three instincts of Border collies that are heritable: clapping (crouching), eye (staring at the sheep) and barking (not done when herding but done in other circumstances). He also noted that studies of the effects of genetics plus environment (heritability) in guarding dogs would help breeders improve their performance.

Possible chemical reasons for these behaviour differences were studied by Arons (1989). She reported livestock guarding dogs to be different from herding dogs and from sledge dogs in both the distribution and amount of neurotransmitters in various sections of the brain. The guarding dogs (several breeds) had low levels of dopamine in the basal ganglia, whereas Border collies and Siberian huskies had higher quantities of this neural transmitter. Dopamine is related to neural activity.

# Herding Dogs

'There are expensively mechanised farms', wrote Longton and Hart (1976), 'where, to move bullocks from one field to another or bring sheep into the yards, involves turning out the entire farm staff. A good sheep dog could do the whole job more easily and economically.'

Brown and Brown (1990) explain that the instinct of a gathering and herding dog such as a Border collie is to circle around the livestock. The dog will instinctively position itself opposite the handler on the other side of the livestock (Fig. 13.3). If the handler moves to the right the dog will tend to move to the left and vice versa. A good dog can be ruined if the handler tries to position him/herself on the same side of the livestock as the dog. The gathering instinct is part of dog predatory behaviour. This behaviour can be modified by training, to enable the handler to send the dog out and have it bring the livestock to him/her.

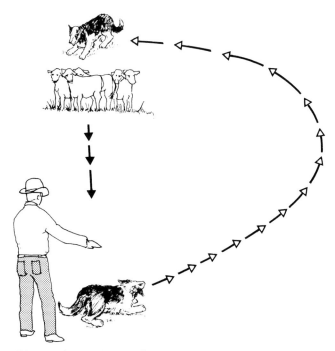

**Fig. 13.3.** Diagram from Brown and Brown (1990) on teaching a young Border collie to circle around the livestock and bring them to the handler.

### Training information

Modern technology has greatly enriched the methods by which livestock producers can acquire needed information about training stock dogs. The training methods may vary, but details are quickly visible when shown on videotapes or explored on the World Wide Web. Several 'standard' books, however, should never be replaced on the shelves of dog trainers. The techniques of Jones and Collins (1987) and Jones (1991) are slightly different from those of Longton and Hart (1976), but the results are similar. The classic, 'old reliable' book on training a sheepdog is *The Farmer's Dog*, by John Holmes (1976). His descriptions of training and using dogs in daily shepherding are based on a broad understanding of dog behaviour, which he applies effectively to explain the process. For example, he describes herding behaviour as inhibited (by the handler, usually) predatory behaviour by the dog. Iley's (1978) *Sheepdogs at Work* includes chapters on the history of the working dog, the pup and its early training, trials and breeding. A right-to-the-point book is by Means (1970), who trains Border collies to work with cattle. The text includes basic training and use of a stock dog, and also what can go wrong and how to prevent or cure it, rather than assuming everything goes right the first time.

Recent books include *Working Sheepdogs: Management and Training*, by John Templeton and Matt Mundell (1992), and *Border Collies: Everything about Purchase, Care, Nutrition, Breeding, Behavior, and Training*, by Michael Devine and David Wenzel (1997). The former is excellent, and written by two of the best working dog trainers in Great Britain. The second appeals more to owners of the Border collie as a pet. *Ranch and Farm Dogs: Herders and Guards*, by Elizabeth Ring (1994), is geared for younger (ages 4–8 years) readers, and provides for youngsters a much-needed connection with these important tools. It has colour photographs of most of the breeds.

The above-mentioned books concentrate on Border collies, mostly with sheep, whereas Taggart (1991) discusses 21 herding breeds, including the Australian cattle dog and others bred to work more with cattle than with sheep. Training methods for all are similar, but Taggart's descriptions allow for the different styles of the different breeds. She, too, discusses common problems and suggests solutions. Diagrams and photographs of the different breeds and of dogs in action help emphasize the fine points.

Training cattle dogs is also addressed by Little (1975), whose positive view of her job as dog trainer is evident from the title, *Happy Herding Handbook*. Little describes how to train for the various commands, and what to do about problems such as biting, grabbing or harassing stock, not obeying commands, not outrunning well, and so on. Charoy's (1989) descriptions of standard selection and training methods are enhanced by the inclusion of techniques relevant to the daily needs of European shepherds, such as training a dog to conduct sheep along a road with cars or crossroads.

Among the new cattle dog books is *The Complete Australian Cattle Dog* (1993) by John Holmes and Mary Holmes, long-term cattle dog trainers. Narelle Robertson's *Australian Cattle Dogs* (1996) is good for beginning owners, providing a history of the breed and a discussion of the dogs' behaviour.

For people desiring to learn how to train a dog, step by step, Vergil S. Holland's *Herding Dogs: Progressive Training* (1994) is the most detailed. He writes mostly about dogs working with sheep, but his depiction of problems and solutions, and the way different dogs approach stock, plus the amusing drawings and photographs, broaden the scope of this book to make it useful to trainers of any stock dogs.

The traits that allow herding dogs to be trained so well to their tasks are the essence of an article by Vines (1981). Fiennes and Fiennes (1968) also noted reasons for sheepdogs being so well suited for studies of how dogs process information and what they do with it.

For a range of techniques and advice, the World Wide Web provides immediate access. Web sites run by experts and other practitioners contain articles, breeder lists, membership applications and links to other dedicated sites, as well as to news groups, whose participants discuss issues and give advice at all levels. On the Internet, search engines such as *Yahoo!* lead to scores of resources. Performing a search for 'stock dog' or 'herding dog' is the best way to start. A main web site is www.stockdog.com.

### Acoustic signals

One of few scientific studies on herding dogs (McConnell and Baylis, 1985) discusses: (i) locomotion and postural behaviour of Border collies when working around sheep (visual communication from dog to sheep); and (ii) acoustic signal systems used by shepherds to control their dogs. Analysis of data on 'mature trained and immature untrained Border collies shows that the [stalking] posture [of a hunting mammalian predator] was innate, but was refined by training and experience'. The signals (whistling) that stimulated the dog into activity (e.g. go fetch, go closer to the sheep, go faster) were short, rapidly repeated notes, tending to rise in frequency. Those that inhibited activity (e.g. slow down, stop) were continuous, long and descending. McConnell (1990) subsequently looked more closely at the acoustic structure and response, and confirmed these observations.

The original question asked by McConnell and Baylis, 'How do individuals of two different species effectively communicate to maneuver a third (and usually unwilling) species?', might well be asked also within the interspecific triangle of guarding dog, livestock and predator. McConnell and Baylis mention that shepherds exploit the interactions between predator and prey species (in order to move their stock), and that ethologists can exploit the shepherds' systems by studying the results of the shepherd-generated interspecific behaviour. Researchers into the behaviour of guarding dogs have begun to explore these interactions.

## Guarding Dogs

> In the two and one half years that we had Ike we didn't have a single incidence of livestock being harmed by dogs or coyotes. Ike literally made it possible for us to continue having sheep and to begin purebred breeding with feelings of security regarding the safety of our animals.
>
> (S. Sorensen, 1990, personal communication)

> Estimates indicate coyotes . . . kill an average of 1–2.5% of the domestic adult sheep and 4–9% of the lambs in the 17 western states . . . Livestock producers reduce losses [mortality] by using various livestock management practices, frightening devices, trapping, snaring, calling and shooting, sodium cyanide guns, denning, aerial gunning, and livestock guarding dogs'.
>
> (Andelt, 1992)

> Only the dogs stopped coyote predation (O'Gara *et al.*, 1983); 'but they also harassed sheep' (Linhart *et al.*, 1979)

The ancient Romans, generations of Old World shepherds and a few 19th-century New World ranchers knew how to protect their livestock from predators with guarding dogs. But, in late 20th century USA, guarding dogs were essentially unknown. In spite of an array of 'sophisticated' poisons and various high-tech devices, sheep and cattle producers were still losing 8–10% of their animals to predators. In the mid-1970s, a 3-month, on-the-road survey of producers in the USA using guarding dogs with sheep and/or goats showed seven dogs out of the 12 located to be successful (Coppinger and Coppinger, 1978). Yet few producers had even heard about guarding dogs, and many were highly sceptical that a carnivore could protect a prey species from other carnivores. 'Why should I pay for my predators when I can get them for free?' one of them asked.

Research began in 1976 at Hampshire College in Amherst, Massachusetts, and at the Denver Wildlife Research Center (US Fish and Wildlife Service) in Colorado. At the Denver project and its successors at the US Sheep Experiment Station in Idaho, researchers studied dog/coyote interactions under controlled conditions and ran field trials with dogs placed on western ranches. At Hampshire, biologists and their students focused on the basic, instinctive behaviour that results in good guardians, and hypothesized that guarding dogs achieve protection by being attentive to the livestock and trustworthy with them. Both projects placed dogs on farms and ranches for on-site trials, and almost immediately reports came back of lower predation. By the end of the 1980s, the data looked good: predation on dog-guarded flocks was reduced by 64–100% (Coppinger *et al.*, 1988), dogs were an economic asset for 82% of the people surveyed (Green and Woodruff, 1988), and dogs in a Colorado survey saved an average of $3216 worth of sheep annually per dog (Andelt, 1992). Green *et al.* (1984) reported that 37 ranchers spent about 9 h a month feeding and maintaining their adult guardians, corroborated by Andelt's (1992) 10 h. These authors agreed on costs of a guarding dog: first-year dollar amounts, including purchase, were $700–900 and subsequent annual costs were $250–290.

Use of guarding dogs may have died out among Spanish and Anglo-American ranchers in the American south-west during the late 19th century, but at least one native American group either learned from the immigrants or reinvented the system. Black (1981) and Black and Green (1985) found Navajos in Arizona using mixed-breed guarding dogs in a practice very like those still seen in Europe and Asia.

> Navajos call their dogs 'sheep dogs' but, unlike sheep dogs used by other ranchers to assist in herding and moving the flocks, Navajo dogs function primarily as guardians of sheep and goats to whom they have developed social bonds. This attraction is a result of raising dogs essentially from birth in visual, olfactory, auditory, and tactile association with sheep and goats. A minimum of handling of pups reduced the likelihood that they will bond strongly to humans. Mixed-breed dogs of the Navajo appear to exhibit all behavioral traits believed to be important in protecting flocks from predators, especially coyotes: they are attentive,

defensive, and trustworthy. If ranchers choose to employ dogs, the rather simple Navajo recipe for training may serve them well. Mixed-breed dogs could be quickly deployed in a variety of ranching situations to help reduce predation on livestock.

## Training guard dogs

From the following recipe, reported in the Black and Green (1985) paper, it is obvious that guarding dogs need much less formal training than do herding dogs.

> Raise or place mixed-breed pups in corrals with sheep, lambs, goats, and kids at 4–5 weeks of age. Feed the pups dog food and table scraps. Provide no particular shelters such as dugouts or doghouses (the pups will sleep among the sheep and will dig their own dirt beds). Minimize handling and petting. Show no overt affection. Return pups that stray to the corral (chase them, scold them, toss objects at them). Allow pups to accompany the herds onto the rangeland as age permits. Punish bad behaviour such as biting or chasing the sheep or goats, and pulling wool by scolding and spanking. Dispose of dogs that persist in chasing, biting, or killing sheep.

In this straightforward approach is the basis of the attentive, trustworthy and protective behaviour identified as characteristic of a good guarding dog (Coppinger *et al.*, 1988). It also contains the wisdom behind the dog's critical socialization periods identified under laboratory conditions by Scott and his colleagues (Scott and Marston, 1950; Scott, 1958). Applying the findings of Scott and other modern ethologists to the training of a livestock guarding dog results in a more formal recipe for what the Navajos (and the ancient Romans) knew all along (Table 13.2).

Making a case against the importation and use of large Old World guardians, Black and Green (1985) wrote that the small, mongrel-type dogs used by Navajos are cheaper, easier to get and keep, of lower liability and easier to dispose of if they show unacceptable behaviour. Coppinger *et al.* (1985) agreed that mongrels can be an efficient dog for wider use in the USA, and better than most pure non-guarding breeds. However, they reiterated that dogs selected specifically for the task of stock-guarding, and bred true are likely to be more successful than non-guarding breeds. Green and Woodruff (1990) also found pure-breds to be more successful than 'others'.

The success of the research in the USA has attracted the attention of wildlife managers and sheep producers back in Europe. In countries where guarding dogs were rarely, if ever, used, such as Norway and Sweden, recent reintroductions of wolves and bears have caused wildlife and agricultural agents to look at these dogs as potential stock protectors (Klaffke, 1999). Researchers in Norway determined that guardians were effective there at some level against bears, although the problem of their tending to chase reindeer needs to be solved (Hansen and Bakken, 1999).

**Table 13.2.** First-year development of livestock guarding dogs.

| Attentive behaviour | | |
| --- | --- | --- |
| Stage 1 | | |
| Neonatal | Transitional | Primary socialization |
| 0–2 weeks | 2–3 weeks | 3–8 weeks (ends at weaning) |
| Pup is insulated from the environment outside the litter. Reflex care-soliciting behaviour: cries, sucks, roots towards warmth. Crawls | Eyes open, teeth appear, walks. Non-reflexive learning behaviours appear. Mother stops responding to pups' cries | Ears and eyes begin to work. Notices other animals at a distance. Begins to form primary social relationships that determine later attachments. Can eat solid food. Food pan dominance begins and wrestling with littermates |

| Stage 2 | Stage 3 |
| --- | --- |
| Early juvenile | Late juvenile |
| 8–16 weeks | 4–6+ months (ends at puberty) |
| Secondary socialization begins; attachments made to other animals and even species. Non-reflexive care soliciting behaviour, such as dominance–submission and food-begging, appear. These become the basis for the complex social behaviours of the adult. The target of these behaviours is determined to some degree by primary socialization. But in guarding dogs this is the period for bonding pups with livestock. By 16 weeks the 'critical period' or window during which social attachments are made is closed | Emerging social behaviours of stage 2 must be reinforced. Pup must be kept with livestock all the time and not be allowed to play or interact extensively with other dogs or people. Exception would be if pup is put in a pasture with another guardian dog, presumably older, which is acting as a 'teacher' dog. Any wandering or other inattentive behaviour should be stopped immediately |

*continued overleaf*

As with herding dogs, guarding dogs are well served by the World Wide Web. A main web page is at www.lgd.org. One of its links is to the livestock guarding dog newsgroup (LGD-L), an active, highly informative source of advice – some of which conflicts, but all of which is of great interest to an owner of a livestock guarding dog.

**Table 13.2.**   *Continued.*

| Trustworthy behaviour | Protective behaviour |
|---|---|
| Stage 4 | Stage 5 |
| Subadult | Adult |
| 6+–12+ months | 12+ months |
| Onset of predatory behaviour patterns and of 'play', which includes the predatory movements of chase, grab-bite, wool pull, ear chew. If this behaviour is allowed to be expressed, which to the pup is a reinforcement of the behaviour, it will become common and be almost impossible to correct. If the behaviours are not reinforced, they will disappear from the pup's repertoire of behaviours. Heat cycles begin in females, sometimes resulting in unexpected behaviour such as wandering or chewing on sheep. Males may stray if attracted by a female on heat | Care-giving and mature sexual behaviours emerge. A dog that has been properly bonded with livestock and not allowed to disrupt them should be an effective guardian at this point. First experience with serious predators must not be overwhelming: dog needs to gain confidence in its ability as it matures |

### Guarding-dog behaviour

To try to understand the underlying behaviours that characterize the guarding breeds, Coppinger and his colleagues at Hampshire College studied three basic guardian behaviours: attentive, trustworthy and protective. The attentiveness to sheep of dogs which had been used in the USA for several years was found to be similar to that of their parent and grandparent generations still working in Italy (Coppinger *et al.*, 1983b). Looking at trustworthy and untrustworthy dogs, Coppinger *et al.* (1987a) quantified differences in motor patterns and found that untrustworthy guarding dogs shared some behaviours with herding dogs. Their protective abilities were severely tested in wild, forested wolf territory in northern Minnesota, where several guardians showed that dogs, if managed correctly, could protect stock against North American wolves (Coppinger and Coppinger, 1995).

McGrew and Blakesley (1982) found that interspecific adjustments to the dog are made by both sheep and coyotes: 'The sheep learned to run to or to stand with the dogs when attacked, and usually bedded with the dog. The coyotes learned to attack the flock when the dog was not present.' Anderson *et al.* (1988) used a trained Border collie as a mock predator in an experiment to discover the advantage of interspecific behaviour in providing protection from predators. The collie's 'aggressive, threatening' approach did result in a closer association between sheep and cows that had been bonded with each other, but non-bonded animals moved in distinctly intraspecific groups when threatened. McGrew (1982) reported that early exposure of pups to sheep (i.e. bonding) is important, and that a dog's value is determined by what breed it is and by its own personality.

### Guard dogs on open ranges

Early doubts that dogs could protect sheep on the huge, unfenced ranges of the western USA were diminished by reports by Coppinger *et al.* (1983a) and Green and Woodruff (1983). Methods of increasing the dogs' effectiveness were studied: by looking at causes of presenescent mortality (Lorenz *et al.*, 1986) and by transferring inattentive or untrustworthy dogs to new ranches (Coppinger *et al.*, 1987b). That guarding dogs are a positive presence in managing livestock is evident in reports from other researchers. Pfeifer and Goos (1982) found them to be effective in North Dakota, and Hagstad *et al.* (1987) noted that the presence of a dog with dairy goats reduced predation in Louisiana.

## Conclusions

Sheepdogs – both herding and guarding types – are highly efficient for moving or protecting livestock. Within the herding types, different breeds tend to specialize in moving different stock, although most breeds are versatile and adaptable to sheep, cattle, llamas – whatever needs moving. Livestock are safer when moved by herding dogs than by people, for dogs are fast at heading off a stampede and deft at aiming reluctant beasts into the pen or changing a maverick's mind. Sheep are safer, too, when protected by guarding dogs, whose senses match those of most predators and which live with the stock day and night. Sheepdogs extend a herder's control over the stock, saving time, energy and animals. Many livestock growers who use dogs – both herding and guarding dogs – confirm, in chorus, 'Without our dogs, we would be out of business.'

## References

Andelt, W.F. (1992) Effectiveness of livestock guarding dogs for reducing predation on domestic sheep. *Wildlife Society Bulletin* 20, 55–62.

Anderson, D.M., Hulet, C.V., Shupe, W.L., Smith, J.N. and Murray, L.W. (1988) Response of bonded and non-bonded sheep to the approach of a trained Border Collie. *Applied Animal Behaviour Science* 21, 251–257.

Anon. (1873) How they train sheep dogs in California. *Forest and Stream* (New York), 1, (11 December), 279.

Anon. (1913) *Roman Farm Management: The Treatises of Cato and Varro* (circa 150 BC) *A Virginia Farmer*, translated by Macmillan, New York.

Arons, C. (1989) Genetic variability within a species: differences in behaviour, development, and neurochemistry among three types of domestic dogs and their F1 hybrids. Unpublished PhD thesis, University of Connecticut.

Baur, J.E. (1982) *Dogs on the Frontier*. Denlinger's, Fairfax, Virginia.

Black, H.L. (1981) Navajo sheep and goat guarding dogs. *Rangelands* 3, 235–237.

Black, H.L. and Green, J.S. (1985) Navajo use of mixed-breed dogs for management of predators. *Journal of Range Management* 38, 11–15.

Brown, L. and Brown, M. (1990) *Stock Dog Training Manual*. Eagle Publishing Company, Ekalaka Montana.

Caius, J. (1576) *Of English Dogges*. Rychard Johnes, London.

Charoy, G. (1989) L'éducation du chien. *Ethnozootechnie* 43, 35–50.

Coppinger, R. and Coppinger, L. (1978) *Livestock Guarding Dogs for US Agriculture*. Hampshire College, Amherst, Massachusetts.

Coppinger, R. and Coppinger, L. (1995) Interactions between livestock guarding dogs and wolves. In: Carbyn, L.N., Fritts, S.H. and Seip, D.R. (eds) *Ecology and Conservation of Wolves in a Changing World*. Occasional Publication number 35, Canadian Circumpolar Institute, University of Alberta, Edmonton, pp. 523–526.

Coppinger, R. and Coppinger, L. (1998) Differences in the behavior of dog breeds. In: Grandin, T. (ed.) *Genetics and the Behavior of Domestic Animals*. Academic Press, San Diego, California, pp.167–202.

Coppinger, R. and Schneider, R. (1995) The evolution of working dog behavior. In: Serpell, J. (ed.) *The Domestic Dog: Its Evolution, Behavior and Interactions with People*. Cambridge University Press, Cambridge, pp. 21–47.

Coppinger, R., Lorenz, J. and Coppinger, L. (1983a) Introducing livestock guarding dogs to sheep and goat producers. In: Decker, D.J. (ed.) *Proceedings of the First Eastern Wildlife Damage Control Conference*. Cornell University, Ithaca, New York, pp. 129–132.

Coppinger, R., Lorenz, J., Glendinning, J. and Pinardi, P. (1983b) Attentiveness of guarding dogs for reducing predation on domestic sheep. *Journal of Range Management* 36, 275–279.

Coppinger, R.P., Smith, C.K., and Miller, L. (1985) Observations on why mongrels may make effective livestock protecting dogs. *Journal of Range Management* 38, 560–561.

Coppinger, R., Glendinning, J., Torop, E., Matthay, C., Sutherland, M. and Smith, C. (1987a) Degree of behavioral neoteny differentiates canid polymorphs. *Ethology* 75, 89–108.

Coppinger, R., Lorenz, J., and Coppinger, L. (1987b) New uses of livestock guarding dogs to reduce agriculture/wildlife conflicts. In: *Proceedings of the Third Eastern Wildlife Damage Control Conference*. Gulf Shores, Alabama, pp. 253–259.

Coppinger, R., Coppinger, L., Langeloh, G., Gettler, L. and Lorenz, J. (1988) A decade of use of livestock guarding dogs. In: Crabb, A.C. and Marsh, R.E. (eds) *Proceedings of the Thirteenth Vertebrate Pest Conference*. University of California, Davis, pp. 209–214.

Dawson, W.M. (1965) *Studies of Inheritance of Intelligence and Temperament in Dogs*. ARS 44–163, Animal Husbandry Research Division, US Department of Agriculture, Beltsville, Maryland.

de Planhol, X. (1969) Le chien de berger: développement et signification géographique d'une technique pastorale. *Bulletin de l'Association des Géographes Français* 370, 351.

Devine, M., and Wenzel, D. (1997). *Border Collies*. Barrons Educational Series, Hauppauge, New York.

Fiennes, R. and Fiennes, A. (1968) *The Natural History of Dogs*. Bonanza Books, New York.

Green, J.S. and Woodruff, R.A. (1983) The use of three breeds of dog to protect rangeland sheep from predators. *Applied Animal Ethology* 11, 141–161.

Green, J.S. and Woodruff, R.A. (1988) Breed comparisons and characteristics of use of livestock guarding dogs. *Journal of Range Management* 41, 249–251.

Green, J.S. and Woodruff, R.A. (1990) *Livestock Guarding Dogs: Protecting Sheep from Predators*. Agriculture Information Bulletin No. 588, US Department of Agriculture.

Green, J.S., Woodruff, R.A. and Tueller, T.T. (1984) Livestock-guarding dogs for predator control: costs, benefits, and practicality. *Wildlife Society Bulletin* 12, 44–49.

Hagstad, H.V., Hubbert, W.T. and Stagg, L.M. (1987) A descriptive study of dairy goat predation in Louisiana. *Canadian Journal of Veterinary Research* 5, 152–155.

Hahn, M.E. and Schanz, N. (1996) Issues in the genetics of social behavior revisited. *Behavior Genetics* 26, 417–422.

Hahn, M.E. and Wright, J.C. (1998) The influence of genes on social behavior of dogs. In: Grandin, T. (ed.), *Genetics and the Behavior of Domestic Animals*. Academic Press, San Diego, California, pp. 299–318.

Hansen, I. and Bakken, M. (1999) Livestock-guarding dogs in Norway: Part I. Interactions. *Journal of Range Management* 52, 2–6.

Holland, V.S. (1994) *Herding Dogs: Progressive Training*. Howell Book House, New York.

Holmes, J. (1976) *The Farmer's Dog*. Popular Dogs, London.

Holmes, J. and Holmes, M. (1993) *The Complete Australian Cattle Dog*. Howell Book House, New York.

Hubbard, C.L.B. (1947) *Working Dogs of the World*. Sidgwick and Jackson, London.

Iley, T. (1978) *Sheepdogs at Work*. Dalesman Books, Clapham, UK.

Jones, H.G. (1991) *Come Bye! and Away!* Farming Press Videos, Ipswich, UK, colour, 45 min.

Jones, H.G. and Collins, B.C. (1987) *A Way of Life: Sheepdog Training, Handling and Trialling*. Farming Press Books, Ipswich, UK.

Klaffke, O. (1999) The company of wolves. *New Scientist*, www.newscientist.com, 6 February 1999.

Kupper, W. (1945) *The Golden Hoof*. Alfred A. Knopf, New York.

Laurans, R. (1975) Chiens de garde et chiens de conduite des moutons. *Ethnozootechnie* 12, 15–18.

Linhart, S.B., Sterner, R.T., Carrigan, T.C. and Henne, D.R. (1979) Komondor guard dogs reduce sheep losses to coyotes: a preliminary evaluation. *Journal of Range Management* 43, 238–241.

Little, M.E. (1975) *Happy Herding Handbook*. Wheel-A-Way Ranch, Riverside, California.

Longton, T. and Hart, E. (1976) *The Sheep Dog: Its Work and Training*. David and Charles, North Pomfret, Vermont.

Lorenz, J., Coppinger, R. and Sutherland, M. (1986) Causes and economic effects of mortality in livestock guarding dogs. *Journal of Range Management* 39, 293–295.

Lory, J. (1989) Le chien de berger, son utilisation. *Ethnozootechnie* 43, 27–34.

McConnell, P.B. (1990) Acoustic structure and receiver response in domestic dogs, *Canis familiaris. Animal Behaviour* 39, 897–904.

McConnell, P.B. and Baylis, J.R. (1985) Interspecific communication in cooperative herding: acoustic and visual signals from human shepherds and herding dogs. *Zeitschrift für Tierpsychologie* 67, 302–328.

McGrew, J.C. (1982) Behavioral correlates of guarding sheep in Komondor dogs. Unpublished PhD thesis, Colorado State University.

McGrew, J.C. and Blakesley, C.S. (1982) How Komondor dogs reduce sheep losses to coyotes. *Journal of Range Management* 35, 693–696.

Means, B. (1970) *The Perfect Stock Dog.* Ben Means, Walnut Grove, Missouri.

O'Gara, B.W., Brawley, K.C., Munoz, J.R. and Henne, D.R. (1983) Predation on domestic sheep on a western Montana ranch. *Wildlife Society Bulletin* 11, 253–264.

Olsen, J.W. (1985) Prehistoric dogs in mainland East Asia. In: Olsen, S.J. (ed.), *Origins of the Domestic Dog: the Fossil Record.* University of Arizona Press, Tucson, pp. 47–70.

Pfeifer, W.K. and Goos, M.W. (1982) Guard dogs and gas exploders as coyote depredation control tools in North Dakota. *Proceedings of the Vertebrate Pest Conference* 10, 55–61.

Ring, E. (1994). *Ranch and Farm Dogs.* Millbrook Press, Brookfield, Connecticut.

Robertson, N. (1996) *Australian Cattle Dogs.* TFH Publications, Neptune City, New Jersey.

Ružić, R. (1988) *The Yugoslav Sheepdog – Šarplaninac.* Jugoslovenska Revija, Belgrade.

Schmitt, R. (1989) Chiens de protection des troupeaux. *Ethnozootechnie* 43, 51–58.

Scott, J.P. (1958) *Animal Behavior.* University of Chicago Press, Chicago.

Scott, J.P. and Marston, M. (1950) Critical periods affecting the development of normal and maladjustive social behavior of puppies. *Journal of Genetic Psychology* 77, 25–60.

Taggart, M. (1991) *Sheepdog Training: An All-breed Approach.* Alpine Publications, Loveland, Colorado.

Templeton, J. and Mundell, M. (1992) *Working Sheepdogs: Management and Training.* Crowood Press, Ramsbury, UK.

Thomas, K. (1983) *Man and the Natural World.* Pantheon, New York.

Vines, G. (1981) Wolves in dogs' clothing. *New Scientist* 91, 648–652.

Willis, M.B. (1992) *Practical Genetics for Dog Breeders.* H.F. & G. Witherby, London.

## Further Reading

Adams, H. (1980) *Diary of Maggie (the Komondor).* Middle Atlantic States Komondor Club, Princeton, New Jersey.

Aurigi, M. (1983) Migration of sheep dogs of Abruzzo (Il cane da pastore abruzzese emigra). *Informatore Zootecnico* (Bologna) 30, 30–33.

Austin, P. (1984) *Working Sheep with Dogs.* Department of Primary Industry, Canberra.

Dalton, C. (1983) Training working dogs: short lessons and simple commands are the key. *New Zealand Journal of Agriculture* 146, 42.

Fytche, E. (1998) *May Safely Graze: Protecting Livestock Against Predators.* Creative Bound, Toronto, Ontario, Canada.

Gordon, J. (1985) More from the fool at the foot of the hill. *New Zealand Journal of Agriculture* 150, 47–49.

Hartley, S.W.G. (1967/1981) *The Shepherd's Dogs: a Practical Book on the Training and Management of Sheepdogs.* Whitcoulls, Sydney.

Henderson, F.R. and Spaeth, C.W. (1980) *Managing Predator Problems: Practices and Procedures for Preventing and Reducing Livestock Losses.* Bulletin No. C-620, Cooperative Extension Service, Kansas State University, Manhattan.

Howe, Ch.E. (1983) Training the farm dog. In: Vidler, P. (ed.), *The Border-Collie in Australasia.* Gotrah Enterprises, Kellyville, New South Wales.

Lorenz, J.R. (1989) *Introducing Livestock-guarding Dogs.* Extension Circular 1224 (rev.), Oregon State University, Corvallis.

Lorenz, J.R. (1990) Diffusion of Eurasian guarding dogs into American agriculture: an alternative method of predator control. Unpublished PhD thesis, Oregon State University.

Lorenz, J.R. and Coppinger, L. (1989) *Raising and Training a Livestock-guarding Dog.* Extension Circular 1238 (rev.), Oregon State University, Corvallis.

Parsons, A.D. (1986) *The Working Kelpie: the Origins and Breeding of a Fair Dinkum Australian.* Nelson, Melbourne, Victoria.

Sherrow, H.M. and Marker, L. (1988) *Livestock Guarding Dogs in Namibia.* Cheetah Conservation Fund, Otjiwarongo, Namibia.

Sims, D.E. and Dawydiak, O. (1990) *Livestock Protection Dogs: Selection, Care and Training.* OTR Publications, Fort Payne, Alabama.

Tatarskii, A.A. (1960) *Herd Dogs in Sheep Farming. (Pastush'i sobaki v ovtsevodstve).* Sel'khozgiz, Moscow.

Von Thüngen, J. and Vogel, K. (1991) *Como Trabajar con Perros Pastores.* Instituto Nacional de Tecnologia Agropecuaria, Bariloche, Rio Negro, Argentina.

Wick, P. (1998) *Le Chien de protection sur troupeau ovin: utilisation et méthode de mise en place.* Editions Artus, St Jean de Braye, France.

Zernova, M.V. (1979a) Working dogs in agriculture: training recommendations. *Tvarynnytstvo-Ukrainy* (Ministerstvo sil's'koho hospodarstva, Kyiv, URSR) 9, 54–55.

Zernova, M.V. (1979b) General training of herd dogs. *Tvarynnytstvo-Ukrainy* (Ministerstvo sil's'koho hospodarstva, Kyiv, URSR) 10, 54–55.

Zernova, M.V. (1980) Rex! To me! Recommendations for training dogs for herding animals. *Tvarynnytstvo-Ukrainy* (Ministerstvo sil's'koho hospodarstva, Kyiv, URSR) 2, 50–51.

Zernova, M.V. (1981) Restraining and keeping in line of a herd: recommendations for training sheep dogs. *Tvarynnytstvo-Ukrainy* ('Urozhai', Kyiv, URSR) 7, 52–53.

# Behavioural Principles of Pig Handling

**14**

## P.H. Hemsworth

*Animal Welfare Centre, Agriculture Victoria and University of Melbourne, Victorian Institute of Animal Science, Private Bag 7, Snedyes Road, Werribee, Victoria 3030, Australia*

## Introduction

Human–animal interactions are a key feature of modern pig production because commercial pigs receive frequent and, at times, close human contact. In supervising pigs and their conditions, stockpeople maintain regular visual contact with their animals and may physically interact with the animals when imposing routine husbandry procedures or when moving them. Stockpeople use tactile, auditory and visual interactions when moving most forms of livestock. Tactile interactions, including a push, slap or hit, are common interactions used by stockpeople when handling pigs, while auditory and visual interactions, such as shouts and waves are also used.

The effects of human–animal interactions on livestock, in general, appear to have been neglected until recently. This lack of interest was presumably due to industry personnel and animal scientists considering that either the intensity and frequency of these interactions were low enough to render the effects of any negative interactions ineffective on the animals or the type of interactions were harmless to the animals. However, commencing in the 1980s, reports appeared in the scientific literature on the practical implications of human–animal interactions in agriculture. Research, particularly on the pig, has now shown that the quality of the relationship that is developed between stockpeople and their animals can have surprising effects on both the animals and the stockpeople. For example, there is good evidence, based on handling studies and observations in the pig industry, that human–animal interactions may markedly affect the growth, reproduction and welfare of pigs (Hemsworth and Coleman, 1998). Recent experiments also indicate that human–animal interactions may affect the meat quality of pigs as a consequence of stress prior to slaughter (D'Souza *et al.*, 1998a, b). Furthermore, by influencing the behavioural response of the animal to humans and, in particular, the ease with which animals can be observed, handled and managed by the stockperson, human–animal interactions may have implications for a number of important work-related characteristics of the stockperson, such as job satisfaction. Although not well documented, anecdotal observations suggest that many stockpeople consider the handling of pigs as one of the most frustrating tasks in a piggery. Indeed, poor handling facilities or fearful pigs may cause frustration for stockpeople, with adverse effects on their job satisfaction, work motivation and thus work performance (English *et al.*, 1992; Hemsworth and Coleman, 1998).

An important objective in handling livestock is to minimize their fear. In situations of high fear, animals behave in a self-protective way by either fleeing or fighting back (Toates, 1980) and are therefore generally difficult to handle (Holmes, 1984). As with other forms of livestock, the two main behavioural responses by pigs during a handling bout in which they are being moved are the responses to the handler and to the environment into which they are being moved. If pigs are highly fearful of both the handler and the environment, unexpected and exaggerated behavioural responses, such as baulking, freezing and fleeing, are likely to occur, hindering ease of handling. Nevertheless, some judicious use of fear-provoking stimuli by handlers is required to move pigs efficiently in the desired direction. Thus, the aim of any handling bout should be to elicit sufficient fear to encourage movement but not to create either immediate handling difficulties or an escalation of fear of humans in the longer term. In terms of improving ease of movement and minimizing injury to the stock and the stockperson, it is clearly important that stockpeople reduce the level of fear in their animals.

Hutson (1993) described sheep as a 'visual, flocking and follower-type animal' that is 'intelligent'. For most livestock, this description is reasonably apt, although there are clearly age, sex and species differences in these characteristics. Describing farm animals in such terms emphasizes that an understanding of the behavioural and sensory characteristics of livestock is important in efficiently handling and controlling livestock. This chapter reviews some of the main principles of pig handling based on current knowledge of the fear and explorative responses, learning capability, sensory characteristics and social behaviour of commercial pigs. The second part of the chapter considers opportunities to improve handling of pigs.

## Principles of Pig Handling

### Fear and exploration

There has been considerable debate over the concept of fear (Hinde, 1970; Murphy, 1978). Gray (1987) defines fear as a hypothetical state of the brain or the neuroendocrine system, arising from certain conditions and eventuating in certain forms of behaviour. Fear is usually listed among the emotions and, as such, fear can be viewed as a form of emotional reaction to the threat of punishment, where punishment refers to a stimulus which the animal works to terminate, escape from or avoid (Gray, 1987). For the purpose of this chapter, fear will be considered as a state of motivation and will be viewed as eliciting escape or avoidance responses. Furthermore, fear-provoking stimuli will include those to which the animal may not have had previous exposure (e.g. novel and so-called 'sign stimuli') and those to which a fear response has become attached through the process of conditioning (McFarland, 1981). Exploratory behaviour has been defined as 'behaviour which serves to acquaint the animal with the topography of the surroundings included in the range' (Shillito, 1963), and the amount of exploration of an object will depend on characteristics of the object, such as its novelty (the time since it was last encountered or its degree of resemblance to other encountered situations), complexity, intensity and contrast, and on the poverty of the preceding environment (Berlyne, 1960). In this chapter, exploration will be considered as a state of motivation and as involving behaviours resulting in close contact with a novel stimulus.

A sudden environmental change will usually elicit the movement of turning towards the source, a response called the orientation response, which may be followed by a startling response and defensive or flight reactions by the animal (Hemsworth and Barnett, 1987). As fear responses wane, the animal will also approach and examine the stimulus (Hinde, 1970). Exploration will be terminated once the animal is somewhat acquainted with the stimulus. Therefore, the responses of animals to novel stimuli can be considered to contain elements of both fear and exploratory responses. Furthermore, in response to a novel stimulus or a stimulus perceived as aversive (with or without previous experience), the initial avoidance and subsequent exploration can be viewed as a consequence of the conflicting motivations of fear and exploration and the waning of fear responses.

## Fear of humans

### Fear and ease of handling

Most of the limited research on fear of humans and ease of handling pigs and other farm animals indicates that animals that are highly fearful of humans are generally the most difficult to handle. As shown in Table 14.1, moderate to large correlations have been found between the behavioural response of pigs to humans and their ease of handling. These correlations indicate that pigs that showed high levels of fear of humans, based on their avoidance behaviour of an experimenter in a standard test, were the most difficult pigs to move along an unfamiliar route. These fearful pigs took longer to move, displayed more baulks and were subjectively scored as the most difficult to move by the handler. Animals are generally wary of entering an unfamiliar location and if

**Table 14.1.** Correlations between the behavioural response of pigs to an experimenter in a standard test to assess fear of humans and the ease of movement of 24 pigs along an unfamiliar route by an unfamiliar handler (Hemsworth *et al.*, 1994b).

| Variables recorded in test to assess fear of humans | Variables recorded in ease-of-movement test | | |
|---|---|---|---|
| | Time to move | Baulks | Score |
| Time to approach experimenter | 0.34 | 0.44* | −0.63** |
| Number of interactions with experimenter | −0.42** | −0.42* | 0.51* |

Correlation coefficients with * = $P < 0.05$ and ** = $P < 0.01$.
Score was given based on ease of movement, with 0 reflecting substantial difficulty and 4 reflecting little or no difficulty in moving the pig.

they are fearful of both the new environment and the handler, they are likely to show exaggerated responses to handling. For example, they may baulk or flee back past the handler, thus requiring more effort and time on the part of the handler to move them.

Other authors have also reported that high levels of fear of humans will decrease the ease of handling pigs (Gonyou *et al.*, 1986; Grandin *et al.*, 1987). In contrast, Hill *et al.* (1998) found no effect of fear of humans on the time taken to move pigs to and from a weighing area. In this study, handling treatments were imposed on pigs in their home pens and fear of humans was assessed in these pens. Therefore, this measure of fear of humans may not have reflected actual differences in the behavioural responses of these pigs to humans in other locations outside the home pen. In other words, the measured fear response may have involved elements of a location-specific response. It is of interest that these authors also reported that genotype affected the time taken to move pigs. A number of authors have commented that genetics may affect ease of handling pigs (Grandin, 1991), but there is little evidence of such effects apart from the experiment by Hill *et al.* (1998).

Handling effects on ease of handling have been found in other livestock. Studies on cattle, sheep and horses have shown that handling, generally involving speaking to and touching the animals, particularly during infancy, improved their subsequent ease of handling (for example, Boissy and Bouissou, 1988; Boivin *et al.*, 1992; Hargreaves and Hutson, 1990; Hemsworth *et al.*, 1996b; Lyons, 1989; Mateo *et al.*, 1991; Waring, 1983).

A number of studies have reported that pigs regularly moved out of their pens prior to slaughter were quicker to move during the early stages of transport, such as moving out of their home pen and into a transport crate or box (Abbott *et al.*, 1997; Geverink *et al.*, 1998). While increased human contact may be implicated, it is likely that increased familiarity with locations early in the transport process is responsible for these effects. However, Eldridge and Knowles (1994) reported that commercial grower pigs that were regularly handled and moved out of their pens to a range of locations were easier to move in an unfamiliar environment.

*Factors affecting the pig's fear of humans*

The pig's response to a stockperson in an intensive farming system may have components of both stimulus-specific fear and general fear. While the initial response of a naïve pig to humans may involve a response to novelty or unfamiliarity (i.e. general fearfulness), with subsequent experience of humans there is the development of a specific response to humans (Hemsworth and Coleman, 1998). The initial response of a naïve farm animal to humans may be similar to the animal's response to an unfamiliar object or to unfamiliar animals of another species. Furthermore, Suarez and Gallup (1982) have suggested that the predominant response of naïve animals to humans may be a response to a predator.

As a consequence of the amount and nature of interactions with humans, commercial pigs will develop a stimulus-specific response to humans. Therefore, although there will be some components of novelty in the response of experienced animals to humans, which will occur with changes in the stimulus property of humans (e.g. changes in behaviour, clothing, location of interaction, etc.), a major component of this response will be experientially determined. There is some evidence that the behavioural response of relatively naïve pigs to humans, which may be predominantly a result of general fearfulness, may be moderately heritable; however subsequent experience with humans appears to dilute the genetic effects (Hemsworth *et al.*, 1990). The behavioural response of experienced pigs to humans only accounted for less than a quarter of the variance of their behavioural response to humans earlier in life, when they were relatively inexperienced with humans.

There is considerable support for this view of the development of a stimulus-specific response of farm animals to humans. For example, numerous handling studies have shown that handling treatments varying in the nature of human contact, but not in the amount of human contact, resulted in rapid changes in the level of fear of humans by pigs (Hemsworth *et al.*, 1981a, 1986b, 1987; Gonyou *et al.*, 1986; Hemsworth and Barnett, 1991). Murphy (1976), in studying two stocks of chickens, termed 'flighty' and 'docile' on the basis of their behavioural responses to humans, found that the so-called docile birds did not necessarily show fewer withdrawal responses to novel stimuli, such as a mechanical scraper and an inflating balloon, than the flighty birds. Jones *et al.* (1991) and Jones and Waddington (1992) examined the effects of regular handling on the behavioural responses of quail and domestic chickens to novel stimuli (such as a blue light) and humans and found that handling predominantly affected the responses of birds to humans, rather than to the novel stimuli. Handled birds showed less avoidance of humans but their responses to novel stimuli were unaffected.

Considerable research has been conducted over the past 10 years on the stockperson behaviour–animal behaviour relationships in the pig industry.

This research has shown some large and consistent correlations between the behaviour of the stockperson towards pigs and the behavioural response of pigs to humans, which generally confirm the predictions of handling studies that have been conducted under experimental conditions (Hemsworth *et al.*, 1981a, 1986b, 1987; Gonyou *et al.*, 1986; Hemsworth and Barnett, 1991). It was found in these studies on commercial breeding pigs that use of what can be termed 'negative tactile interactions' by stockpeople was predictive of the level of fear of humans by pigs. Negative tactile interactions by stockpeople include mild to forceful slaps, hits, kicks and pushes, while the positive tactile interactions include pats, strokes and the hand resting on the pig's back. In these studies, fear of humans was assessed by the measuring the time spent by pigs near a stationary experimenter in a standard test. It was consistently found that the percentage of negative tactile interactions to the total tactile interactions by the stockperson was highly correlated with the level of fear of humans by pigs (Hemsworth *et al.*, 1989; Coleman *et al.*, 1998): high fear levels were observed where stockpeople displayed a high percentage of negative tactile interactions. Surprisingly, high levels of fear of humans were best predicted when the classification of negative behaviours included not only forceful kicks, hits, slaps and pushes, but also negative behaviours used with less force, such as mild and moderate slaps and pushes. This finding indicates the sensitivity of pigs to mild and moderate negative interactions by humans, something that is not intuitively obvious to most of us.

The results of these studies in the industry (Hemsworth *et al.*, 1989; Coleman *et al.*, 1998), together with a number of handling studies (Hemsworth *et al.*, 1981a, 1986b, 1987; Gonyou *et al.*, 1986; Paterson and Pearce, 1989; Pearce *et al.*, 1989; Hemsworth and Barnett, 1991), indicate that conditioned approach–avoidance responses develop as a consequence of associations between the stockperson and aversive and rewarding elements of the handling bouts. The main aversive properties of humans include hits, slaps and kicks by the stockperson, while the rewarding properties include pats, strokes and the hand of the stockperson resting on the back of the animal. It is the percentage of these negative tactile interactions to the total tactile interactions that appear to determine the commercial animal's fear of humans. While auditory interactions by stockpeople may not be highly important in regulating these fear responses (Hemsworth *et al.*, 1986a), visual interactions by the stockperson, such as speed of movement and unexpected movement, may affect fear of humans. Furthermore, pigs with limited experience with humans may habituate to the regular presence of humans and thus may perceive humans as part of the environment without any particular significance. Habituation will occur over time as the animal's fear of humans is gradually reduced by repeated exposure to humans in a neutral context; that is, the human's presence has neither rewarding nor punishing elements for the animal.

Evidence from a number of handling studies on pigs supports the view that the animal's response to a single human might extend to include all humans through the process of stimulus generalization (Hemsworth and Coleman, 1998). For example, pigs which previously were briefly but regularly handled by either a handler in a predominantly negative manner or two handlers who differed markedly in the nature of their behaviour towards pigs, showed similar behavioural responses to familiar and unfamiliar handlers (Hemsworth *et al.*, 1994b). Similar evidence is also available from studies with poultry and sheep (Barnett *et al.*, 1993; Jones, 1993; Bouissou and Vandenheede, 1995).

Such results suggest that, in commercial situations, the behavioural response of pigs to one handler may extend to other humans. However, it is possible that there are handling situations in which pigs may not exhibit stimulus generalization. In situations in which there is intense handling, animals may learn to discriminate between this handler and other handlers to which the animals may be subsequently exposed. Following an extensive period of intense human contact, Tanida *et al.* (1995) found that young pigs showed greater approach to the familiar handler than to an unfamiliar handler, even though both handlers wore similar clothing. Furthermore, in situations in which the physical characteristics of the handlers may differ markedly, farm animals may learn to discriminate between the handlers. For example, in a series of experiments, de Passille *et al.* (1996) found that dairy calves exhibited clear avoidance of a handler that had previously handled them in a negative manner in comparison with handlers wearing different-colour clothing who were either unfamiliar to the calves or had previously handled them in a positive manner. Initially, there was a generalization of the aversive handling, with calves showing increasing avoidance of all handlers, but, with repeated treatment, calves discriminated between handlers and, in particular, between the 'negative' and 'positive' handlers. It is of interest that discrimination was greatest when tested in the area in which handling had previously occurred rather than in a novel location. In fact, in one experiment when animals were tested in a location where handling had not been performed, 40% of the calves actually approached and interacted with the negative handler. These data on calves indicate that discrimination between people by farm animals will be easier if the animals have some distinct cues on which they can discriminate, such as colour of clothing or location of handling.

Although several species of farm animals, including pigs, are capable of discriminating between stockpeople, they do not appear to do so under normal commercial circumstances. Nevertheless, even when farm animals learn to discriminate between humans, fear responses to humans in general are likely to increase in response to the most aversive handler (Hemsworth *et al.*, 1994b; de Passille *et al.*, 1996). Such a finding has important implications in situations in which several stockpeople may interact with pigs.

An important finding in terms of training stockpeople to improve human–animal interactions in the pig industry is that the attitudes of stockpeople towards interacting with their animals are predictive of the behaviour of the stockpeople towards their animals. Hemsworth *et al.* (1989)

and Coleman *et al.* (1998) used questionnaires to assess attitudes of the stockpeople on the basis of the stockpeople's beliefs about their behaviour and the behaviour of their pigs. Positive attitudes to the use of petting and the use of verbal and physical effort to handle pigs were negatively correlated with the use of negative tactile interactions, such as slaps, pushes and hits. These correlations indicate that stockpeople are likely to use a lower percentage of negative interactions when handling their pigs if they believed that: (i) petting should be frequently used; and (ii) verbal and physical effort should be infrequently used when interacting with pigs.

## Fear of novelty or unfamiliarity

Pigs, like other animals, are initially fearful of strange objects and locations and will generally baulk. Therefore, features such as floor surfaces, floor levels and wall types should be as consistent as possible throughout a race or corridor to reduce baulking. It is probably particularly important to minimize such changes at critical points in the route, such as pen exits, corners and entrances to corridors or races, where unfamiliarity is likely to have a greater effect on ease of handling. Novel objects in the race or moving and flapping objects will also cause baulking.

If pigs become fearful in an unfamiliar location, it is preferable to allow them some time to familiarize themselves with the environment. Trying to move pigs quickly in this situation may be costly in terms of time and effort, as well as risking injury to both pigs and the handler. Habituation to novel stimuli of moderate intensity will occur over time, as the animal's fear is gradually reduced by continuous exposure in a neutral context.

Grandin (1982/83) reported that a smooth concrete floor with a wet slippery surface inhibits pig movement. Furthermore, Grandin (1988) observed that fattening pigs reared on metal mesh floors were difficult to move on concrete floors. Lack of confidence in gaining a firm footing on an unfamiliar surface may be responsible for these effects.

There are conflicting data in the literature on the behaviour and performance of livestock provided with additional complexity in their environments. Such manipulations, which is often called 'environmental enrichment', may be similar to the phenomenon of 'infantile stimulation' seen in laboratory animals. Infantile stimulation, which involves handling of young animals, has at times been shown to advance behavioural and physiological maturation (Schaefer, 1968; Hinde, 1970). One notable effect of infantile stimulation is decreased general fearfulness or fear of novelty. It has been proposed that these subsequent effects on maturation act via an acute stress response early in life (Schaefer, 1968). Grandin *et al.* (1987) found that fattening pigs given novel objects to manipulate, such as rubber hoses, were easier to handle and Pearce *et al.* (1989) found that young pigs housed in pens with novel objects, such as chains and tyres, showed less avoidance of humans than those reared in barren pens. In contrast, Hill *et al.* (1998) found no effect of the provision of such novel objects on either the behavioural response of pigs to humans in their pens or their ease of movement. It has been reported that pigs are subsequently more easily startled and more difficult to handle when reared in darkness (Grandin, 1991) or semi-darkness (Warriss *et al.*, 1983). In the latter study, control pigs were reared outdoors, while in the former study the control pigs were presumably reared indoors with greater illumination. It is possible that these effects observed on general fearfulness and ease of handling may have been a consequence of increased environmental stimulation.

## Learning ability

Contrary to what many people may believe, farm animals will readily learn a variety of tasks. Pigs are generally considered easy to classically condition: they are capable of quickly learning to show a range of conditioned or associative responses, such as salivary and cardiac responses, to a range of stimuli, including auditory stimuli (Houpt and Wolski, 1982; Kilgour, 1987). Pigs can be trained quite easily to perform operant responses and will respond, for example, by pushing or manipulating levers with their snouts to receive sensory rewards, such as lighting or temperature (Houpt and Wolski, 1982), or jumping across an obstacle to avoid the punishment of an electric shock (Craig, 1981). They perform well in maze learning tests, but often perform poorly in visual discrimination tests (Kilgour, 1987).

Pigs have good short- and long-term memories (Houpt and Wolski, 1982) and this can be used to develop handling routines. With breeding pigs, there is an excellent opportunity for pigs to learn to move easily to and from commonly used locations by regular exposure. By allowing the pigs to initially move at their own pace and thus by minimizing aversive experiences, pigs are likely to move more easily on subsequent introductions.

## Sensory characteristics

Pigs have good colour vision (Grandin, 1987) and thus may respond to the novelty of a change in the colour of the routine clothing of handlers. Pigs have a wide angle of vision (310°) (Prince, 1977) and therefore walls of corridors, pen fronts and gates should be solid (at least up to pig height) to prevent the pigs that are being moved from becoming distracted by what they see, such as other people or pigs (Grandin, 1980). Stockpeople should follow behind and slightly to one side, and use a solid board to prevent the pigs from turning back (Fig. 14.1). Livestock have sensitive hearing and thus may avoid excessive and unfamiliar noise (Grandin, 1990). While livestock appear to move more easily

**Fig. 14.1.** The use of a solid panel for moving pigs will prevent them from attempting to turn back.

on a level surface, excessively steep ramps were avoided by pigs in a preference test: 20–24° ramps were preferred to 28–32° ramps (Grandin, 1990).

Pigs, like other animals, have a tendency to move towards a more brightly lit area (Van Putten and Elshof, 1978). Tanida *et al.* (1996) found that piglets preferred to move from dark to lighter areas and were encouraged to move to darker areas with the provision of lighting. Experiments indicate that the light should be even and diffuse (Grandin, 1980). Thus lights can be used to encourage movement into poorly illuminated areas, such as races and dark corridors. Although Grandin (1987) suggests that shadows will cause baulking, Tanida *et al.* (1996) found no effect on the movement of piglets of either shadows or lines on the floor. While the light intensity of a flashlight (160 lux) did not affect pig movement (Tanida *et al.*, 1996), excessive light (1200 lux) caused avoidance (Grandin, 1990). Furthermore, pigs avoid black and white patterns on the floor (Tanida *et al.*, 1996) and, since pigs are only moderate judges of distance (Grandin, 1980), they are reluctant to cross changing light patterns, drain grates, steps, puddles of water, gutters and other high-contrast objects. Batching gates and pig boards should be solid to block the vision of pigs in order to encourage their movement away from the gates or boards.

## Social behaviour

Several aspects of the pigs' social behaviour will affect ease of handling. Herding or flocking, in which social spacing and orientation are maintained, is most pronounced in sheep but is also evident in other livestock, including pigs. Following behaviour, in which there is synchrony of behaviour, such as walking, running, feeding and lying, is commonly seen in pigs and other livestock. Pigs show pronounced herding and following behaviour (Van Putten and Elshof, 1978).

This motivation of pigs to follow other pigs and maintain body and visual contact with other pigs obviously can and should be utilized in moving pigs. For example, the walls of corridors, pen fronts and gates should be solid (at least up to pig height) to prevent the pigs being moved from becoming distracted by adjacent pigs; however, race design should utilize the attraction of pigs to the sight of others moving ahead (Fig. 14.2; Grandin, 1982). Thus corridors or races should be wide enough to provide the animals with a clear view ahead and of other animals moving ahead; Grandin (1990) suggests that

**Fig. 14.2.** A 'see-through' partition promotes following in this loading ramp. The outer fences are solid.

corridors on farms for pigs should be 1 m wide. If one animal needs to be isolated from the group, it may be preferable to move it within a small group to a location where the animal can easily be drafted from the rest or to use a large pig board in the pen to direct the pig out of the pen or away from the group.

## Main Recommendations on Handling Pigs Arising from an Understanding of Pig Behaviour and Sensory Capacity

### Desirable human contact

Modern pig production involves several levels of interaction between stockpeople and their animals. Many interactions are associated with regular observation of the animals and their condition and thus this type of interaction often involves only visual contact between the stockperson and the pigs, perhaps without the stockperson entering the animals' pen. Pigs in most production systems have to be moved by the stockperson and this often also involves tactile and auditory interaction with the animals. Growing pigs are occasionally moved from pen to pen, in order to provide accommodation suitable to their stage of growth, and breeding pigs are regularly moved according to their stage of the breeding cycle. It is during these situations that human–animal interactions have considerable potential to influence animal behaviour.

There are some basic principles in handling pigs, which involve an understanding of their behavioural response to humans. Pigs can be moved in the desired direction by entering the flight zone of the animal at 45–60° from directly behind the animal. Handlers should work on the edge of this zone to avoid an extreme reaction. The often widely held view that high fear responses to humans is desirable to effectively use the animal's flight distance is a dangerous one, since high fear responses may actually exacerbate handling problems.

During handling bouts, the risk of eliciting high fear of humans can be reduced by stockpeople minimizing their negative interactions towards pigs, while increasing their positive ones (Hemsworth *et al.*, 1989, 1994a). This can be achieved by using negative interactions, such as hits, slaps and pushes, only when necessary and using, when the opportunity arises, positive interactions, such as pats, strokes and the hand of the stockperson resting on the pig's back. For example, these positive interactions can be used when animals are moving, have arrived at the destination or are feeding. Stockpeople also interact with pigs when they inspect them and their equipment, such as feeders and waterers. Although these interactions do not necessarily involve tactile contact with the animals, visual interactions at the time can also be potentially fear-provoking. For example, fast speed of movement and unexpected movement or appearance by the stockperson may be fear-provoking, particularly if frequently occurring.

There are occasions when negative behaviours have to be used to move pigs and, in these situations, the elicitation of a fear response (i.e. withdrawal from the stockperson) is used to encourage the animal to move in the desired direction. Such situations may include moving the pig out of the home pen or into an unfamiliar area and, in these situations, it is normal for the animal to be wary or hesitant when moving. Thus stockpeople should only use these negative behaviours when necessary. When pigs baulk or are difficult to move, stockpeople should first examine the features of the environment for factors which may be fear-provoking and which could be eliminated, before resorting to the use of negative behaviours. Therefore, there will be occasions when negative behaviours have to be used; however, the stockperson's objective should be to use negative interactions only when necessary and, when the opportunity arises, to use positive behaviours.

Human–animal interactions also occur in situations in which animals must be restrained and subjected to management or health procedures. Pigs are restrained for procedures such as blood sampling, castration, injections and ear tagging. The association of fear and pain from these husbandry procedures with the humans performing them will increase the fear of humans which animals exhibit in other situations, such as during routine inspections. The effect that these procedures have on the human–animal relationship is a consequence of both the aversiveness of the procedure and the association of people with that aversion. Rewarding experiences, such as provision of a preferred feed or even positive handling, around the time of the procedure, may ameliorate the aversiveness of the procedure and reduce the chances that animals associate the punishment of the procedure with humans or the location. In comparison with positive handling and handling involving minimal human contact, Hemsworth *et al.* (1996) judged a treatment involving daily injections over a 3-week period to be, at the worst, moderately aversive to young pigs. It was concluded by the authors that there may have been some rewarding components in this daily injection treatment, such as the presence of the experimenter and the opportunity to closely investigate and interact with this somewhat novel stimulus, which may mask or overcome any aversive components of the treatment. Similarly, feed rewards, such as barley, can reduce the avoidance shown by sheep to a location in which an aversive procedure has been previously carried out (Hutson, 1985).

## Training stockpeople to improve human–animal interactions

The significant interrelationships between stockperson attitudes and behaviour and pig fear (behaviour) and performance (Hemsworth *et al.*, 1981b, 1989; Coleman *et al.*, 1998) indicate opportunities to improve pig behaviour and productivity by improving the attitudes and behaviour of stockpeople. Indeed, studies in the pig industry have shown that it is possible, first, to improve the attitudinal and behavioural profiles of stockpeople towards pigs and, secondly, to reduce the level of fear and improve the productivity of their pigs (G.C. Coleman *et al.*, unpublished data; Hemsworth *et al.*, 1994a).

This approach in improving the attitudes and behaviour of stockpeople has been described in detail by Hemsworth and Coleman (1998). Basically, such a training programme for stockpeople has to specifically target the key attitudes and behaviours that regulate the pig's fear of humans. It should be appreciated that simply presenting stockpeople with information on the effects of human–animal interactions on the behaviour and productivity of commercial pigs is unlikely to have long-lasting benefits on their attitudes and behaviour. Most people, when presented with this information, simply deny problems with their attitudes and behaviour. This is quite common, particularly when many do not fully appreciate the definitions of positive and negative behaviours which regulate the fear responses of pigs and also the need to use positive behaviours to reduce fear in pigs. For example, many stockpeople are surprised to learn that the use of moderate slaps and pushes can result in fearful animals.

Cognitive-behavioural modification techniques, which have been success-fully used to modify human behaviour, involve retraining people in terms of their behaviour, as well as changing their attitudes and beliefs. Because of the reciprocal relationship between the attitudes and behaviour of the stockperson and the equally strong relationships between the stockperson's attitude and behaviour and the animal's fear and productivity, any behavioural modification procedure should target both the attitudes and behaviour of stockpeople. Furthermore, inducing attitudinal and behavioural change involves processes somewhat different from those used in the normal classroom situation; it involves not only imparting knowledge and skills, but also changing established habits, altering well-established attitudes and beliefs and preparing the person to handle reactions from others towards the individual following change. The process of inducing behavioural change is really a comprehensive procedure, in which all of the personal and external factors which are relevant to the behavioural situation are explicitly targeted. Therefore achieving change requires the presentation of factual material on the key beliefs affecting stockperson behaviour, the key stockperson behaviours regulating the human–animal relationship, the consequences of this relationship for both stockpeople and their animals, recommendations on the use of these stockperson behaviours and advice on maintaining attitudinal and behavioural change. The cognitive-behavioural intervention procedure used as an experimental tool in the above studies has recently been offered as a training package to the Australian pig industry by the Australian Pig Research and Development Corporation, and this development may soon be available to the international pig industry.

An opportunity may also be available for the pig industry to use attitude questionnaires to assist in the selection of stockpeople with desirable attitudinal and behavioural profiles, and work is proceeding on this aspect.

Furthermore, some recent research in the pig industry (Coleman *et al.*, 1998) has indicated relationships between the stockperson's attitudes and a number of job-related variables. It was found that some measures of work motivation of stockpeople were correlated with attitudes towards characteristics of pigs and towards most aspects of working with pigs. Job enjoyment and opinions about working conditions showed similar relationships with attitudes. Thus, the stockperson's attitudes may be related to aspects of work apart from handling of animals. Significant relationships have also been found in the pig industry between personality types of stockpeople, based on an evaluation using the '16 personality-factor questionnaire', and productivity in farrowing units of both small independent and larger integrated units (Ravel *et al.*, 1996). Thus selection tools may be developed in the near future to assist the industry in selecting stockpeople with desirable characteristics, including some that have important implications for pig handling.

## Utilization of the characteristics of pigs to handle and control them

An understanding of the learning and sensory ability and social behaviour of pigs can be effectively utilized in handling pigs and designing features of handling facilities. The scientific and popular literature on livestock handling describes many of the features of livestock that are relevant to livestock handling and the main features that should be considered in handling and controlling pigs include the following:

- Familiarity with direction of flow and route will help the pigs to learn where to go.
- The sight of stationary pigs adjacent to the race will slow movement and thus race walls adjacent to other animals should be covered.
- Wide, clear, well-lit areas will promote movement.
- Lighting can be used to promote movement. For example, at the time of handling, darkening the area in which the animals are held and brightening the area in which they are to move will promote movement.
- Races with a clear, unobstructed view towards the exit or where the animals are meant to move will promote movement.
- Pigs will be attracted by the sight of others moving ahead and thus visual contact with these animals needs to be maintained and not obstructed.
- Changes in race construction material or changes in floor type (slats to concrete) will inhibit flow.
- Walls painted one colour to avoid contrasts will promote flow.
- The use of covered and open panels can direct movement and vision.
- Ramps with covered sides will not allow animals to see the elevation.
- It is easier to move pigs as a group, rather than individually, to a holding facility where individual animals can be separated and treated.

- The direction of movement inside a shed should be across the direction of the grating or slats to improve footing and to reduce the ability of pigs to see through the floor or perceive heights.
- Reducing excessive noise, such as banging gates and engines, will facilitate movement.
- Reducing the aversiveness of handling and treatment in a location will promote subsequent entry and movement in the location. Thus it is useful to consider the following: minimize or reduce the duration of restraint and severity of handling and treatment, provide rewards after any treatment (opportunity for exploration, feed, interaction with handler, etc.), allow habituation to location and handler if possible (through repeated exposure) and apply all aversive husbandry treatments in a location other than those in which animals are routinely introduced or handled.

For detailed information on the designs of races and loading and unloading facilities for livestock, including pigs, readers are referred to the review by Grandin (1990).

## Conclusion

Human–animal interactions are a key feature of modern pig production. Research has shown that the quality of the relationship that is developed between stockpeople and their animals can have surprising effects on both the animals and the stockpeople. Handling studies on pigs and observations in the pig industry show that human–animal interactions may markedly affect the behaviour, productivity and welfare of pigs. Furthermore, by influencing the behavioural response of the pigs to humans, these interactions can affect the ease with which pigs can be observed, handled and managed by the stockperson. In addition to human contact, physical features of the environment will also influence animal movement and thus animal handling. Therefore an understanding of the behavioural and sensory characteristics of pigs is also important in effectively handling and controlling pigs. There is a clear ongoing need for the pig industry to train their personnel to effectively handle and move their stock, as well as ensuring that current knowledge on the characteristics of pigs are utilized in the design of handling facilities. Such improvements will not only have implications for the behaviour, stress, productivity and welfare of commercial pigs, but may also have implications for a number of important work-related characteristics of the stockperson, such as job satisfaction and work performance.

# References

Abbott, T.A., Hunter, E.J., Guise, H.J. and Penny, R.H.C. (1997) The effect of experience of handling pigs on willingness to move. *Applied Animal Behaviour Science* 54, 371–375.

Barnett, J.L. Hemsworth, P.H. and Jones, R.B. (1993) Behavioural responses of commercial farmed laying hens to humans: evidence of stimulus generalization. *Applied Animal Behaviour Science* 37, 139–146.

Berlyne, D.E. (1960) *Conflict, Arousal and Curiosity*. McGraw-Hill, New York.

Boissy, A. and Bouissou, M.F. (1988) Effects of early handling on heifers' subsequent reactivity to humans and to unfamiliar situations. *Applied Animal Behaviour Science* 20, 259–273.

Boivin, X., Le Neindre, P. and Chupin, J.M. (1992) Establishment of cattle–human relationships. *Applied Animal Behaviour Science* 32, 325–335.

Bouissou, M.F. and Vandenheede, M. (1995) Fear reactions of domestic sheep confronted with either a human or a human-like model. *Behavioural Processes* 43, 81–92.

Coleman, G.C., Hemsworth, P.H., Hay, M. and Cox, M. (1998) Predicting stockperson behaviour towards pigs from attitudinal and job-related variables and empathy. *Applied Animal Behaviour Science* 58, 63–75.

Craig, V.C. (1981) *Domestic Animal Behaviour: Causes and Implications for Animal Care and Management*. Prentice-Hall, Englewood Cliffs, New Jersey.

de Passille, A.M., Rushen, J., Ladewig, J. and Petherick, C. (1996) Dairy calves' discrimination of people based on previous handling. *Journal of Animal Science* 74, 969–974.

D'Souza, D.N., Dunshea, F.R., Warner, R.D. and Leury, B.J. (1998a) The effects of handling pre-slaughter and carcass processing rate post-slaughter on pork quality. *Meat Science* 50, 429–437.

D'Souza, D.N., Warner, R.D., Dunshea, F.R. and Leury, B.J. (1998b) Effects of on-farm and pre-slaughter handling of pigs on meat quality. *Australian Journal of Agricultural Research* 49,1021–1025.

Eldridge, G.A. and Knowles, H.M. (1994) *Improving Meat Quality Through Improved Pre-slaughter Management*. Research Report DAV91P, Pig Research and Development Corporation, Victorian Institute of Animal Science, Werribee, Australia.

English, P., Burgess, G., Segundo, R. and Dunne, J. (1992) *Stockmanship: Improving the Care of the Pig and Other Livestock*. Farming Press Books, Ipswich, UK.

Geverink, N.A., Kappers, A., van de Burgwal, J.A., Lambooij, E., Blokhuis, H.J. and Wiegant, V.M. (1998) Effects of regular moving and handling on the behavioural and physiological responses of pigs to preslaughter treatment and consequences for subsequent meat quality. *Journal of Animal Science* 76, 2080–2085.

Gonyou, H.W., Hemsworth, P.H. and Barnett, J.L. (1986) Effects of frequent interactions with humans on growing pigs. *Applied Animal Behaviour Science* 16, 269–278.

Grandin, T. (1980) Livestock behaviour as related to handling facilities design. *International Journal for the Study of Animal Problems* 1, 33–52.

Grandin, T. (1982/83) Pig behaviour studies applied to slaughter-plant design. *Applied Animal Ethology* 9, 141–151.

Grandin, T. (1987) Animal handling. In: Price, E.O. (ed.) *The Veterinary Clinics of North America: Food Animal Practice*, vol. 3. W.B. Saunders, Philadelphia, pp. 323–338.

Grandin, T. (1988) Livestock handling pre-slaughter. In: *Proceedings of the 34th International Congress of Meat Science and Technology*. Copenhagen, Denmark, pp. 41–45.

Grandin, T. (1990) Design of loading facilities and holding pens. *Applied Animal Behaviour Science* 28, 187–201.

Grandin, T. (1991) Handling problems caused by excitable pigs. In: *Proceedings of the 37th International Congress of Meat Science and Technology*, vol. 1. Kulmbach, Germany.

Grandin, T., Curtis, S.E. and Taylor, I.A. (1987) Toys, mingling and driving reduce excitability in pigs. *Journal of Animal Science* 65 (Suppl. 1), 230 (abstract).

Gray J.A. (1987) *The Psychology of Fear and Stress*. 2nd edn. Cambridge University Press, Cambridge.

Hargreaves, A.L. and Hutson, G.D. (1990) The effect of gentling on heart rate, flight distance and aversion of sheep to a handling procedure. *Applied Animal Behaviour Science* 26, 243–252.

Hemsworth, P.H. and Barnett, J.L. (1987) Human–animal interactions. In: Price, E.O. (ed.) *The Veterinary Clinics of North America: Food Animal Practice*, vol. 3. W.B. Saunders, Philadelphia, pp. 339–356.

Hemsworth, P.H. and Barnett, J.L. (1991) The effects of aversively handling pigs either individually or in groups on their behaviour, growth and corticosteroids. *Applied Animal Behaviour Science* 30, 61–72.

Hemsworth, P.H., and Coleman, G.J. (1998) *Human–Livestock Interactions: the Stockperson and the Productivity and Welfare of Intensively Farmed Animals*. CAB International, Wallingford, UK.

Hemsworth, P.H., Barnett, J.L. and Hansen, C. (1981a) The influence of handling by humans on the behaviour, growth and corticosteroids in the juvenile female pig. *Hormones and Behavior* 15, 396–403.

Hemsworth, P.H., Brand, A. and Willems, P. (1981b) The behavioural response of sows to the presence of human beings and its relation to productivity. *Livestock Production Science* 8, 67–74.

Hemsworth, P.H., Gonyou, H.W. and Dzuik, P.J. (1986a) Human communication with pigs: the behavioural response of pigs to specific human signals. *Applied Animal Behaviour Science* 15, 45–54.

Hemsworth, P.H., Barnett, J.L. and Hansen, C. (1986b) The influence of handling by humans on the behaviour, reproduction and corticosteroids of male and female pigs. *Applied Animal Behaviour Science* 15, 303–314.

Hemsworth, P.H., Barnett, J.L. and Hansen, C. (1987) The influence of inconsistent handling on the behaviour, growth and corticosteroids of young pigs. *Applied Animal Behaviour Science* 17, 245–252.

Hemsworth, P.H., Barnett, J.L., Coleman, G.J. and Hansen, C. (1989) A study of the relationships between the attitudinal and behavioural profiles of stockpersons and the level of fear of humans and reproductive performance of commercial pigs. *Applied Animal Behaviour Science* 23, 310–314.

Hemsworth, P.H., Barnett, J.L., Treacy, D. and Madgwick, P. (1990) The heritability of the trait fear of humans and the association between this trait and the subsequent reproductive performance of gilts. *Applied Animal Behavioural Science* 25, 85–95.

Hemsworth, P.H., Coleman, G.J. and Barnett, J.L. (1991) Reproductive performance of pigs and the influence of human–animal interactions. *Pig News and Information* 12, 563–566.

Hemsworth, P.H., Coleman, G.J. and Barnett, J.L. (1994a) Improving the attitude and behaviour of stockpeople towards pigs and the consequences on the behaviour and reproductive performance of commercial pigs. *Applied Animal Behaviour Science* 39, 349–362.

Hemsworth, P.H., Coleman, G.J. Cox, M. and Barnett, J.L. (1994b) Stimulus generalisation: the inability of pigs to discriminate between humans on the basis of their previous handling experience. *Applied Animal Behaviour Science* 40, 129–142.

Hemsworth, P.H., Barnett, J.L. and Campbell, R.G. (1996a) A study of the relative aversiveness of a new daily injection procedure for pigs. *Applied Animal Behaviour Science* 49, 389–401.

Hemsworth, P.H., Price, E.O. and Bogward, R. (1996b) Behavioural responses of domestic pigs and cattle to humans and novel stimuli. *Applied Animal Behavioural Science* 50, 43–56.

Hill, J.D., McGlone, J.J., Fullwood, S.D. and Miller, M.F. (1998) Environmental enrichment influences on pig behaviour, performance and meat quality. *Applied Animal Behaviour Science* 57, 51–68.

Hinde, R.A. (1970) *Behaviour: a Synthesis of Ethology and Comparative Psychology.* McGraw-Hill Kogakusha, Japan.

Holmes, R.J. (1984) *Sheep and Cattle Handling Skills.* Accident Compensation Corporation, Wellington, New Zealand.

Houpt, K.A. and Wolski, T.R. (1982) *Domestic Animal Behavior for Veterinarians and Animal Scientists.* Iowa State University Press, Ames, Iowa.

Hutson, G.D. (1985) The influence of barley food rewards on sheep movement through a handling system. *Applied Animal Behaviour Science* 14, 263–273.

Hutson, G.D. (1993) Behavioural principles of sheep handling. In: Grandin, T. (ed.) *Livestock Handling and Transport.* CAB International, Wallingford, UK, pp. 127–146.

Jones, R.B. (1993) Reduction of the domestic chick's fear of humans by regular handling and related treatments. *Animal Behaviour* 46, 991–998.

Jones, R.B. and Waddington, D. (1992) Modification of fear in domestic chicks, *Gallus gallus domesticus* via regular handling and early environmental enrichment. *Animal Behaviour* 43, 1021–1033.

Jones, R.B., Mills, A.D. and Faure, J.M. (1991) Genetic and experimental manipulation of fear-related behaviour in Japanese Quail chicks *(Coturnix coturnix japonica).* *Journal of Comparative Psychology* 105, 15–24.

Kilgour, R. (1987) Learning and training of farm animals. In Price, E.O. (ed.) *The Veterinary Clinics of North America: Food Animal Practice,* vol. 3. W.B. Saunders, Philadelphia, pp. 269–284.

Lyons, D.M. (1989) Individual differences in temperament of dairy goats and the inhibition of milk ejection. *Applied Animal Behaviour Science* 22, 269–282.

McFarland, D. (1981) *The Oxford Companion to Animal Behaviour.* Oxford University Press, Oxford, UK.

Mateo, J.M., Estep, D.Q. and McCann, J.S. (1991) Effects of differential handling on the behaviour of domestic ewes *(Ovis aries).* *Applied Animal Behaviour Science* 32, 45–54.

Murphy, L.B. (1976) A study of the behavioural expression of fear and exploration in two stocks of domestic fowl. PhD dissertation, Edinburgh University, UK.

Murphy, L.B. (1978) The practical problems of recognizing and measuring fear and exploration behaviour in the domestic fowl. *Animal Behaviour* 26, 422–431.

Paterson, A.M. and Pearce, G.P. (1989) Boar-induced puberty in gilts handled pleasantly or unpleasantly during rearing. *Applied Animal Behaviour Science* 22, 225–233.

Pearce, G.P., Paterson, A.M. and Pearce, A.N. (1989) The influence of pleasant and unpleasant handling and the provision of toys on the growth and behaviour of male pigs. *Applied Animal Behaviour Science* 23, 27–37.

Prince, J.H. (1977) The eye and vision. In: Swenson, M.J. (ed.) *Dukes Physiology of Domestic Animals*, Cornell, New York, pp. 696–712.

Ravel, A., D'Allaire, S.D., Bigras-Poulin, M. and Ward, R. (1996) Personality traits of stockpeople working in farrowing units on two types of farms in Quebec. In: *Proceedings of 14th Congress of the International Pig Veterinary Society, 7–10 July, Bologna, Italy*. Faculty of Veterinary Medicine, University of Bologna, Bologna, p. 514.

Schaefer, T. (1968) Some methodological implications of the research on early handling in the rat. In: Newton, G. and Levine, S. (eds) *Psychobiology of Development*, Charles Thomas, Springfield, Illinois, pp. 102–141.

Shillito, E.E. (1963) Exploratory behaviour in the short-tailed vole *Microtus agrestis*. *Behaviour* 21, 145–154.

Suarez, S.D. and Gallup, G.G., Jr (1982) Open-field behaviour in chickens: the experimenter is a predator. *Journal of Comparative Physiology and Psychology* 96, 432–439.

Tanida, H., Miura, A., Tanaka, T. and Yoshimoto, T. (1995) Behavioural response to humans in individually handled weanling pigs. *Applied Animal Behaviour Science* 42, 249–259.

Tanida, H., Miura, A., Tanaka, T. and Yoshimoto, T. (1996) Behavioural responses of piglets to darkness and shadows. *Applied Animal Behaviour Science* 49, 173–183.

Toates, F.M. (1980) *Animal Behaviour – a Systems Approach*. John Wiley & Sons, Chichester, UK.

Van Putten, G. and Elshof, W.J. (1978) Observations on the effect of transport on the well-being and lean quality of slaughter pigs. *Animal Regulation Studies* 1, 247–271.

Walker, S. (1987) *Animal Learning: an Introduction*. Routledge and Kegan, London.

Waring, G.H. (1983) *Horse Behaviour. The Behavioural Traits and Adaptations of Domestic and Wild Horses, Including Ponies*. Noyes Publications, Park Ridge, New Jersey.

Warriss, P.D., Kestin, S.C. and Robinson, J.M. (1983) A note on the influence of rearing environment on meat quality in pigs. *Meat Science* 9, 271–279.

# Transport of Pigs

15

## E. Lambooij

*Institute for Animal Science and Health, PO Box 65,
8200 AB Lelystad, The Netherlands*

## Introduction

Several centuries ago, the increasing numbers of inhabitants in towns made it necessary to transport animals to slaughterhouses. In Europe, the farmer walked to the municipal slaughterhouse, leading one or a few animals. Cowboys or shepherds drove large numbers of animals to the towns on the prairies of the USA. Occasionally, animals were transported by ship to new areas and ships would also carry animals as a food store, which, after slaughter, could be used as food for the crew. In the 20th century, walking to market or slaughter was gradually replaced by transport. This ranged from large numbers of cattle transported by train in the West to, in the Far East, single pigs tied and transported by bicycle or trailer to the slaughterhouse.

At present, transport by train is not common, because the animals have to be transported to a station and reloaded, so increasing the adverse effects of loading and lengthening some journeys. However, conditions by train can be very good. The use of aircraft is limited to breeding animals and day-old chicks, because it is expensive. A large number of sheep are shipped from Australia to the Middle East. Cattle are also sometimes transported by ship. The most common means of transport for all farm animals is the road vehicle, even though it is generally found that road transport is worse for the animals than rail, sea or air transport. Pigs are usually transported in large trucks, which may hold over 200 animals. In the European Union (EU) most of these trucks are equipped with a loading lift.

Nowadays, transport distances of farm animals by road to another farm or to the slaughterhouse are expanding, because of the economic consequences of greater opportunities for long-distance and international trade, improved infrastructure and increased demand for live animals for fattening and

©CAB *International* 2000. *Livestock Handling and Transport,* 2nd edn
(ed. T. Grandin)

slaughtering. Within the EU, free movement of animals from one member state to another and more uniformity have resulted in more long-distance travel to slaughter. Regulations to protect animals during transport are laid down in a European Community (EC) Council Directive (1991) and in governmental legislation. Consumers are now demanding better treatment of animals in the whole production chain, including transport. The conditions during transport and the welfare of transported animals are more and more the subject of discussion.

## Welfare

Transport and associated handling always have some adverse affects on the welfare of pigs (van Putten and Lambooij, 1982). Adverse effects are related to psychological, physical, environmental, metabolic and treatment factors (Fig. 15.1). Indicators of poor welfare include behavioural responses indicative of coping ability. Where control systems are overtaxed, the term stress is used (Broom and Johnson, 1993). A well-known disease related to transport is the porcine stress syndrome (PSS) (Tarrant, 1989). This syndrome is the acute reaction to stress, mediated by the sympathetic nervous system, which can cause severe distress and even death. The affected animal shows severe signs of dyspnoea, cyanosis and hyperthermia and may develop rigor in the muscles before death occurs.

During road transport, weather conditions (temperature, air velocity, humidity), loading density and duration of the trip are important factors influencing the condition of the animals (Augustini, 1976; Hails, 1978).

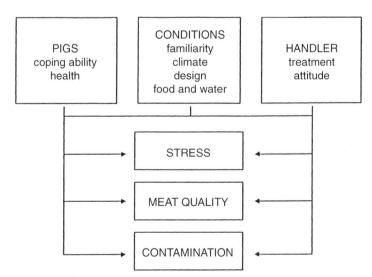

**Fig. 15.1.** Factors that affect stress, meat quality and contamination.

On long trips of 2 or 3 days, pigs are exposed to wide variations in weather conditions. In general, live-weight losses during transport of 1 or 2 days are 40–60 g kg$^{-1}$, whereas the mortality is 0.1–0.4% (Hails, 1978; Holloway, 1980; Grandin, 1981; Markov, 1981; Lambooij, 1983, 1988). For pigs being transported from the farm to a nearby slaughterhouse, death losses ranged from 0.1% to 1.0% (Fabianson *et al.*, 1979; Allen *et al.*, 1980; Warriss, 1998a). Losses increase during hot weather conditions (Smith and Allen, 1976; van Logtestijn *et al.*, 1982). Clark (1979) found that 70% of the Canadian losses occurred on the truck. When the temperature is over 35°C, death losses in 120 kg pigs may rise to 0.27%–0.3% (T. Grandin, 1999, unpublished data).

Vehicle motion and vibration are known to have effects on humans. As well as generating motion sickness, health, comfort, postural stability and ability to perform a task can be severely compromised. It is likely that similar responses occur in pigs; however, the relevant ranges of frequencies could be different (Randall, 1992). Vibration magnitudes are highest on a small, towed, twin-axle trailer as used by farmers to transport up to ten pigs. A small and fixed-body truck provides conditions which could be classified as only a little uncomfortable on the best roads to fairly uncomfortable on minor roads. A large fixed-body truck with air suspension provides a very smooth ride, classified as not uncomfortable to a little uncomfortable (Randall *et al.*, 1996). Low-frequency vibrations may cause pigs to vomit during transport (Randall and Bradshaw, 1998). In piglets a decreasing effect is observed with increasing frequency for a given acceleration, an independence of acceleration at higher frequencies and a large between-animal variability (Perremans *et al.*, 1996). Further information on the effects of vertical vibration on pigs can be found in Perremans *et al.*, (1998).

## Coping Style

Pigs are kept under specific housing conditions for several months, which vary according to the particular production system. After this period, they have to be transported either to another farm or to a slaughterhouse. Individual pigs respond in different ways to stress factors. The response is dependent on the genotype, coping style, treatment and experience of the animal. On the basis of their response, pigs can be divided into groups in which different behavioural and physiological characteristics may be noted. Pigs may be divided into active and passive individuals based on behavioural tests. The normal cortisol value in blood is higher in passive animals compared with active ones. However, after an adrenocorticotrophic hormone (ACTH) challenge, the value is equal in both groups. When piglets are placed on their back ('back test') the active animals show a higher increase in heartbeats than passive animals (Hessing, 1994). However, it should be noticed that behavioural characteristics vary in a population (Jensen *et al.*, 1995). Based on agonistic interactions, pregnant sows may be divided into three groups of animals with high success in winning, low success and no success. The normal cortisol value is highest in animals with low success, and these animals respond highest on the ACTH challenge test (Mendle *et al.*, 1992).

Coping style may correlate with response to transport and associated conditions. Transport conditions involve exposure to social stress (e.g. mixing with unfamiliar pigs) and non-social stress (rough handling). Individual differences in behaviour in home-pen conditions and during mild challenge tests may be related to subsequent reaction during transport, driving and mixing (Geverink, 1998; Geverink *et al.*, 1998). During fattening, when the pigs are aged between 14 and 20 weeks, the piglets can be tested for social status by scoring the agonistic interactions or by using a 'food competition test'. In the first test, social ranking of individuals may be determined by allowing a total of 4 h of focally sampled data on agonistic interactions (bite, head knock, threat, displace, avoid). In the second test, the pigs have no access to food for 29 h prior to the test. At the start of the test, a fixed amount of food is placed in the trough. For 15 min, interactions and frequency and duration of eating are scored. Finally, an 'open door test' may be used to test the individual activity and exploration. In this test, the door of a pen is opened and the reaction of pigs is scored. Pigs that were regularly given the opportunity to leave their pen and in addition were accustomed to transport showed increased willingness to move during preslaughter treatment. Pigs that are easier to move are less likely to be subject to rough handling, which implies improved welfare, while the workload for stockmen is reduced (Geverink, 1998)

The most detrimental effect of coping style is death, which normally follows a period of very poor welfare. Stress activates hormones via the pituitary–adrenal system (glucocorticoids) and the sympathetic–adrenal medullary system (cathecholamines), resulting in behavioural (overreaction to normal stimuli) and clinical (increase in heart and respiration rate) deviations from normal functioning, followed by exhaustion. The death rate following transport varies between 0.1 and 1% (Warriss, 1998a).

## Thermoregulation

The variation in temperatures encountered by the pigs during transport may increase up to approximately 20°C. This variation in temperature within the vehicle is related to variation in temperature outside (Lambooij, 1988). Therefore ventilation rate during transport should be adapted to the inside temperature, which is the resultant of heat flowing from outside to inside and heat produced by the animals. Data of heat production at climatic conditions that occur during transport of slaughter pigs are not known. Little quantitative information is available on thermal thresholds during transport (Schrama *et al.*, 1996).

Pigs are homoeothermic animals and maintain a constant body temp-erature by balancing heat loss and heat production, as presented in Fig. 15.2 and Table 15.1. Animals may maintain a constant body temperature in zone AC. Heat loss is kept constant by regulation of both sensible and evaporative heat losses in the thermoneutral zone BC. Factors such as feeding level, physical activity and stress determine the heat production. Mechanisms to reduce and control heat loss are depleted in zone AB. In order to maintain homoeothermia, the animal has to increase its heat production. Climatic conditions in the thermoneutral zone are optimal for the animal (Schrama *et al.*, 1996). Heat production at maintenance can be assumed normally at about 420 kJ kg$^{-0.75}$ day$^{-1}$ (Holmes and Close, 1975). At feeding time, heat production will increase with about 30% (van der Hel *et al.*, 1986). Lambooij *et al.* (1987) simulated a 2-day transport of slaughter pigs, which were non-fed and held at a loading density of 225 kg m$^{-2}$. It was calculated that the animals

**Table 15.1.** Calculated effect of age on the thermoneutral zone of individually housed pigs fed at maintenance and under standard environmental conditions (Verstegen, 1987).

|  | Live-weight (kg) | Temperature (°C) |
|---|---|---|
| Piglets | 2 | 31–33 |
| Growing piglets | 20 | 26–33 |
| Feeder pigs | 60 | 24–32 |
| Slaughter pigs | 100 | 23–31 |

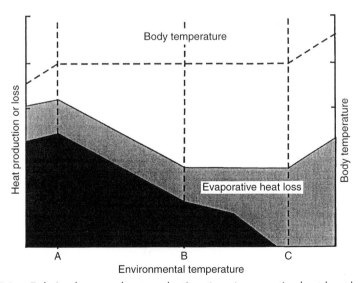

**Fig. 15.2.** Relation between heat production, (non-)evaporative heat loss, body temperature and environmental temperature (from Mount, 1974).

lose 824–944 g body fat during exposure. The loss at 16°C was lowest. The metabolic rate was, on average, above the maintenance requirement as normally assumed. The mean heat production at 16°C was 551 kJ kg$^{-0.75}$ day$^{-1}$. The animals have produced this heat as a result of their maintenance and response to their environment. The heat production increased during light and decreased during darkness, with a minimum early in the morning (Fig. 15.3). Heat production values during environmental temperatures of 8 and 24°C tended to be higher than during 16°C. This extra heat production may be due to some extra activity (Fig. 15.3). It appeared that 8°C is below thermoneutrality, which agrees with data derived from literature (Holmes and Close, 1975). A higher weight loss during an air velocity of 0.8 m s$^{-1}$ compared with 0.2 m s$^{-1}$ may be related to a lower water consumption and a higher heat production. It is assumed that 16°C and an air velocity of 0.2 m s$^{-1}$ are at thermoneutrality (Lambooij et al., 1987).

During transport the ventilation rate cannot be altered, in general. Therefore a solution might be the use of adjusted vents or an artificial ventilation system (Lambooij, 1988). At the moment, different ventilation systems are in development (Barton Gade et al., 1996b).

## Meat Quality

During loading and unloading transport, injuries and bruising commonly occur in all animal species (Grandin, 1990). These defects occur by forceful contacts in passageways, compartments and containers, by fighting between

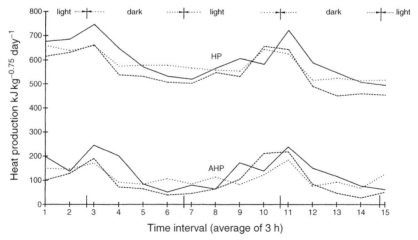

**Fig. 15.3.**   Heat production (HP) and activity-related heat production (AHP) of non-fed slaughter pigs in a calorimeter at an environmental temperature of 8°C (———), 16°C (– – –), and 24°C (·····) (from Lambooij et al., 1987).

animals and by mounting (Connell, 1984). Skin blemish is a serious commercial problem. The skin blemish score reflects the amount of fighting which pigs have indulged in preslaughter (Barton Gade *et al.*, 1996a). It is observed that 63% of 5484 carcasses from pigs in the EU had some damage; however, only in about 10% of carcasses was this moderate to severe. Based on the association between the level of skin blemish and ultimate increased muscle pH values, a probable factor contributing to this is fighting between mixed groups of unfamiliar animals (Warriss *et al.*, 1998a).

Loss of live- and carcass weight is the result of excretion, evaporation and respiratory exchange (Dantzer, 1982; Warriss, 1993) and is a normal physiological reaction. However, when food and water are withdrawn for a longer time, dehydration and mobilization of fat and muscle glycogen may occur (Warriss and Bevis, 1987; Tarrant, 1989; Warriss *et al.*, 1989).

Transport conditions may affect post-mortem meat quality, either via adrenal or other stress responses or by fatigue of the animals. Important meat quality parameters related to stress and exhaustion before slaughter are pH, rigor mortis, temperature, colour and water-binding capacity. Acidity of the meat after slaughter is caused by a breakdown of glycogen to lactate. The rigor mortis value is indirectly caused by a decrease of the energy store. Temperature is increased by chemical metabolism (Sybesma and van Logtestijn, 1967). Colour and water-binding capacity are determined by protein denaturation, caused by a rapid acidification after death (Tarrant, 1989). The rate of acidification after death is controlled by the degree of hormonal and contractile stimulation of muscle immediately before and during slaughter, while muscle temperature at death and rate of cooling are also important (Warriss, 1987; Tarrant, 1989; Monin and Ouali, 1992).

It is assumed that stress before slaughter leads to an increased breakdown of glycogen and a greater decrease in the energy store and thus a rapid acidification, an earlier and increased rigor mortis value and an increased post-mortem muscle temperature. In pigs, this results in pale, soft and exudative (PSE) meat (Tarrant, 1989). This explanation is too simply postulated, because the physiological response to stress factors from the environment is partly influenced by the genotype (coping ability) of the animal (see Fig. 15.1). Slaughter pigs of different genotypes from the same production unit that are subjected to identical preslaughter handling may show different values of meat quality parameters (Tarrant, 1989; Klont *et al.*, 1993; Klont and Lambooij, 1995a,b) and stress-resistant Hampshire pigs can have a low water-binding capacity (Monin and Sellier, 1985). Information from about 5500 pigs killed in the EU has shown large differences in the prevalence of potentially PSE meat. There are no observed apparent relationships between indices of stress and characteristics associated with PSE meat. In contrast, greater stress tended to be reflected in more dark, firm and dry (DFD) meat (Warriss *et al.*, 1998). The occurrence of DFD meat is more readily attributable to effects of the transport environment and is less variable amongst genetic lines. It occurs when the animals are fatigued. In this case, the glycogen energy store is exhausted at slaughter, resulting in no acidification, an increase in rigor mortis value and dark-coloured meat. In well-fed rested animals, meat pH falls to about 6.0–6.5, the rigor mortis value is below 10 and the temperature is below 40°C about 45 min post-mortem (Sybesma and van Logtestijn, 1967; Klont et al., 1993).

Meat quality is influenced by an increased muscle temperature (Klont and Lambooij, 1995a). The normal range in resting pigs varies between 37.0 and 39.6°C (Hannon et al., 1990). An increase to the upper level of the normal range may already cause an increased incidence of PSE meat, especially in stress-susceptible pigs (Klont and Lambooij, 1995a). Preslaughter stress factors, such as exercise and high ambient temperatures, might easily elevate the body temperature up to values between 39.0 and 41.0°C, with the highest increases in stress-susceptible pigs compared with stress-resistant pigs (Aberle et al., 1974; Lundström, 1976; Gariepy et al., 1989; Geers et al., 1992). Higher post-mortem muscle temperatures, in combination with an increased lactate formation after normal slaughter conditions, will lead to a greater incidence of PSE meat. Showering and resting pigs for 2–4 hours will help to reduce PSE (Malmfors, 1982; Smulders et al., 1983). When the period is longer, the percentage of carcasses with DFD may increase (Verdijk, 1974; Culau et al., 1991; Warriss, 1993). Stress before slaughter also affects the microbiological contamination in the live animal by influencing the meat quality, which may result in a more contaminated carcass. In PSE and DFD carcasses, micro-organisms can grow better to a great extent.

## Contamination

Animals require proper preparation before transport. This means that, in pigs, feed should be withheld for 16–24 h (Eikelenboom et al., 1990; Warriss, 1993). Other advantages of feed withdrawal are less labour at slaughter, less contamination of the carcass and a lower percentage of PSE meat (Eikelenboom et al., 1990, Warriss, 1993). It is thought that, after great physical and psychological labour in clinically healthy animals carrying Salmonella and other pathogenic microorganisms, the excretion pattern from the intestinal tract may be changed from intermittent to constant shedding. This disturbance may also lower the immunological response and facilitate the spreading of intestinal bacteria. Feeding and environmental conditions during transport and lairage, including the total time involved and mixing animals from several herds, have been shown to be the main factors.

Experiments show that from pigs to be delivered to the slaughterhouse no Salmonella were isolated, but after delivering to the abattoir 0.1% of the samples were positive, while after slaughter this percentage had increased to 0.7%. It was concluded that stress factors were responsible for the increase in the carrier percentage (Slavkov et al., 1974). When pigs stayed in lairage for a longer time, in larger pens and in worse hygienic conditions (cross-)contamination also increased. Carcass contamination was caused by *Salmonella* of intestinal origin, as demonstrated by the *Salmonella* recovery rate and the *Salmonella* serotypes from caecal contents and the carcass surface (Morgan *et al.*, 1987). As carcass contamination was determined by intestinal *Salmonella* entering the slaughterhouse with the pigs, a very important strategy may be reduction of contamination, by improved preslaughter handling avoiding any form of multiplication of *Salmonella* in the live animals (Huis in't Veld *et al.*, 1994).

The mechanism of spreading microorganisms by stressed carrier animals is not clear. However, it is known that preslaughter conditions affect the contamination rate of the product after slaughter. Thus, it is recommended to pay more attention to the procedures before transport, while loading during transport and in lairage (Mulder, 1995).

Cleaning and sanitizing trucks reduced *Salmonella* from 41.5% of the samples collected from truck floors to 2.77% (Rajkowski *et al.*, 1998). There was no significant difference in the number of *Salmonella*- or *Escherichia coli*-positive trailers attributed to the distance travelled or the season of the year (Rajkowski *et al.*, 1998).

## Procedures at Loading and Unloading

Farm animals are kept under specific housing conditions for several months, the exact time depending on the species and production system. After that period, the animals have to be transported to a slaughterhouse. Transport causes physical and behavioural problems, because the animals are not accustomed to transport conditions and procedures. The loading procedures, the design and the other animals are unfamiliar to the animals and will frighten them. The drivers may not treat the animals in a proper way to minimize stress and sometimes animals are mixed or regrouped, thus increasing stress and resulting in fighting amongst the animals to determine social order (Fig. 15.4; Connell, 1984).

Climbing a loading ramp is easy for horses, cattle and sheep when ramp design and handling procedures are good. For pigs, climbing a loading ramp is difficult, since the situation is often psychologically disturbing. The animals may simply refuse to try and even turn their sides towards the ramps. As a result, the heart rate may increase to a level where the heart starts to lose synchronization. The angle of the loading ramp should not be greater than 15–20° (van Putten and Elshof, 1978; Phillips *et al.*, 1988; Fraser and Broom, 1990). Descending a loading ramp steeper than 20° is difficult for all animals and should be avoided (Grandin, 1981).

Correct treatment by the driver is detailed below:

- Load and unload the animals quietly and do not use electric goads. Let the animals observe their environment and let them go from reduced light levels to better-illuminated areas (Fig. 15.4).

**Fig. 15.4.**   Unloading of slaughter pigs quietly and without use of electric goads.

- The passageways should be of sufficient width and solid. Steel projections and channels in the walls are not acceptable. The slope of the ramp should not exceed 15–20°.
- Groups of animals must be kept stable and limited to avoid fighting and stress when put with unfamiliar animals. This means keeping animals from one rearing pen together and not mixing them, even with others from the same farm. The number of farms where the animals have to be fetched from should be as low as possible.
- The loading density should be correct (Table 15.2), while the microclimate should be adapted to the species and ambient weather conditions.
- Food and water should be available at appropriate times.
- The vehicle should be driven carefully so that there is no sudden acceleration, braking or sideways movement, which causes animals in compartments or containers to be thrown around or unduly disturbed. Transport on highways or bigger roads is preferred to urban traffic.

# Treatment During Transport

## Feeding and watering

Animals should be fit for the intended journey irrespective of the purpose of the journey. Pigs are likely to suffer motion sickness during road transport, which may result in vomiting after eating 4 h before transport (Bradshaw *et al.*, 1996a,b). For this reason, pigs require careful preparation before transport and comprehensive plans for the journey should be made. This means that pigs' feed should be withheld for 16–24 h before slaughter (Warriss and Brown, 1983; Eikelenboom *et al.*, 1990). Depending on the distance to the slaughterhouse or transit station, feeding should be stopped the night before transport, but water should be available. It is suggested that pigs suffer motion sickness as a result of low-frequency vibrations (Randall, 1992) and may vomit when the stomach is full. The pigs may die by choking and by inhalation of their own vomit (Guise, 1987). Other advantages of feed withdrawal are less labour at slaughter, less contamination of the carcass, a decreased weight loss of the carcass during chilling and a lower percentage of PSE meat (Eikelenboom *et al.*, 1990; Warriss, 1993).

During stops in transport of long duration, piglets were observed to drink from bite nipples (Barton-Gade and Vorup, 1991). Observations of slaughter pigs during such journeys showed that pigs drank only 0.65 or 1.6 litres water per pig when available via bite nipples during motion or in a trough after unloading (Lambooij, 1983, 1984). If feed was supplied, vomiting was observed. There were no physiological differences between watered and non-watered animals, due to mobilization of water within the fat cells (Lambooij *et al.*, 1985). However, an increased blood protein concentration was found in pigs after 6 h transport (Warriss *et al.*, 1983).

The stress of handling, transport and fasting lowered blood glucose level; however, additional energy was obtained from fat breakdown. Most liver glycogen was utilized in the first 18 h of transport and fat was broken down 9 h after feed withdrawal (Warriss and Brown, 1983). Because of the physiological changes, the EC Working Group (1992) recommended that pigs should be given water every 8 hours and the transport should be limited to 24 h.

## Loading density

Animals must be able to stand in their natural position and all must be able to lie down at the same time. For animals which may stand during the journey, the roof must be well above the heads of all animals when they are standing with their heads up in a natural position. This height will ensure adequate freedom of movement and ventilation and will depend on the species and breed concerned (EC Working Group, 1992). Loading density has a major effect on animal welfare and post-mortem meat quality. At loading densities of over 200 kg m$^{-2}$ pigs showed increased body temperature, heart rate and breathing frequency after a short journey and a high frequency of PSE meat post-mortem (Heuking, 1988; von Mickwitz, 1989). When the loading density is higher than 235 kg m$^{-2}$, not all pigs are able to lie down, hence there is a continual changing of positions and the pigs cannot rest (Fig. 15.5). The consequences are more skin blemishes and rectal prolapses and bad meat quality (Lambooij et al., 1985; Guise and Penny, 1989; Lambooij and Engel, 1991). A loading density for slaughter pigs of 235 kg m$^{-2}$ is suggested as a compromise between animal welfare, meat quality and the economics of transport (Fig. 15.6). The loading densities for different weight groups and species are presented in Table 15.2.

Recent studies on pig loading density indicates that a density of 321 kg m$^{-2}$ results in clear evidence of physical stress and a density of 250 kg m$^{-2}$ is the minimum requirement for pigs to rest in sternal recumbancy (Warriss et al., 1998). There may be a need to differentiate loading densities for long and short journeys. On longer journeys, pig welfare will be severely compromised if they are unable to lie down, but on short trips during cool weather they can remain standing. Guise et al. (1998) reported that pigs remain standing during short journeys of 3 hours. For such short journeys, there was relatively little evidence of adverse effects with a loading density of 281 kg m$^{-2}$, which is 0.35 m$^2$ per pig of 100 kg. On short journeys, the source of the pigs had a greater effect on carcass quality and welfare than the loading

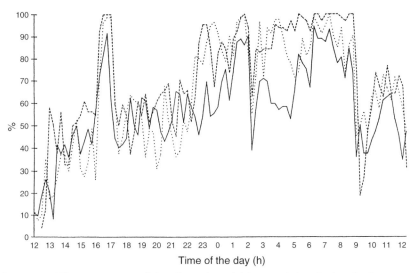

**Fig. 15.5.** The percentage of slaughter pigs which were lying during the journey. The loading densities were 186 (– – –), 232 (······) and 278 kg m$^{-2}$. Stops in transport were between 16.00 and 17.00 h, 1.00 and 2.00 h and 6.00 and 9.00 h, while the driver was changed at 21.00 h (from Lambooij and Engel, 1991).

**Fig. 15.6.** During transport of long duration pigs sit or lie down when the conditions are suitable.

density. The source farm had a significant effect on measurements of meat quality.

Practical experience in the USA has shown that during hot summer weather with high temperatures and humidity, pigs must be loaded according to the recommendations in Table 15.2 to prevent death losses from heat stress. Temperatures in the southern and midwestern USA are much hotter than the temperatures in most regions of the EU. Differences in the results between different studies are likely to be partly due to differences in climate.

## Microclimate

The effect of climatic conditions is difficult to measure. The weather conditions are dependent on the location, the time of the day and the season. During hot weather conditions, the number of transport deaths is increased in pigs (van Logtestijn *et al.*, 1982). Experiments in climate-controlled calorimeters showed that the lowest heat production ante-mortem and best meat quality post-mortem in pigs occurred at an environmental temperature of 16°C and an air velocity of 0.2 m s$^{-1}$ (Lambooij *et al.*, 1987).

The most common method of ventilating compartments and containers is via vents positioned at the upper part of the left and right sides. During the journey, the correlation between the outside and inside temperature was positive and significant, but for humidity this was not the case. The temperature during stops increased by 1–4°C in the compartments (Lambooij,

**Table 15.2.** Loading densities recommended by the EC Working Group (1992).

|                | Live-weight (kg) | $m^2$ animal$^{-1}$ | Animals $m^{-2}$ |
| -------------- | ---------------- | ------------------- | ---------------- |
| Piglets        | < 25             | 0.15                | 6.60             |
| Feeder pigs    | 60               | 0.35                | 2.80             |
| Slaughter pigs | 100              | 0.42                | 2.35             |
| Heavy pigs     | 120              | 0.51                | 1.96             |

1988). Thus the microclimate depends on the ambient weather conditions. Placing covers at the holes, which can be opened or closed, can vary the ventilation. In pigs this variable ventilation only improved meat quality when it was combined with showering during the journey (Lambooij and Engel, 1991).

Artificial ventilation in compartments and containers may improve conditions during transport. When pigs were ventilated artificially during transport, the rigor mortis value post-mortem was decreased, which pointed to a decreased energy loss in the muscles (Lambooij, 1988). For breeding pigs, air conditioning during transport is developed and applied during short and long journeys by road. The air velocity is low and the temperature is held at 16°C. At the moment, different ventilation systems for piglets and slaughter pigs are in development (Barton Gade *et al.*, 1996b).

### Air, railway and ship transport

Specially designed containers are used to ship young breeding pigs by air. Each container holds 25–30 piglets. The floor of the containers is covered with a thick layer of sawdust (approx. 10 cm). Each container should have at least two drinking nipples with sufficient water supply. It is usual to tranquillize piglets during transport to calm them down; however, this practice is not recommended. Pigs in stressful conditions are calmed by the presence of their pen mates. Another comforting factor is a familiar well-lighted environment. To prevent fighting during air transport, pigs should be grouped into groups that will fly together in a container at the farm of origin. Dust from dry sawdust might cause frequent sneezing. Heat stress occurs when the plane is on the ground. To prevent death losses at destination, prompt unloading of the containers is essential. Upon arrival at the destination airport, the pigs have to be transported to the farm of destination by truck. The piglets may suffer from jet lag. Their diurnal rhythm is completely out of phase. They gradually become accustomed to a new rhythm. To reduce death losses during air transport shippers should: (i) avoid use of tranquillizers; (ii) premix pig groups; (iii) provide an adequate water-supply; (iv) keep a dim light on during the flight; (v) control dust; and (vi) avoid heat stress (G. van Putten, personal communication).

Transport by rail is often condemned because it takes much more time than transport by road. However, rail transport is considered to have positive effects on animal welfare. Since the trip is longer, there has to be an accompanying attendant to look after the animals and provide them with water and feed. Jackson (1973) reported that pigs eat and drink during transport and they gain weight instead of losing it. Railway vans have to be provided with artificial ventilation, because they may stand in sunshine for hours. In the USA, there is one slaughterhouse to which slaughter pigs are transported almost 2000 km by rail. The pigs are fed and watered from large troughs in the car. The animals arrive at their destination in good condition (T. Grandin, personal communication).

Transport by ship is not feasible, because it takes too long and it is difficult to deliver the pigs to inland destinations. It is sometimes used in a roll-on roll-off situation, where the truck with pigs is placed on a ferry to transport them to an island. The truck is placed on the upper deck to guarantee ventilation.

## Handling During Transit or Lairage

### Passageways

After arrival at the slaughterhouse or transit station, the animals need to be unloaded carefully and as soon as possible, because ventilation in stationary vehicles is often not good. Loading ramps at the height of the deck are necessary because all species have problems with descending loading ramps. Another possibility is the use of a deck lift, in which the whole deck moves upwards or downwards. The passageways need to be solid, while projections and channels should be avoided. The floors must not be allowed to be slippery (Grandin, 1990). Different colours and shadows may frighten the animals and they walk most easily from a dark to a lighter place (van Putten and Elshof, 1978; Grandin, 1990). Electric goads are in common use in lairage and in transit. However, these goads cause stress to the animals and should be banned (van Putten and Elshof, 1978; Grandin, 1988). Moreover, petechial haemorrhages can then be reduced (Grandin, 1988). In pigs, the width of the passageway is sufficient when four to five animals can walk side by side. Groups of approximately 15 animals should be driven. With such a group, the use of force, such as an electric goad on the last animals, is not effective. Unwanted returning of animals can be reduced by using gates (Grandin, 1990). A fully automatic system, which is controlled by a computer, has been developed in Denmark (Barton-Gade *et al.*, 1992). Special designs of race to drive pigs to the stunning place have been developed. The pigs are brought into a single file step by step (Hoenderken, 1976) or via a curved crowd pen to a double race ramp with an entrance restrictor to prevent jamming (Grandin, 1990).

**Waiting pens**

In transit stations, pigs have to be showered and to rest a few hours. Before long journeys, they can be fed beforehand with a thin porridge consisting of one part feed (high sugar content) and three parts water in order to reduce weight loss. A resting period of 2–4 h before slaughter of pigs is recommended. This may result in a lower percentage of carcasses with PSE meat. When the period is longer, the percentage of carcasses with DFD may increase (Verdijk, 1974; Culau *et al.*, 1991; Warriss, 1993).

During transit or lairage the pigs should be showered intermittently. Showering has the following advantages: (i) the stress reaction as a result of the transport is reduced, due to the calming and cooling effect; (ii) fighting will be reduced, resulting in less skin damage; and (iii) the animals are cleaned (van Putten *et al.*, 1983; Tarrant, 1989; Warriss, 1993). Practical experience and experiments have shown that showering is most effective during hot weather conditions and that it should not be used when the temperature drops below 11°C (T. Grandin, personal communication; Santos *et al.*, 1997). Showering should cease if pigs are seen to be shivering (Knowles *et al.*, 1998). A loading density of two pigs m$^{-2}$ is recommended, because higher densities diminish meat quality (van Putten *et al.*, 1983). Larger pigs of 120 kg will require more space. All pigs should have room to lie down.

## Summary

- Transport and associated handling of animals always have some adverse effects on animal welfare and health, due to inadequate human behaviour, environmental conditions and the inability of the animal to cope with the situation,
- Governments, transport organizations and research institutes have their programmes for improvement of transport conditions. However, more effort is needed to stimulate this process, which is too slow.
- Truck drivers should be aware that the quality of transport of animals has to be improved because of consumers' demands. Correct facilities for loading, unloading, design of compartments and environmental conditions should be available. Drivers should be educated, trained and licensed. Courses should be taught in each country.
- A minimal acceptable level of effects of these procedures on animals should be specified for each species and laid down in regulations. It is necessary that government inspectors control the transport procedures during loading, movement, unloading, transit and lairage, independent of borders.

# References

Aberle, E.D., Merkel, R.A., Forrest, J.C. and Alliston, C.W. (1974) Physiological responses of stress susceptible and stress resistant pigs to heat stress. *Journal of Animal Science* 38, 954–959.

Allen, W.M., Herbert, C.N. and Smith, L.P. (1980) Death during and after transportation of pigs in Great Britain. *Veterinary Record* 94, 212–214.

Augustini, C. (1976) ECG – und Körper Temperature Messungen an Schweinen während der Mast und auf dem Transport. *Fleischwirtschaft* 56, 1133–1137.

Barton Gade, P. and Vorup, P. (1991) *Investigation into Various Aspects of Long Distance Transport in Pigs.* Danish Meat Research Institute Report 02.487/93.

Barton Gade, P., Blaajberg, L. and Christensen, I. (1992) New lairage system for slaughter pigs: effect on behaviour and quality characteristics. In: *Proceedings 38th International Congress of Meat Science and Technology* Clermont-Ferrand. pp. 161–164.

Barton Gade, P., Warriss, P.D., Brown, S.N. and Lambooij, B. (1996a) Methods of improving pig welfare and meat quality by reducing stress and discomfort before slaughter – methods of assessing meat quality. In: *Proceedings EU Seminar: New Information on Welfare and Meat Quality of Pigs as Related to Handling, Transport and Lairage Conditions.* FAL-Sonderheft 166, Völkenrode, pp. 23–34.

Barton Gade, P., Christensen, L., Brown, S.N. and Warriss, P.D. (1996b) Effect of tier and ventilation during transport on blood parameters and meat quality in slaughter pigs. In: *Proceedings EU Seminar: New Information on Welfare and Meat Quality of Pigs as Related to Handling, Transport and Lairage Conditions.* FAL-Sonderheft 166, Völkenrode, pp. 101–116.

Bradshaw, R.H., Parrott, R.F., Goode, J.A., Lloyd, D.M., Rodway, R.G. and Broom, D.M. (1996a) Behavioural and hormonal responses of pigs during transport: effect of mixing and duration of journey. *Animal Science* 62, 547–554.

Bradshaw, R.H., Parrott, R.F., Forsling, M.L., Goode, J.A., Lloyd, D.M., Rodway, R.G. and Broom, D.M. (1996b) Stress and travel sickness in pigs: effects of road transport on plasma concentrations of cortisol, beta-endorphin and lysine vasopressin. *Animal Science* 63, 507–516

Broom, D.M. and Johnson, K.G. (1993) *Stress and Animal Welfare.* Chapman and Hall, London.

Clark, E.G. (1979) A post mortem survey of transport deaths in Saskatchewan market hogs. *Western Hog L.* 1, 34–36.

Connell, J. (1984) *International Transport of Farm Animals Intended for Slaughter.* Commission of the European Community, Brussels.

Culau, P.O.V., Ourique, J.M. and Nicolaiewsky, S. (1991) The effect of transportation distance and pre-slaughter lairage time on the pig meat quality. In: *Proceedings 37th International Congress on Meat Science Technology.* Kulmbach. pp. 224–228.

Dantzer, R. (1982) Research on farm animal transport in France: A survey. In: Moss, R. (ed.) *Transport of Animals Intended for Breeding, Production and Slaughter.* Current Topics Veterinary Medicine and Animal Science, vol. 18, Martinus Nijhoff, The Hague, pp. 218–231.

EC Council Directive (1991) On the protection of animals during transport. *Official Journal of the European Community,* 1. 340/17.

EC Working Group on Transport of Farm Animals (1992) *Report of the Scientific Veterinary Commission on Animal Welfare Section.* VI/3404/92., Brussels.

Eikelenboom, G., Bolink, A.H. and Sybesma, W. (1990) Effects of feed withdrawal before delivery on pork quality and carcass yield. *Meat Science* 29, 25–30.

Fabianson, S. Lundström, K. and Hansson, I. (1979) Mortality among pigs during transport and waiting time before slaughter in Sweden. *Swedish Journal of Agricultural Research* 9, 25–28.

Fraser, A.F. and Broom, D.M. (1990) *Farm Animal Behaviour and Welfare.* Baillière Tindall, London.

Gariepy, C., Amiot, J. and Nadai, S. (1989) Ante-mortem detection of PSE and DFD by infrared thermography of pigs before stunning. *Meat Science* 25, 37–41.

Geers, R., Ville, H., Janssens, S., Goedseels, V., Goosens, K., Parduyns, G., van Bael, J., Bosschaerts, L. and Heylen, L. (1992) Changes of body temperatures of piglets as related to halothane sensivity and treadmill exercise. *Journal Thermal Biology* 17, 125–130.

Geverink, N.A. (1998) Preslaughter treatment of pigs; consequences for welfare and meat quality. Thesis, Wageningen.

Geverink, N.A., Bradshaw, R.H., Lambooij, E., Wiegant, V.M. and Broom, D.M. (1998) Effects of simulated lairage conditions on the physiology and behaviour of pigs. *Veterinary Record* 143, 241–244.

Grandin, T. (1981) *Livestock Trucking Guide.* Livestock Conservation Institute, Madison, Wisconsin.

Grandin, T. (1988) Effect of temperature fluctuation on petechial haemorrhages in pigs. In: *Proceedings Workshop on Stunning of Livestock, 39th International Congress on Meat Science and Technology.* Brisbane, pp 23–26.

Grandin, T. (1990) Design of loading facilities and holding pens. *Applied Animal Behaviour Science* 28, 187–201.

Guise, H.J. (1987) Moving pigs from farm to factory. *Pig International* December, 8–12.

Guise, H.J. and Penny, R.H. (1989) Factors influencing the welfare and carcass meat quality of pigs. *Animal Production* 49, 511–515.

Guise, H.J., Riches, H.L., Hunter, E.J., Jones, T.A., Warriss, P.D. and Kettlewell, P.J. (1998) The effect of stocking density in transit on the carcass quality and welfare of slaughter pigs: 1. Carcass measurements. *Meat Science* 50(4), 439–446.

Hails, M.H. (1978) Transport stress in animals: a review. *Animal Regulation Studies* 1, 289–343.

Hannon, J.P., Bossone, C.A. and Wade, C.E. (1990) Normal physiological values for conscious pigs used in biomedical research. *Laboratory Animal Science* 40, 293–298.

Hessing, M.J.C. (1994) Individual behavioural characteristics in pigs and their consequences for pig husbandry. Thesis, Utrecht.

Heuking, L. (1988) Die Beurteilung des Verhaltens von Schlachtschweinen bei LKW-Transporte in Abhängigkeit von der Ladedichte mit Berücksichtigung des Blutbildes, der Herzfrequenz und der Körpertemperatur zur Erfassung Tierschutzwidriger Transportbedingungen. Thesis, Berlin

Hoenderken, R. (1976) Ein verbessertes System für das Treiben von Schlachtschweinen zum Restrainer. *Fleischwirtschaft* 56, 838–839.

Holloway, L. (1980) The Alberta Pork Producers Marketing Board Transit Indemnity Fund. In: *Proceedings Livestock Conservation Institute,* Madison, Wisconsin.

Holmes, C.W. and Close, W.H. (1995) The influence of climatic variables on energy metabolism and associated aspects of productivity in the pig. In: Haresign, W.,

Swan, H. and Lewis, D. (eds) *Nutrition and the Climatic Environment*. Butterworths, London, pp. 51–74.

Huis in't Veld, J.H.J., Mulder, R.W.A.W. and Snijders, J.M.A. (1994) Impact of animal husbandry and slaughter technologies on microbial contamination of meat: monitoring and control. *Meat Science* 36, 123–154.

Jackson, W.T. (1973) To Italy – with 120 pigs. *Veterinary Record* 92, 121–122.

Jensen, P., Rushen, J. and Forkman, B. (1995) Behavioural strategies or just individual variation in behaviour? – A lack of evidence for active and passive piglets. *Applied Animal Behaviour Science* 43, 135–139.

Klont, R.E. and Lambooij, E. (1995a) Influence of preslaughter muscle temperature on muscle metabolism and meat quality in anesthetized pigs of different halothane genotypes. *Journal Animal Science* 73 (1), 96–107.

Klont, R.E. and Lambooij, E. (1995b). Effects of preslaughter muscle exercise on muscle metabolism and meat quality studied in anesthetized pigs of different halothane genotypes. *Journal Animal Science* 73(1), 108–117.

Klont, R.E., Lambooij, E. and van Logtestijn J.G. (1993) Effect of pre-slaughter anesthesia on muscle metabolism and meat quality of pigs of different halothane genotypes. *Journal Animal Science* 71, 1477–1485.

Knowles, T.G., Brown, S.N., Edwards, J.E. and Warris, P.D. (1998) Ambient temperature below which pigs should not be continuously showered in lairage. *Veterinary Record* 143(21), 575–578.

Lambooij, E. (1983) Watering pigs during 30 hours road transport through Europe. *Fleischwirtschaft* 63, 1456–1458.

Lambooij, E. (1984) Watering and feeding pigs during road transport for 24 hours. In: *Proceedings 30th European Meat and Meat Research Workers*. Bristol. pp. 6–7.

Lambooij, E. (1988) Road transport of pigs over a long distance: some aspects of behaviour, temperature and humidity during transport and some effects of the last two factors. *Animal Production* 46, 257–263.

Lambooij, E. and Engel, B. (1991) Transport of slaughter pigs by truck over a long distance: some aspects of loading density and ventilation. *Livestock Production Science* 28, 163–174.

Lambooij, E., Garssen, G.J., Walstra, P., Mateman, G. and Merkus, G.S.M. (1985) Transport by car for two days: some aspects of watering and loading density. *Livestock Production Science* 13, 289–299.

Lambooij, E., van der Hel, W., Hulsegge, B. and Brandsma, H.A. (1987) Effect of temperature on air velocity two days pre-slaughtering on heat production, weight loss and meat quality in non-fed pigs. In: Verstegen, M.W.A. and Henken, A.M. (eds)*Energy Metabolism in Farm Animals: Effect of Housing, Stress and Disease*. Martinus Nijhoff, The Hague, pp. 57–71.

Lundström, K. (1976) Repeatability of response to heat treatment in pigs and the correlation between the response and meat quality after slaughter. *Swedish Journal Agricultural Research* 6, 163.

Malmfors, E. (1982) Studies on some factors affecting pig meat quality. In: *Proceedings 28th European Meat and Meat Research Workers*. Madrid, Spain, pp. 21–23.

Markov, E. (1981) Studies on weight losses and death rate in pigs transported over a long distance. *Meat Industry Bulletin* 14, 5.

Mendle, M., Zanella, A.J. and Broom, D.M. (1992) Physiological and reproductive correlates of behavioural strategies in female domestic pigs. *Animal Behaviour* 44, 1107–1121.

Monin, G. and Ouali, A. (1992) Muscle differentiation and meat quality. In: Lawrie, R.A. (ed.) *Developments in Meat Science.* Elsevier Applied Science, London, pp. 89–159.

Monin, G. and Sellier, P. (1985) Pork of low technological quality with a normal rate of muscle pH fall in the intermediate post mortem period: the case for the Hampshire breed. *Meat Science* 13, 49–63.

Morgan, J.R., Krautil, F.L. and Craven, J.A. (1987) Effect of time and lairage on caecal and carcass *Salmonella* contamination of slaughter pigs. *Epidemiology and Infections* 98, 323–330.

Mount, L.E. (1974) The concept of thermal neutrality. In: Monteith, J.L. and Mount, L.E. (eds) *Heat Loss from Animals and Man.* Butterworths, London, pp. 425–439.

Mulder, R.W.A.W. (1995) Impact of transport and related stresses on the incidence and extent of human pathogens in pig meat and poultry. *International Journal of Food Safety* 15, 239–246

Perremans, S., Randall, J., Ville, H., Stiles, M., Duchateau, W. and Geers, R. (1996) Quantification of pig response to vibration during vertical motion. In: *Proceedings EU-Seminar: New Information on Welfare and Meat Quality of Pigs as Related to Handling, Transport and Lairage Conditions.* FAL-Sonderheft 166, Völkenrode, pp. 135–141.

Perremans, S., Randall, J.M., Allegaert, L., Stiles, M.A., Rombouts, G. and Geers, R. (1998) Influence of vertical vibration on heart rate of pigs. *Journal of Animal Science* 76(2), 416–420.

Phillips, R.A., Thompson, B.K. and Fraser, D. (1988) Preference tests of ramp designs for young pigs. *Canadian Journal of Animal Science* 68, 41–48.

Rajkowski, K.T., Eblen, S., and Laubauch, C. (1998) Efficacy of washing and sanitizing trailers used for swine transport in reduction of *Salmonella* and *Escherichia coli. Journal of Food Protection* 61(1), 31–35.

Randall, J.M. (1992) Human subjective response to lorry vibration: implications for farm animal transport. *Journal Agricultural Engineering Research* 52, 295–307.

Randall, J.M. and Bradshaw, R.H. (1998) Vehicle motion and motion sickness in pigs. *Animal Science* 66, 239–248.

Randall, J.M., Stiles, M.A., Geers, R., Schütte, A., Christensen, L. and Bradshaw, R.H. (1996) Vibrations on pig transporters: implications for reducing stress. In: *Proceedings EU-Seminar: New Information on Welfare and Meat Quality of Pigs as Related to Handling, Transport and Lairage Conditions.* FAL-Sonderheft 166, Völkenrode, pp. 143–159.

Santos, C., Almeida, J.M., Matias, E.C., Fraqueza, M.J., Roseiro, C. and Sardina, L. (1997) Influence of lairage environmental conditions and resting time on meat quality in pigs. *Meat Science* 45(2), 253–262.

Schrama, J.W., van der Hel, W., Gorssen, J., Henken, A.M., Verstegen, M.W. and Noordhuizen, J.P. (1996) Required thermal thresholds during transport of animals. *Veterinary Quarterly* 18(3), 90–95.

Slavkov, I., Iodanov, I., Milev, M. and Danov, V. (1974) Study of factors capable of increasing the number of *Salmonella* carriers in clinically normal pigs before slaughter. *Veterinary Medicine Nauki, Bulgaria* 11, 88–91.

Smith, L.P. and Allen, W.M. (1976) A study of the weather conditions related to the death of pigs during and after their transportation in England. *Agricultural Meteorology* 16, 115–124.

Smulders, F.J.M., Romme, A.M.T.S., Woolthuis, C.H.J., de Kruijf, J.M., Eikelenboom, G. and Corstiaensen, G.P. (1983) Pre-stunning treatment during lairage and pork quality. In: Eikelenboom, G. (ed.) *Stunning of Animals for Slaughter*. Martinus Nijhoff, The Hague, pp. 90–95.

Sybesma, W. and van Logtestijn, J.G. (1967) Rigor mortis und Fleischqualität. *Fleischwirtschaft* 4, 408–410.

Tarrant, P.V. (1989) The effects of handling, transport, slaughter and chilling on meat quality and yield in pigs – a review. *Irish Journal of Food Science and Technology* 13, 79–107.

van der Hel, W., Duijghuisen, R. and Verstegen, M.W.A. (1986) The effect of ambient temperature and activity on the daily variation in heat production of growing pigs kept in groups. *Netherlands Journal of Agricultural Science* 34, 173–184.

van Logtestijn, J.G., Romme, A.M.T.C. and Eikelenboom, G. (1982) Losses caused by transport of slaughter pigs in the Netherlands. In: Moss, R. (ed.) *Transport of animals intended for breeding, production and slaughter*. Martinus Nijhoff, The Hague, pp 105–114.

van Putten, G. and Elshof, W.J. (1978) Observations on the effect of transport on the well-being and lean quality of slaughter pigs. *Animal Regulation Studies* 1, 247–271.

van Putten, G. and Lambooij, E. (1982) The international transport of pigs. In: *Proceedings 2nd European Conference on the Protection of Farm Animals*. Strasburg. pp. 92–103.

van Putten, G., Corstiaensen, G.P., van Logtestijn, J.G. and Zuidhof, S. (1983) Het douchen van slachtvarkens. *Tijdschrift voor Diergeneeskunde* 108, 645–652.

Verdijk, A.T.M. (1974) Oorzaken van afwijkende vleeskwaliteit bij stress-gevoelige varkens. Thesis, Utrecht.

Verstegen, M.W.A. (1987) Swine. In: Johnson, H.D. (ed.) *World Animal Science, B5: Bioclimatology and the Adaptation of Livestock*. Elsevier, Amsterdam, pp. 245–258.

von Mickwitz, G. von (1989) Lowest possible requirements for dealing with slaughter pigs beginning with the loading–transport–resting time, and ending with stunning seen from the animal welfare and meat quality point of view. In: *Proceedings European Association for Animal Production*. Dublin.

Warriss, P.D. (1987) The effect of time and conditions of transport and lairage on pig meat quality. In: Tarrant, P.B., Eikelenboom, G. and Monin, G. (eds). *Evaluation and control of meat quality in pigs*. Martinus Nijhoff, Dordrecht, pp. 245–264.

Warriss, P.D. (1993) Ante mortem factors which influence carcass shrinkage and meat quality. In: *Proceedings 39th International Congress on Meat Science and Technology*. Calgary, Canada, pp. 51–56.

Warriss, P.D. (1998a) The welfare of slaughter pigs during transport. *Animal Welfare* 7, 365–381.

Warriss, P.D. (1998b) Choosing appropriate space allowances for slaughter pigs transported by road: a review. *Veterinary Record* 142(17), 449–454.

Warriss, P.D. and Bevis, E.A. (1987) Liver glycogen in slaughtered pigs and estimated time of fasting before slaughter. *British Veterinary Journal* 143, 254–260.

Warriss P.D. and Brown, S.N. (1983) The influence of pre-slaughter fasting on carcass and liver yield in pigs. *Livestock Production Science* 10, 273–282.

Warriss, P.D., Dudley, C.P. and Brown, S.N. (1983) Reduction in carcass yield in transported pigs. *Journal of the Science of Food Agriculture* 34, 351–356.

Warriss, P.D., Bevis, E.A. and Ekins, P.J. (1989) The relationships between glycogen stores and muscle ultimate pH in commercially slaughtered pigs. *British Veterinary Journal* 145, 378–383.

Warriss, P.D., Brown, S.N., Barton Gade, P., Santos, C., Nanni Costa, L., Lambooij, E. and Geers, R. (1998a) An analysis of data relating to pig carcass quality and indices of stress collected in the European Union. *Meat Science* 49(2), 137–144.

Warriss, P.D., Brown, S.N., Knowles, T.G., Edwards, J.E., Kettlewell, P.J. and Guise, H.J. (1998b) The effect of stocking density in transit on the carcass quality and welfare of slaughter pigs: 1. Results from the analysis of blood and meat samples. *Meat Science* 50(4), 447–456.

# Horse Handling and Transport | **16**

## K.A. Houpt[1] and S. Lieb[2]

[1]*Department of Physiology, College of Veterinary Medicine, Cornell University, Ithaca, NY 14853–6401, USA;*
[2]*Department of Animal Science, University of Florida, Gainesville, FL 32611, USA*

## Handling

Livestock species, with the exception of dairy cows, are handled rarely. Most livestock are not trained, nor is their value dependent on their trainability. In contrast, most horses are handled daily as individuals. Their ability to be handled safely and trained to respond to physical and verbal cues are the reasons for their continued maintenance and the basis of their value. For these reasons, most of the handling methods described will be for individual animals.

Handling methods to be covered include 'imprint training' of the newborn foal, training the young horse to lead, round pen or tackless training, methods of catching and methods of restraint. Retraining of problem horses will also be addressed.

### Early handling

The easiest way to ensure that horses will be tractable is to handle the foal from birth. One method of handling foals was termed 'imprint training' by Miller (1991). The term itself is probably something of a misnomer because the foal does not become imprinted on to humans, as a duckling would. The foal does not follow people nor does it show sexual behaviour toward people. Imprint training involves handling the foal as soon after birth as possible. The goal is to accustom the foal to manipulations that will be done to it in the course of saddling, shoeing and medicating it. For that reason, the handler puts his/her arms around the foal's chest and squeezes as if a girth were tightening. The hoofs are each tapped with the hand 50 times to simulate shoeing and hoof care. Foals seem to resist manipulation of their hooves more than the other handling. The foal is rubbed all over. Fingers are inserted into its mouth, ears and (while wearing a rubber glove) the rectum so that the foal will not be frightened by bits, otoscopes or thermometers. Clippers are held, while running, against its head, neck and body so that it will not be afraid of their sound or vibration.

There is some controversy about how soon after birth the foals should be handled. If the foal is manipulated before it can stand, the phenomenon of 'learned helplessness' is probably involved. Learned helplessness is the phenomenon whereby an animal that cannot escape a situation will not even try to escape or avoid later when escape or avoidance is possible; the animal simply gives up (Maier and Seligman, 1976). The neonatal foal cannot escape and so in later life may not try to escape from a farrier or a pair of clippers. Handling the foal a few hours or days later may not be as effective, because the foal can move vigorously. This would indicate that early handling is desirable. However, care must be taken to avoid disturbing the mare. Some mares will become so agitated at the presence of a person that they will either attack the person or redirect their aggression toward the foal. The odour of the handler may interfere with the mare's recognition of the foal, but in some cases the smell of the handler may reassure the mare.

Handling later in the foal's life may be as important or more important than handling at birth. The Japan Racing Association (Ryo Kusunose, personal communication) tested the behavioural characteristics of 270 foals on 25 Thoroughbred breeding farms and compared the rearing methods used on each farm. This is the most extensive test of foal behaviour in relation to handling method.

The test was the foal's response to six manipulations. The foal's head was touched and the circumference of the front and hind cannon-bone was measured. The height at the withers, the heart girth and the hip width were also measured. Foals were tested at 50, 90 and 140 days of age and the scores were added for each foal. The reaction of each foal while a handler touched its head and took each measurement was scored. If the foal stood still during measuring, its score was 4. If the foal moved a little, its score was 3 points. If the handler could not measure the foal without applying a twitch or having someone help hold the foal or if the foal reared, its score was only 1 point. The maximum, if the foal always stood still, motionless whatever the handler did to the foal, was 24 points. The minimum was 6 points if the foal rejected all of the handlers' manipulations.

The average score increased with the age of the foals, indicating that they became calmer as they matured, but differed between farms. On one stud farm (A), all of the foals scored 24 or 23 points. These foals were very calm and gentle. However, on farm Y, there were far fewer gentle foals than at A. The behavioural characteristics of the foals reared on the same farm resembled each other. The differences among the farms were not great when the foals were 50 days old, but by 90 days, the behavioural characteristics of foals reared on the same farm were found to resemble each other. There was no further change at 140 days.

The farm's size was not related to the behaviour of foals, but the number of mares per worker was inversely related to the mean scores of foals. The top three rankings of farms had 3.7, 5.3 and 3.1 mares per handler. In contrast, the bottom three farms had 7.0, 7.5 and 7.5 mares per handler. The fewer mares and foals each handler was responsible for, the more obedient the foals were. There were marked differences in foal handling at these farms. The handlers on the high-score farm groomed their foals twice a day, every morning and evening.

One of the most important differences was in the manner of moving the foals from stable to pasture. The two methods were leading and driving. When leading a mare and a foal to pasture, the handler walked along between the mare and the foal with a lead on each animal. There was a lot of contact between the human and the foal, and the leading method apparently makes the foal develop confidence in humans. The further the mare and foal were led per day, the quieter the foal was during testing. On other farms, where the mares and foals were driven to pasture, the scores were lower.

On some farms, the halter was taken off at the gate of the pasture every day for security because the foal might catch its halter on a tree or fence and injure itself. Foals from farms on which the halters were taken off at the pasture gate every day and replaced in the evening were easier to handle.

Another indication that handling extending over the first few weeks of the life is important is the study of Mal and McCall (1996), in which foals were handled for 10 min a day, 5 days a week for either the first or the second 7 weeks of their lives. The foals handled for the first 42 days were easier to halter-break than those handled later. The handling involved rubbing the foals all over their bodies. In an earlier study, Mal *et al.* (1994) had found that foals handled for 10 min twice a day for the first week were no more easy to manage at 120 days than unhandled foals.

Jezierski *et al.* (1999) tested the effects of environment and handling on Konik (Polish primitive) horses. There were four groups: stable-reared, intensively handled (SIH), stable-reared, not handled (SNH), reserve-reared, intensively handled (RIH) and reserve-reared, not handled (RNH). Stable-reared foals were raised under typical domestic conditions in a stable. Reserve-raised foals were born and lived in a harem group within a semi-reserve and were not handled at all until they were 10 months old. At that time, they were divided into intensively handled (RIH) and non-handled (RNH) groups. Handling occurred 10 min a day, 5 days a week, from 2 weeks of age until 2 years for SIH foals and from 10 months until 2 years for RIH foals. Handling consisted of haltering, rubbing the foal all over its body and picking up the hooves. The two non-handled (SNH and RNH) groups were not handled except for routine management and veterinary care. The test, given at 6 months (for the SIH and SNH foals only) and at 12, 18 and 24 months for all groups, consisted of three people catching the horse in a paddock, leading it to and from the stable, picking up its feet and holding it while a strange person approached. The horses were given a score for each test, with high being easy to handle and low being difficult. The intensively handled foals were easier to mange than the non-handled foals and the stable-reared horses were easier to manage than the reserve horses at 12 months, but, by 18 months, handling was more important than early rearing environment. Fillies had higher heart rates than colts, but their test scores were similar.

## Stroking and massaging

A method of handling horses that has gained considerable popularity is that of Linda Tellington-Jones (Tellington-Jones and Bruns, 1985). The basis of this method is no different from most training in that the horse is rewarded for correct or desirable behaviour and reprimanded for incorrect or undesirable behaviour, but emphasis is on positive methods, particularly pleasurable tactile stimuli. The equipment needed is a 4-foot (90 cm) long wand or dressage whip without a lash and lead shank with a long (30 in ~ 1 m) chain. The chain is threaded through the lower near (left) side and right halter ring and fastened to the off (right) side upper halter ring. This serves to control the horse if it should try to evade or react aggressively to the handler. The wand is used for two purposes: (i) the thickened base is used to rap the horse on the nose if it plunges ahead when led; and (ii) the tapered end is used as an extension of the handler's arm to caress the horse or, more rarely, to block the horse from moving forward. The first step is usually to stroke the horse with the wand on the back and withers, then down the rump and the hind limbs. The wand is also used to stroke the forelegs and belly. This approach – stroking the body without even attempting to touch the horse directly – is an excellent way to overcome resistance to handling of the tail, the feet, the flanks, etc., and accustoms the horse to touches that can later be used as cues. When the horse relaxes and accepts the wand strokes without moving away or flinching, it is ready for direct manual contact.

Another use of the wand is to reprimand the horse for invading the handler's personal space, by knocking it on the forehead or nose. The horse can be taught to halt by applying slight pressure on the lead line and tapping the line with the whip while standing directly in front of the horse.

The manual contact is in the form of light digital pressure. The digital pressure is applied in a circular motion. A small (3 cm) circle is made, usually in the clockwise direction from 6 o'clock, 360° and beyond to 9 o'clock. The amount of pressure is no more than that which is comfortable when applied to the person's eyelid, i.e. a light touch. The massage can be done anywhere on the body. A good place to start is on the shoulders because this is where horses most frequently mutually groom one another. When horses groom, they pick up a fold of skin with their incisors and pull back allowing the skin to slip out of their teeth. Mutual grooming is believed to serve two purposes: to groom areas, such as the back and withers, that the horse cannot reach itself and to strengthen bonds between the horses. Fillies are more likely than colts to mutually groom and, as adults, mares groom their preferred associates and those close in social rank (Waring, 1983). Feh and de Mazières (1993) found that scratching a horse on the withers, but not on the chest, reduces heart rate, i.e., is calming

The massage method, commercially termed the TTouch, is good for calming an anxious horse. It is a useful adjunct to veterinary examination. Massaging the area around the tail can make a horse much more relaxed and willing to accept a thermometer. Head shyness can be improved by massaging the head closer and closer to the ears or other sensitive area. The circular motion should not be applied again and again at the same spot during one session or the horse will be irritated. Instead, work across a given area. Pretreatment with massage can relax a horse sufficiently for it to accept a thorough veterinary examination without becoming tense. Other exercises consist of manipulating and rotating the horse's ears and massaging its gums. One can tell whether the technique is relaxing the horse by its posture. The horse's head will drop and its upper lip lengthen when it is relaxed, especially when pleasurable sensations are being perceived. This technique is also useful for reassuring a horse on first acquaintance.

## Greeting

Blowing in a horse's nostrils has been recommended (Woodhouse, 1984) as a technique for greeting a strange horse. That is, in fact, the usual behaviour when two horses meet (Houpt, 1991). Usually, however, one or both horses squeal, and they may also strike. Therefore, when a person blows in a horse's nostril it may threaten in a similar manner. Rubbing the horse is a much safer approach.

# Physical and Chemical Restraint

## Tying

Horses are restrained by tying or cross-tying. Descriptions of the methods used to train a horse to be tied can be found in Wright (1973) and Miller (1975). Ideally, a horse should be taught to lead and be fully accepting of the pressure of a halter before tying is attempted. Then, the first tie session should be accomplished in a safe, non-confining place, with a halter and lead (including snaps) that will not break. Few horses that respect the halter will do more than 'test' the tie and the few that do usually give up quickly if they don't 'escape' the tie when they pull hard the first few times. There are several important points to remember. Quick-release knots or safety fasteners should always be used so that the horse can be freed if it becomes entangled. Ideally, the point to which the horse is to be fastened should be higher than the horse's eyes. Although well-trained horses may be tied to hitching rails, fences, etc., it is not recommended because a downward pull exerts pressure on the poll and is more frightening to the horse than an upward pull. These tying situations are not recommended by 4-H horse manuals, even for quiet horses.

Several methods of restraining horses that attempt to break ropes have been suggested (McBane, 1987). Both methods involve use of ropes around the horse's body and both have the rope passing through, but not attached to, the halter. In one method, a long rope is attached around the heart girth using a bowline (non-slip) knot and the free end of the rope is run between its forelegs, through the centre halter ring and then to a post or wall. If the horse pulls back, the rope presses down on the withers and pulls up on the chest. It should be tied at the horse's eye level or above. The second method is similar but involves an even longer rope that is looped around the horse from in front, passing under its tail. The two lengths of rope are knotted twice, at the croup and at the withers. The loose ends pass through the lower halter rings on each side and are attached to the post or wall. When the horse pulls back, pressure will be exerted under its tail. For some horses, a bungee tether can solve tying problems. When the horse pulls back, it gives and doesn't exert pressure on the poll; however, if the tether breaks, the snap may fly back and injure the horse.

## Hand restraint

Most movement of horses is done by means of a halter and lead. Descriptions of training to lead can be found in Wright (1973) and Miller (1975). A trained gelding or mare can usually be handled by means of a rope and snap fastened to the ring in the noseband of the halter. For stallions and fractious horses, a little more control can be exerted using a lead shank with a chain. There are various techniques of restraint using a chain shank, the effectiveness of which depends on the amount of leverage afforded the handler and the sensitivity of the area contacted by the chain. In all methods using a chain on the face, the chain should be threaded through the rings with the snap opening facing down toward the horse's face. For minor restraint in this method, the chain is threaded from one side ring to the other across the horse's nose. The reason for placing the chain across the nose is that pressure applied to the nose is painful and will cause the horse to stop or back up. A properly fitted and properly used chain should release quickly and not apply painful pressure for more than a second or two. The principle is that punishment must be applied instantly, preferably as the horse misbehaves, and should be removed as soon as the misbehaviour ceases. This is the same principle as that used in dog training, in which leash corrections are applied with a choke chain collar. In both cases, the quick application and release is much more effective than a steady pull because the animal quickly habituates to a steady pull or begins to fight it if too painful. Another method has the chain running through the lower right ring behind the horse's jaw to fasten on the upper left ring (Fig. 16.1). This gives leverage without the danger of twisting the halter. The chain can be threaded through the lower left halter ring, and fastened to the upper right ring. The chain can then be tightened to pass through the horse's mouth. The chain applies pressure to the gums and the commissure of the lip, causing the horse to stop and raise its head.

Placing the chain behind the chin applies pressure to the sensitive soft tissue there. It causes the horse to raise the head and move forward rather than stop.

The use of a chain to treat head-shy horses has been demonstrated in a video, *Influencing the Horse's Mind*, by Robert Miller, DVM (BandB Equestrian Films, 1325 Thousand Oaks Blvd, #102, Thousand Oaks, CA 91366). The principle is that the horse is rewarded for desirable behaviour and punished for undesirable behaviour. The reward is scratching the horse with flexed fingers while clucking (cha chah cha chah). The scratching is directed to the horse's face if head shyness is the problem, but can as easily be applied to any part of the horse.

The head-shy horse is stroked on the head, but each time it throws its head, the lead shank is jerked. The scratches and clucking must stop as soon as the head goes up and resume as soon as it comes down, but there are no verbal reprimands. If the goal is to bridle the horse, the scratching should progress up toward the horse's ears and any head tossing should be punished. If the goal is

**Fig. 16.1.** Chain shank across nose and under middle of lower jaw for control of a fractious horse.

to place a crupper on a horse that resents handling of its tail, the scratching should be applied to the neck, then the withers, then the back and then the rump and finally the area under the tail should be rubbed. If the horse picks up a hind foot or moves away, the lead shank should be jerked. These activities can be tried first with a chain just over the nose, and the lip chain, which is very severe, should be a last resort. To use the severe method, the chain is run through the left lower ring and attached to the right lower or upper ring and is worked under the upper lip with the left hand (Fig. 16.2). It should only be used to make the horse stand still.

## The twitch

Another method of restraint used almost exclusively on horses is the twitch. The twitch is made of a loop of rope or chain attached to a pole. The horse's upper lip is grasped and the loop slipped over the lip and twisted. This applies pressure and, presumably, produces pain. Most horses will stand quietly when the twitch is applied. It was assumed that this was to avoid more pain that would ensue if the horse moved against the restraint. Lagerweij *et al.* (1984) found that administration of the opiate blocker, naloxone, greatly reduced the effectiveness of twitching, indicating that the horses are being sedated by endogenous opiates. Apparently the pain sensed when the twitch is first applied leads to release of opiates, which sedate the horse. The calm facial

**Fig. 16.2.** Chain positioned against upper gums. This is the harshest use of a chain and should be used only when necessary.

expression and hanging head of many horses when twitched indicates that they are somewhat sedated.

Recently, a bridle, the Stableizer®, has been developed that makes use of a strap beneath the lip and over the poll which can be tightened. This is supposed to calm the horse and may operate on the same principle as the twitch.

### Stocks and chutes

Although most handling of horses can be done by tying or holding the horse, there are occasions when it is necessary to further restrain the animal. The occasions are usually for veterinary care or any painful procedure. Stocks, consisting of pipes at the level of the horse's chest, serve the purpose well when the horse is fairly tractable. The vertical pipes should be 80–90 cm apart, attached to the ceiling and provided with cleats, so that ropes can be attached to cross-tie the horse. The horizontal pipe should be approximately 120 cm from the floor. The purpose is to prevent lateral movement. If more restraint is needed, solid panels can form the front and back of the stocks. This type of arrangement is ideal for rectal palpation, ultrasonic examination and other reproductive procedures. If an unhandled horse must be restrained, solid sides are necessary. This is the manner in which halters can be placed on feral horses that have never before encountered humans. Typical livestock chutes with metal panels separated from the next panel by spaces are very dangerous for horses. They can easily thrust a limb through the space between the panels and injure themselves.

### Chemical restraint

The methods discussed above are physical. It is also possible to use chemical restraint. The most commonly used drugs are the tranquillizer acepromazine ($0.04$–$0.1$ mg kg$^{-1}$ intravenously) or the alpha-adrenergic agonist xylazine ($0.4$–$2.2$ mg kg$^{-1}$ intramuscularly or $0.2$–$1.1$ mg kg$^{-1}$ intravenously). More recently, a more potent $\alpha$-adrenergic agonist, detomidine, has been developed and is useful for restraint and pain control. The dose is $20$–$40$ µg kg$^{-1}$ ($0.02$–$0.04$ mg kg$^{-1}$ intravenously or intramuscularly) (Robinson, 1992). These should be used under veterinary supervision.

## Social Facilitation

When horses have to be moved, one can take advantage of the phenomenon of social facilitation. Social facilitation is the technical term for a behaviour that is influenced by the behaviour of other animals. Because horses are herd animals, there is considerable social facilitation of their behaviour. They graze and rest at the same time (Tyler, 1972; Arnold, 1984/85) and they usually move together to water and away from danger. This behaviour that motivates one horse to do what the other horses are doing can be a nuisance if one is trying to take one horse away from a group, but the astute handler can use it to advantage. If a horse will not enter a barn, cross a creek or walk into a trailer alone, it may be perfectly willing to follow another horse into those areas.

## Catching

Horses are prey animals who have evolved flight as the best means of defence against predators. Therefore, it is not surprising that horses quickly learn to escape, especially if they are not too eager to be ridden.

One of the easiest ways to catch a horse is to teach it to come. In general, horses will usually not come when called simply for attention. The exception is the single horse, especially a young horse, which may want companionship. Food is the best reward for most horses, although even grain may not be more attractive than a field of young grass. Approach the fence and call the horse. Give it grain, apples or carrots for approaching. Try to accustom the horse to being caught by repeating the feed and catch routine many times. The ratio of catches to catches followed by something unpleasant should be low, so do not catch the horse only to work it. Catch the horse and let it go or catch the horse and take it out of the pasture only to graze or to be groomed and massaged.

The problem with using food is that some horses at all times and most horses at some time are more motivated to avoid capture than to eat. In that case, another technique must be used, but even then it is best to have some food to reward the horse when it is caught. To catch the horse, avoid direct eye contact because that is threatening to the horse and avoid running at it from behind because that will stimulate the animal to flee. Move with the horse at the level of its shoulder, but stop when it stops. Leaning away from the horse will encourage it to approach, whereas leaning forward or crouching down will encourage it to move away. Decrease the distance by walking at an angle. When a difficult horse is to be caught, plan to take several hours. Horses would rather not be in continual motion so most will eventually allow themselves to be approached. When the horse is caught, stroke him and reward him. No matter how frustrating the process may have been, do not reprimand or yell at the horse.

## Herd Behaviour and Fear

When handling horses, one must always bear in mind that they are herd animals. This is reflected in the behavioural (Houpt and Houpt, 1988) and physiological (Mal *et al.*, 1991) response to isolation. The handler should take advantage of this to move horses and to keep horses calm in a strange environment. Conversely, extra care must be taken to prevent the escape or injury of a horse that is isolated.

When moving horses, whether as a group or while riding, driving or leading a single horse, one should be aware of those things that frighten horses. Horses are reluctant to enter a dark area from a brightly lit one. They are also reluctant to step on anything that sounds or feels unusual. They often shy when crossing a drain, presumably because of the hollow sound made by their hooves. Many horses are reluctant to cross a stream or even a puddle. The unfamiliar sound of rushing water probably frightens it and then the feel of wet water only serves to increase its fear. One can desensitize a horse to water by starting with a trickle of water from a hose in a familiar situation. When the horse will cross the trickle, the trickle can be increased to a stream or to create a puddle.

Windy days are particularly apt to result in problems. Horses will shy from bits of paper or light-coloured leaves blowing near them. Pennants or flags can also frighten them. Horses may be reluctant to pass through gates with flapping flags. This can be especially dangerous if the horse is pulling a wagon, which can catch on the gate as the horse swerves to avoid the flag.

## Learning Ability of Horses

There have been many studies of learning in horses; this subject has been reviewed by McCall (1990). Unfortunately, few of these studies dealt with learning tasks of interest to riders or drivers. The tasks most frequently studied were visual discrimination and maze learning. Some information which is relevant to handling is found in the study by Heird *et al.* (1981) who found that horses handled moderately (once a week) learned more quickly than either those handled extensively (daily) or who had not been handled at all. Trainers could predict the trainability of a horse after working with it for 10 days. In a later study, Heird *et al.* (1986) found that unhandled horses learned more slowly than handled horses. Mader and Price (1980) found that Quarter-horses learned a visual discrimination more rapidly than Thoroughbreds; they concluded that the Thoroughbreds were more distractable. Others working with one breed (Quarter-horses) have also found that the more emotional the horse, the slower it is to learn (Fiske and Potter, 1979; Heird *et al.*, 1986). Haag *et al.* (1980) found that there was no correlation between position in the equine dominance hierarchy and learning, but there was a correlation between avoidance and maze learning, that is horses that learn one task quickly learn another quickly.

Horses can form concepts. The experiment was an operant conditioning task. The horse simply had to push one of two hinged panels. The correct panel was unlocked, allowing the horse access to a bowl of grain. The incorrect panel was locked, but there was a bowl of grain behind it to ensure the horse did not choose the panel because it could smell the grain. The correct choice was not always on the same side. The first problem was a simple discrimination between a black panel and a white panel. Next, it had to discriminate between a cross and a circle. The third problem was to distinguish a triangle from a rectangle and the next to discriminate triangles from half circles and various other patterns. In the true test of conceptual learning, the horse had to choose between two shapes it had never seen before, one triangular and the other non-triangular (Sappington and Goldman, 1994).

Horses have also been shown to be capable of conditional discrimination – in this case, identity matching to sample. The horse had to touch the two identical cards (marked with Xs or with circles) rather than a third, distractor card, which was marked with a square or star. The horses could learn with a 73–83% accuracy (Flannery, 1997).

The ability of horses to perceive objects is influenced by their size and by environmental conditions that affect contrast, such as overcast skies. Horses responded by hesitating at a distance of 15.2 cm from a narrow stripe (1.27 cm) when contrast was poor or hesitating at a distance of 2.3 m for a wide 10.2 cm stripe when contrast was good (Saslow, 1999). This is important to consider because horses may shy more on an overcast day.

There have been four experiments attempting to prove that horses can learn by observation (Baer et al., 1983; Baker and Crawford, 1986; Clarke et al., 1996; Lindberg, 1999). None proved that they could, but the tasks – visual discriminations – are probably not as relevant to equine ecology as learning which food is safe to eat or which place to avoid. It is probably not a good idea to let a young horse watch another horse misbehave, despite the lack of scientific evidence for observation learning in horses.

Ability to make a visual discrimination (a black from a white bucket) and then a reversal (the formerly incorrect choice is now correct) is not correlated with ability to learn to jump a hurdle or to cross a wooden bridge, despite the fact that food was the reward in all three tasks. Furthermore, those horses who learned to jump quickly were not those who learned to cross the bridge quickly (Sappington et al., 1997). Performance can't be predicted from a discrimination learning task or from a different type of performance.

Emotionality or nervousness has been shown to interfere with learning. Furthermore, injuries to itself and handlers are often the result of the horse's response to a stimulus it perceives as frightening. Therefore, it is useful to be able to predict which horses are likely to shy, bolt or baulk in fear. McCann et al. (1988a,b) have developed a temperament or emotionality test for horses, based on their response to being herded, being isolated and being approached by people. The subjective rating of the observers correlated well with the heart-rate response. Handling or reserpine treatment did not improve the later response of the horses to frightening situations.

One of the most difficult training concepts is distinguishing negative reinforcement from punishment. Negative reinforcement is not punishment. Punishment follows behaviour or misbehaviour. The behaviour will be performed less often when the animal learns that the punishment is contingent on the behaviour. Negative reinforcement proceeds until the horse performs the desired behaviour. The reason it is so important to make the technical distinction is that most horse training uses pressure cues that are negative reinforcement. For example, the untrained horse is 'nudged' with the rider's heels or tapped with a whip until it moves. Gradually, subtle cues, such as a slight pressure by the rider's legs, can replace the more forceful ones. When the rider pulls on the reins, he or she is applying negative reinforcement – pressure or pain on the gums or lips – until the horse slows down.

Positive reinforcement is a reward when the desired behaviour is performed. The horse is rubbed when it walks up to the trailer or given a piece of sugar for coming when called. Positive reinforcement is probably not used enough in horse training. The tackless training and free-lunging methods do use subtle forms of reinforcement – the negative reinforcement is chasing the horse by approaching it from behind and removing that stimulus by backing off when the horse speeds up. Circus and trick horse trainers make much more use of positive reinforcement. The horse is rewarded with a food treat for bowing, for example.

Dougherty and Lewis (1993) studied stimulus generalization. They taught horses that they would receive a reward if they pressed a panel after having felt a touch in a particular spot on their back. Then the touch was moved away from the original location. The horse's response decreased with distance from the original spot. This is certainly applicable to horse training. If a horse is taught to move sideways to a touch on one location on its flank, but the next rider cues it in a slightly different location, it may not respond.

The secret of a good trainer is excellent timing. The negative reinforcement should be removed as soon as the horse begins to perform the desired behaviour and the positive reward also should be given as soon as the horse performs the desired behaviour (both within a few seconds' time). Most important is timing of punishment. Punishment is a punishment only if it reduces the frequency of a behaviour. The frequency will be reduced only if the horse realizes which behaviour is being punished. The punishment must occur within a second of the misbehaviour. If the horse kicks and is immediately struck with a whip, it will be less inclined to kick, but, if it kicks and the handler must walk across the barn to get a whip and then strikes the horse, the horse will not be punished for kicking. The horse will be punished for allowing someone to approach with a whip.

There has been a change in equine training methods over the past 50 years. Gentling has replaced 'breaking', so that inductive methods rather than intimidation and physical force are used. Perhaps the most popular of the spokespersons for these methods is Monty Roberts, whose book, *Listening to Horses*, achieved renown with a population much larger than that of equestrians. Roberts was by no means the first of these people. The Jeffrey method of horse training (Wright, 1973), first used in Australia on their feral horses (brumbies), uses an approach and withdrawal technique with a horse in a paddock wearing a halter and lead rope. The trainer approaches when the horse is calm, but withdraws if the horse is even slightly evasive. Using that method, a horse can be saddled, bridled and ridden within a few hours.

Round-pen training is favoured in the USA. The first trainer to use this method was Archibald and later Ray Hunt, Pat Parelli and John Lyons toured the country giving demonstrations and clinics and writing in the popular equine press. The best description is given in MacKenzie's book, *Fundamentals of Free Lungeing* (1994). The main thrust is to move the horse around the pen, at first simply in a circle, but later inducing the horse to turn toward the outside of the pen and reverse, and to turn toward the inside of the pen and reverse according to the body movements (mostly head and hip) of the person in the centre of the ring. Many trainers use a whip or a rope to keep the horse at the perimeter of the circle and to keep him moving fast. At least part of the technique is simply to tire the horse, but what may be more important is that it teaches the horse to watch the human and to attend to his or her movements. The theory is that the human is assuming a dominant or leadership role over the horse. There may also be an element of predator avoidance because, to push the horse forward, the person leans forward assuming an almost quadrupedal stand and 'nips' at the horse with rope or whip. To encourage the horse to approach him or her, the person stands up straight and leans back, avoiding eye contact, which is threatening to the horse just as sheep avoid Border collies who give them 'the eye', i.e. stare at them. One of the most interesting aspects of round-pen training is that the horses will make a mouth movement, opening their mouths enough to run their tongues in and out. Although Roberts calls this a 'signal meaning I am a herbivore', the same mouth movement is used by horses when anticipating food or when giving in to a farrier's demands. It may be a care-soliciting expression or simply subordination. It is not the 'snapping' behaviour of immature horses, who open and close their mouths but also show their teeth. Snapping is given in situations where a young horse is ambivalent, wishing to stay but frightened enough to flee. It is given when a colt approaches a stallion; he need not approach, but the positive aspects of approach outweigh the negative ones. He signals that he is young by snapping, perhaps to discourage aggression The mouthing movements seem more subordinate without the ambivalence of snapping.

The principle of round-pen training seems to be that horses would rather not run and they are rewarded by being allowed to stop. They appear to be attracted to the trainer when he/she allows them to stop. The horse is then rewarded further by stroking. Some trainers stroke the horse before they begin the circling manoeuvre. The next steps are bridling and saddling the horse. Once the saddle is in place, the horse is again induced to circle until it stops bucking or, if it does not buck, until it has become accustomed to the feel of the saddle and of the stirrups flapping against its sides.

Skilled trainers, such as Lyon and Roberts, can induce a horse to approach them – 'join up' – within a few minutes. The less skilled rely on exhaustion to control the horse. When a horse has been run for hours around a pen the owners are usually not happy. The difference between those who master the technique and those who do not is the ability to observe subtle changes in the horse's demeanour – its ears, tail and muscle tension – and an excellent sense of timing, so that the horse can be rewarded instantly for obeying or for approaching.

There is one impediment to the application of round-pen training or free lungeing: it requires a round pen. If a conventional square or rectangular enclosure is used, the horse will run into a corner. Good practitioners of free lungeing can use a cornered area, but it is much easier, especially for a neophyte, to use a round pen. The best round pens have solid walls so that the horse is not distracted by anything outside the pen. A solid-walled round pen is also much safer than either a rectangular pen or one with rails, because the horse does not have so many opportunities to injure itself.

Handling of horses requires both an understanding of horse behaviour and a willingness to patiently apply the principles of learning.

# Transport of Horses

Horses were probably transported first across seas as the earliest reports have warhorses disembarking near Carthage in the 9th century BC being fit enough to proceed immediately to battle (Cregier, 1989).

### Transportation needs of the horse industry

The use to which a horse is put – work, meat or recreation – has a large influence on the frequency, distance and method of transport to which the horse is subjected. Horses used for livestock ranch work, farm work and hauling of produce and passengers in extensive agricultural enterprises or developing countries are unlikely to travel very far from the area where they were born and used. Horses intended for slaughter are usually cull animals from the work and recreational groups and are purchased from widely dispersed areas and usually loaded in loose groups on to large livestock transports for movement to distant processing plants. There is increasing public concern about the methods used and care given to horses during transit for slaughter and, within the last couple of years, a few states and the Federal government of the USA have initiated research into or controls over the transport of these animals. In 1998, the state of California passed a proposal to prevent the sale of horses for the purpose of slaughter for human consumption.

The last use group, recreational horses, include a very large, diverse group of individuals, which vary greatly in the frequency, distance and method of movement. The simple pleasure mount is likely to be transported less distance and less frequently than a competitive show or racehorse. Most transport of recreational horses is accomplished using a private, towed, horse trailer or large commercial-type van, and, in nearly all cases, these horses have individual stalls and care while in transit. An increasing number of the international- and some national-level competitors and imported breeding stock are transported by air, a method which significantly reduces travel time for long distances. Transport via sea shipping may be used and, as with air shipment, the horses are normally placed in individual stalls in custom containerized freight type crates (Doyle, 1988). Some welfare groups have mentioned inhumane conditions in the shipment of slaughter horses via sea.

It is difficult to establish the number of times or distance that an average horse may be moved in its lifetime. Certainly it would be much greater than other large livestock, such as cattle or pigs, simply because of the use and longer lifespan. Casual observation of today's equine competitions indicates that, during the most active portion of the competition season, a horse might easily be transported every 2–3 weeks to some type of competition within an 80 km radius of its home. Race horses, especially harness horses in the north-east and midwest portions of the USA, are vanned every week to a different track during their summer race circuits, while in Europe, the prominent Thoroughbred racehorses may be flown from one country to another to compete in stake races. Summer or winter horse show circuits for a variety of breeds have been in existence for decades in the USA. Currently, hundreds of hunter–jumper and dressage horses travel from as far as California and Canada to Florida for 8 to 12 weeks of showing in the winter. In addition to a possible 5000+ km two-way round trip from home (requiring several days), these horses compete in shows several times per week and may be moved 50–150 km to a different show facility each week during their stay. Also, many brood-mares are annual commuters from their home farms to the stallion's location, covering distances of a few to over 1500 km, while heavily pregnant or with a foal at their side. Acceptance and use of shipped semen by the breed registries will greatly reduce the necessity for transporting of breeding animals.

Recreational horses are often handled by non-professionals who have minimal experience. This can contribute to behavioural problems and injuries acquired during both loading/unloading and transit.

Some of the horses subjected to extremely long transport, especially when combined with pre- or post-stresses from infectious diseases, contract serious illnesses, such as pleuropneumonia.

### Requirements for transport

Reasonably safe, low-stress and humane transport of horses from point A to point B can be accomplished if the following points are addressed.

**1.** Use of a transport vehicle suitable to the type of horse(s) being moved and its proper maintenance and operation.

**2.** Preconditioning of the animals to be transported, both behaviourally and medically.

**3.** Completion of required governmental medical vaccinations, tests and quarantine specifications necessary for leaving and entering controlled areas (country/state).

**4.** Careful loading, movement and offloading that avoids traumatic injury to the animals.

**5.** Proper care while in transit to ensure that the horses arrive in a healthy condition, free of long-term stress and ready to perform.

**6.** Monitoring of the stress accumulation of horses subjected to repeated and/or prolonged transportation situations, especially when additional stressful factors precede or follow.

**7.** Trained handling and medical personnel to properly accomplish the above.

## Transport vehicles and stalls

Most of the currently used methods and practices for the transportation of horses have been established over a period of time by the demands of the industry with few governmental or industry standards. This has resulted in a wide variety of road-transport vehicles. Conveyances with individual stalls (boxes), where the head and hindquarters of the horse are restrained, have been the norm for many years. This prevents interaction and possible injury due to kicking and biting between horses, but the space given to each horse, the convenience and safety of the stall for horse and handler and the loading designs vary greatly. A discussion of necessities for transport stalls and some advantages/disadvantages to other design features follow.

### Styles and sizes of transport vehicles

Vehicles, usually called vans, are motorized trucks where the bed is enclosed and fitted with individual stalls to accommodate two or more horses. Trailers (also called boxes or floats) are beds made to be separable from the motorized portion; they are enclosed and may or may not be fitted with individual stalls. Trailers which are 'loose' or not fitted with individual stalls are called stock trailers. Trailers which connect over the pulling vehicle's rear axles (goose-neck and tractor-trailer) are more stable for towing than trailers connected to the rear of the vehicle (bumper pull or tag-along). Space and engineered weight restrictions (axle carrying weight, etc.) determine the number of horses which may be carried, and this may range from one to over ten animals for individual stall vehicles and trailers. Large stock-type trailers may carry more. Double-decker or pot-bellied livestock trailers designed to transport cattle are not well suited to transport horses because they lack sufficient height and their internal ramps are difficult for horses to negotiate (Grandin *et al.*, 1999). For example, the upper compartment in a two-level cattle trailer may be only

164 cm (64.5 inches) in height (Wilson Trailer Company, Sioux City, Iowa). Abrasion and laceration-type injuries (most to the head and face) were found by Stull (1998) to occur to 29.2% of the horses transported in double-decker (pot-bellied) trailers, compared with only 8% of horses hauled in straight-deck livestock trailers. Grandin *et al.* (1999) observed that taller horses (those over 162 cm height at the withers) were especially prone to injuries to the heads and toplines (withers, back and croup) and recommended that they should not be transported in these trailers.

### Individual stall construction

An average-size horse (ht = 158 cm, wt = 550 kg) should have approximately 90 cm width and 2.4 m in both length and height of standing space. This allows the horse to use its head and legs to balance during the motion of the transport vehicle. The standing space should be void of projections, edges and protrusions that might cause injury or be contacted by the horse's head or legs. Padding can be added around the walls to prevent bruising when the horse bumps the walls. The construction of the box should be exceedingly strong and include the use of durable material capable of withstanding the forces of weight, pressure, striking, chewing and excrement of the horse. This usually means strong wood (> 5 cm thick) and heavy-gauge metal, or both in combination. Construction of the sides, floor and ceiling should be mostly solid. Openings that could trap head or limbs should be avoided.

Other important factors that need to be considered include adequate ventilation and light. The horse in one stall should be able to see other horses nearby to facilitate ease of loading and calm behaviour. Adaptation of the stall to prevent extreme hot (> 25°C) or cold (< 10°C) temperatures is critical to the horse's life and general health (Leadon *et al.*, 1989). Construction of rear-facing (opposite to the direction of travel) transport stalls should be considered, as there is evidence that horses maintain balance better and show less muscle fatigue (Cregier, 1982; Clark *et al.*, 1988; Kusunose and Torikai, 1996), have lower heart rates and a lower-stress postural stance (Waran *et al.*, 1996) and have a preference for this position when transported untethered in large boxes (Smith *et al.*, 1994; Kusunose and Torikai, 1996). Slant-stalled trailers are now popular (Fig. 16.3) because they provide for more horses and storage space in an overall shorter length of trailer and may increase ease of loading for some horses. Also, designing the front of the stall to allow the horse to lower its head to shoulder height has been shown to be important to normal respiratory function during transits of more than a few hours (Racklyeft and Love, 1990). Most transport stalls have head ties and chest bars that require the horse to maintain its head at 0.25 m or more above the shoulder. Researchers recommended that feed, including hay, be maintained at shoulder level or below.

Floors need special consideration in both construction and maintenance. Wooden boards of 5 × 15 cm size or larger are usually used in smaller trailers, but heavy metal can be used. Rubber mats and/or bedding are almost a

**Fig. 16.3.**   Slant-load three-horse trailer.

necessity to prevent slipping on most floors once they get wet. Bedding also encourages the horse to eliminate wastes during long transits, which is desirable. Cleaning, maintenance and inspection for wear, rot, rust and weakness is a constant requirement. Unfortunately, horses can and do take fatal falls through improperly maintained floors while in transit.

### Stall and vehicle entrance

Horses' heart rates increase dramatically during the loading and unloading process (Waran *et al.*, 1996); therefore horses' behavioural requirements should be considered in the construction of both the transport stall and the entrance to the trailer or van. Because the horse is by instinct afraid of confinement and its eyes neither adapt rapidly to light changes nor see items clearly at close range, the ideal entrance should be wide, well lit and uncomplicated (no steps or ramps) to traverse. A level funnelled walk into the transport vehicle, such as that provided by the special ground loading ramps seen at most large horse facilities are the ideal. However, loading in less than ideal situations is often necessary.

The two most common types of entrances are the step-up, where the horse steps upward directly from ground level into the trailer (Fig. 16.4), and the ramp type (Fig. 16.5), where the horse walks up a sloped platform into the trailer. In both cases, the horse should be allowed to go slowly and look down at its footing during loading. The step-up design, which should have a rubber bumper guard on the leading edge of the floor to prevent injury to the horse's shins, usually allows the horse to get closer to scrutinize the inside of the stall

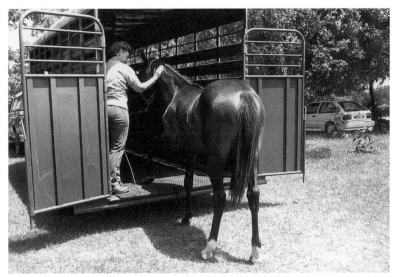

**Fig. 16.4.**   This yearling is standing quietly with one leg in a step-up 16-foot stock-type trailer and being rewarded with a pat on the neck.

before entering. The step-up design requires the horse to learn to pick its feet up to enter, whereas the platform of the ramp type, which solves the stepping up problem, may be yielding, slippery and hollow sounding to the horse's weight and step, all of which may frighten the horse. The disadvantages of both designs need to be minimized and combined with careful preconditioning and patience to teach the horse to accept. Doors which swing out and sides to ramps create a funnel effect to help the horse gradually enter the more confining area of the stall. Also, stall construction that allows one of the side walls to be swung wider for entry and then back into place after the horse enters is very advantageous for young and large horses. Entrances with low step-ups or very short sturdy ramps are adapted to by most horses very quickly. The longer, steeper, loading ramps required by high bedded vans should always have sides for guiding and safety.

*Horse restraint in the transport stall*

As the horse enters the stall, a prompt, quiet and easy method must be available to solidly close off the rear (or front) of the stall. Be sure that open latch handles do not stick out where the horse can be impaled on them and injured during the loading process. Many trailers and vans have very cumbersome or dangerous closures, including having to raise a tail gate from the ground. A simple to latch, but strong, side swinging door that closes off one stall at a time is probably the best. The latching mechanism should be quick and positive to avoid injury to handlers should the horse bump into the door if it comes out prematurely. Where butt or chest bars are used, they must be

**Fig. 16.5.** Two-horse trailer with ramp entrance.

adjustable to the height of the horse to prevent the horse from slipping under them. Head restraint must always be accomplished after exit from the stall is closed off by putting in place the chest or rump bar. When unloading the horse this sequence should be reversed, i.e. the horse's head restraint should be removed before the chest or butt bars are released. Head ties may be installed but should be adjustable and fitted with quick-release snaps. Total stall size and restraints should allow 0.3 m or so of movement forward and back and the head tie should remain slightly slack when the horse's rear is touching the butt bar or back wall. The purpose of the head tie(s) is to prevent the horse from turning around in an individual stall. It is questionable whether a horse's head should be restrained under conditions where there is no hindquarters restraint. The horse inexperienced in tying or one that becomes afraid during transit is likely to pull against a head tie and be injured when it is the only restraint used. If a horse is not accustomed to restraint by the halter, as is often the case with youngsters, they should not be tied, but preferably be transported in a large loose box of approximately 1.5 × 2.0 m. Obviously, vehicles hauling loose horses should not have openings in any cross-gates or the entrance door large enough to encourage the horse to attempt escape.

All transport stalls should allow safe and easy access to the horse's head for tying and care purposes. Shipping crates for air and sea transport may additionally require access to the horse's hindquarters, since it would not be possible to remove the horse in an emergency. In the case of road trailers, extreme care must be taken by both manufacturers and horse owners to properly construct and use escape doors meant for humans only, so that horses do not attempt to use them for exit and thereby become trapped or injured.

## Transport equipment choice and operation

All transports used for livestock, especially those used on the roadways, should be properly engineered for stability and weight-carrying requirements. Transport equipment soundness is paramount to safety on the road. There are few governmental standards regarding the transport equipment, especially the pull-type trailers, currently being used to transport the majority of recreational horses.

Bumper-pull or tag-along single-axle or tandem-axle trailers can be difficult to attach to the pulling vehicle and tow safely, because they are more predisposed to attachment problems, load imbalance, weaving while being towed and jackknifing during the braking phase of driving. The ball-and-hitch arrangement and braking system must be heavy-duty enough for the weight of the trailer plus load to be pulled. For balance reasons, it is important that the trailer is level or slightly uphill at its front when hooked to the pulling vehicle. When a single horse is loaded into a two-horse or larger trailer, the horse should be placed with most of its weight towards the front (over or in front of axles) and driver's side of the trailer, because weaving problems may arise when most of the horse's weight is behind the trailer's axles. Putting the horse on the non-driver's side places it on the low part of the road crown causing the trailer to drift to the outer road edge. Low or uneven tyre air pressure on either the pulling vehicle or trailer can cause swaying and weaving of a moving trailer. Various trailer equipment attachments are available to help adjust the tongue weight of bumper-pull loads and to help reduce swaying, thereby aiding the stability of the rig. Most states in the USA require a safety chain near where the trailer is attached to the pulling vehicle and some, in addition, require a breakaway system which automatically sets the brakes on the trailer should the trailer become disconnected during travel. It is highly recommended that the driver make a last careful inspection of the entire rig just prior to the start of a journey to catch hook-up and other safety problems.

Little research in trailer design has been done, but recently Smith *et al.* (1996a,b) found that, while a two-horse bumper-pull trailer having leaf-spring suspension with bias-ply tyres produced a smoother ride than the same trailer having torsion-bar suspension with normal-pressure radial tyres, no difference in the horses' well-being was found for each type after a 24-h haul.

It is up to the purchaser to check for sound engineering design and purchase only equipment made for the purpose intended. The hitch, axle and brake capacity for trailers and pulling capacity and weight ratings of pulling vehicles and vans should be acquired from the vehicle and trailer manufacturers to determine their suitability for the purpose intended (i.e. total weight). Try to include a safety margin in the equipment. Horse owners must consider that they are transporting an animal that is large, top heavy and, at inopportune times, fractious. These factors readily contribute to sway and weaving of the vehicle/trailer at high speeds. Also, uniform tyre pressure, load-weight distribution and sound undercarriage structure are other factors under the control of the driver that contribute to trailer and vehicular stability on the road. Poor driving technique, neglect of maintenance, improperly used equipment and transport of valuable horses do not mix and, more often than necessary, result in a tragic accident.

Careful maintenance of the running soundness of the motor and wearable parts will reduce the chances of being stranded on the roadside with animals. Extreme cold or hot weather conditions and difficult horses may contribute to the seriousness of a road breakdown. Horse owners should choose the transportation method, design and size best suited to their horse's behaviour and size.

## Training the horse to load, haul and unload

Horses to be transported can be separated into two groups: halter-trained and untrained (or insufficiently trained). There have been no reports in the literature of behavioural responses to loading for transit or of the easiest way to train a horse to accept loading and transport. Recreational horse users and trainers have been dealing with the situation by a great variety of methods. What follows are some common-sense suggestions based upon the horse's normal behavioural patterns.

Horses that have been well trained to be individually handled by halter should be taught to properly load well before the day of an anticipated transport, in order to reduce the stress of the first haul. Loading training may take as little as a couple of days to as long as several weeks. Generally, a longer training period is required for teaching horses to enter and stand in two-horse trailers than in more spacious-type transports (stock trailer or vans). Loading training should be slow and deliberate and without scaring the horse or the use of excess force. If the animal panics and fights or falls, the procedures are in error and need rethinking. Frequently, loading a horse that is already a good loader first and in the presence of the horse learning and/or careful use of food rewards along with asking for only one step at a time, will load most horses. Mares and foals are loaded best by holding the mare at the trailer entrance and loading the foal using two people, one on each side, that lock arms behind the foal and nudge it gently forward straight into one side of the trailer. One handler restrains the foal in the trailer while the mare is loaded. This method has an added advantage of enticing a poor-loading mare to load in order to stay with her foal. Loading the mare first usually causes the mare to lose sight of her foal which may cause her to struggle in the trailer. If there is insufficient time for the loading training then these horses should be treated like untrained horses when loaded and transported. Tranquillization sometimes helps with problem or inexperienced loaders when there is insufficient time for proper training. Horses such as suckling foals and youngsters, which may lead using the halter but do not tie, back up or stand well, need special treatment when loading and hauling in individual stalls.

## Medical preconditioning, case in transit and stress monitoring

Several weeks prior to moving a horse, a licensed veterinarian should be consulted to determine the proper disease testing, vaccinations and paperwork required for its intrastate, interstate or international transit. Requirements vary from state to state and country to country, and 2 or more weeks may be needed to accomplish the disease vaccinations and required tests. The veterinary-issued interstate health certificate is usually acquired a few hours or days before transit.

Transits of 4–12 h or longer tend to be measurably stressful and horses will reduce their water and feed intakes significantly while in a moving vehicle or trailer (Anderson *et al.*, 1985; Traub-Dargatz *et al.*, 1988; Mars *et al.*, 1991; Smith *et al.*, 1996b; Friend *et al.*, 1998). Therefore consideration should be given to body temperature monitoring (pre-, post- and during transit) and giving immunostimulants, prophylactic antibiotics and fluids (pre- and during transit) (Nestved, 1996). These precautions aid the immune system and functions of the kidneys and gastrointestinal tract.

The transport personnel should stop for brief rest periods and offer water every 3–6 h of transit to encourage the consumption of water and hay, and, after every 16–24 h of transit, a complete offloading of the horse for an extended 12-h rest period is recommended. Research has shown that even normal, healthy horses offered water *en route* will dehydrate, and that the consumption of food and water is greatly increased when the van/trailer is standing rather than moving (Kusunose and Torikai, 1996). Transits of up to 24 h were tolerated by healthy horses that were rested in a stopped transport and offered water for 15 min out of every 4 h *en route*. They made an uneventful recovery from the mild dehydration (Smith *et al.*, 1996b). However, some healthy horses, after 24 h of transit without water and in hot environments, became severely dehydrated, fatigued and unsuited to continue travel. Some horses refuse all food and drink while being transported and may need special care during long trips (Friend *et al.*, 1998). Very cold weather may require that horses, especially in more open trailers, be blanketed while *en route*; however, in hot environments, care should be taken not to leave a horse in a parked trailer/van in the sun, as heatstroke can occur from extreme temperature rises within it. The use of leg wraps to protect the lower limbs and head bumper guards are not recommended unless the horse is completely accustomed to them, they are properly applied and checked on periodically *en route*; and they are not kept on for overly long periods of time. Horses which have elevated temperatures and show signs of infectious diseases should not be transported, especially long distances. A horse that begins to show a fever while in transit should be offloaded at the earliest opportunity and receive medical supervision and rest until healthy enough to resume travel.

## Transport stress and post-transport performance

Limited research indicates that short (~1 h) transits just prior to a submaximal exercise performance are not detrimental to the horse (Beaunoyer and Chapman, 1987; Covalesky *et al.*, 1991; Russoniello *et al.*, 1991). For normal healthy horses, longer transits (4–24 h), even though they produce some measurable changes in weight loss, heart rates, dehydration and some metabolites, hormones and other blood factors, including cortisol (Clark *et al.*, 1988, 1993; White *et al.*, 1991; Smith, 1996b; Friend *et al.*, 1998) do not appear to be injurious to a horse's general well-being or health and require only 1–2 days of recovery for most horses. Inexperienced show horses, young horses being hauled for the first time and horses experiencing very rough transport conditions where they were exposed to quick starts and abrupt stops were found to be moderately to highly stressed (Covalesky *et al.*, 1991; Russoniello *et al.*, 1991; Kusunose and Torikai, 1996). Continuous transport of breeding mares for 9–12 h, which produced measurable stress, did not interfere with the normal reproductive functions of the oestrus cycle and early gestation (Baucus *et al.*, 1990a,b).

Mild to severe respiratory dysfunction and disease changes, even pneumonia, can occur during or following transportation by ground and air and tend to remain for several days to weeks after clinical signs of disease disappear (Anderson *et al.*, 1985; Traub-Dargatz *et al.*, 1988). These respiratory problems may be related to high-head-position restraint during transit (McClintock *et al.*, 1986; Racklyeft and Love, 1990). However, Smith *et al.* (1996b), who studied horses during and after 24 h of transit, speculated that exposure to pathogenic agents that initiate injury to the respiratory epithelium before or during transport and trailers that are not well ventilated or horses that have a greater individual stress response may be responsible for the development of respiratory disease post-transport.

## Transport of injured and rescued horses

Many racetracks and large horse facilities now have specially designed horse trailers for the transport of severely injured, non-ambulatory horses. After sedation and stabilization of its injury, either the horse is loaded into the special trailer by a very low mobile floor on to which the horse is initially slid or rolled, or the horse is raised and moved in a harness using an overhead hoist system suspended from a beam which extends from inside the trailer. Horses have even been rescue-airlifted out of inaccessible areas via helicopter, using a special overhead support device and body sling developed by C.D. Anderson (Madigan, 1993).

## Current research related to transportation

There is still little research addressing horse trailer and van design in the scientific literature. However, Smith *et al.* (1996a) evaluated several combinations of vehicle suspensions (lead-spring and torsion bar), tyre types, inflation rate and shock-absorber use for smoothness of ride and common frequencies of vibration using a two-horse, bumper-pull, tandem-axle, forward-facing trailer. The leaf-spring suspension with low-pressure radial tyres (or bias-ply tyres) and without shock absorbers provided the smoothest ride. The torsion-bar suspension combined with normal-pressure radial tyres was the roughest. Shock absorbers did not improve the ride quality. Horses travelling on the right side of the trailer experienced more vibration than horses on the left side which the researchers thought might be caused by the poor conditions of asphalt roads near the shoulders. The International Air Transport Association has standards for horse stalls carried on aircraft in their Live Animals Regulations (Doyle, 1988).

Several researchers have looked at the relationship of the horse's position in the trailer and others have worked to evaluate well-being during longer transits and transport conditions for slaughter horses. The expectation of moving a number of the world's leading performance horses to Sydney, Australia, for the 2000 Olympics brought researchers together in early 1999 for a workshop on equine transportation stress (Miguarese, 1999). This may result in the initiation of more research and new guidelines for equine transport. The need for such research was brought to the fore when several horses transported from the USA to Dubai in the Middle East for the 1998 World Endurance Championships experienced mild tying-up to full-blown, massive myositis (muscle damage) when exercise was initiated post-shipping. The horses had been confined to their shipping boxes for up to 58 h during their flight and layover in Europe (Teeter, 1999). The latest advance in horse transport appears to be a report that you can now Federal Express your horse across the USA via their air-transport system (Bryant, 1999).

Further evidence is accumulating to indicate that horses do have a preference for body position when in transit. Waran *et al.* (1996) found lower heart rates in horses facing rearwards vs. forward in the direction of travel. Horses facing backwards tended to rest on their rumps more. When facing forward, horses tended to move more frequently, hold their necks higher than normal and vocalize more frequently. The authors postulated that rearward-facing horses could use their front limbs more effectively than their hind limbs to balance for lateral trailer movements, placed more weight on the front limbs and protected their heads better. Even loading methods and entrances may make a difference to the horse, as walking into a van from a platform at the same level and backing into a stall produced lower peak heart rates than walking up a sloped ramp directly into the stall. Kusunose and Torikai (1996) observed the behaviour of six pairs of untethered yearling Thoroughbred horses (who had been transported only once previously) in a horse-carrying vehicle and found the time spent feeding was only 10.5% of the total behaviour activity when the vehicle was moving, compared with 67% and 64% when it was parked or parked idling. Half of their driven pairs were exposed to five repeated abrupt stops during each driving trial and the other half to normal slow acceleration and deceleration driving conditions without the abrupt stops. The normal-driving pairs increased their amount of backward facing behaviour (to the direction of travel) and decreased the number of body position changes with each succeeding driving trial, while the abrupt-stop pairs maintained a high incidence of changing position and could not seem to settle into a favoured standing position when driven. The researchers felt that, since the number of body direction changes did not decrease with the number of trials in either the parking or the idling periods, it appeared that the abrupt stops created constant and tremendous stress for the horses. Smith *et al.* (1994) found that mature horses spent more time facing backward to the direction of travel when the trailer was in motion but not when it was parked. Several horses displayed strong individual preferences for the directions they faced during road transport. They also found higher heart rates when the trailer was moving but no difference in heart rate when the horse was tethered facing forward or backward vs. untethered.

In partial disagreement with past research, Smith *et al.* (1996b) studied four horses during transport for 24 h in a two-horse trailer and could detect no decrease in pulmonary (respiratory) function, using aerosol clearance rates, but changes were observed in the red blood cell count, packed-cell volume, haemoglobin, plasma protein and cortisol. The horses lost weight and were slightly dehydrated and water and hay intake rates were lower during transport than pretransport. Heart rates were higher only during the first 120 min of travel. Smith *et al.* (1996b) also looked at trailer environment during transit and found that ammonia and carbon monoxide concentrations in the trailer during transport were within acceptable limits for human exposure; however, respirable articulates (dust particles, probably from the hay) in the atmosphere were too high. The authors did not address head-restraint position–respiratory interactions found by other researchers, but instead suggested that exposure to pathogens either pre- or during shipping and poorly ventilated transports may be the cause of respiratory illnesses. Nestved (1996) has further substantiated the seriousness of respiratory illnesses during transit by finding a 60.9% illness rate (upper respiratory diseases requiring treatment) in non-treated controls (142/233 horses);

however, he managed to reduce this to an 18.4% incidence of disease by preshipment administration of the immunostimulant, *Propionibacterium acnes* (EQSTIM™ Immunostimulate) (40/217 horses) in horses being transported from 390–2300 miles with transit times of 8–50 h.

A few well-done recent studies looking at long-distance shipping stress and slaughter transport have been conducted. Friend *et al.* (1998) made extensive physiological measures of 30 untethered horses on four treatment groups penned with water, penned without water, transported with water and transported without water. Twice as many horses were assigned to the transported as the penned groups. The transported groups contained in one 16-m-long open-top livestock trailer and tractor rig were driven for 4 h and offloaded for 1 h in each 5-h period continuously, until, after five trips (24 h), it was determined that three horses (all in the transport-without-water group) were not fit enough to continue. Two of these three horses had elevated body temperatures (39.6 and 40.6° C (104–105° F)) and the third was classified as too weak to continue. The transported-with-water group drank less water (offered during the last 10 min of each 1-h break) than the penned-with-water group (20.9 vs. 38.2 litres, respectively). Physical measures, such as capillary refill time, mucous membrane score and skin turgor, were not useful predictors of dehydration or welfare assessment. Body temperature, serum sodium and total serum protein concentrations were more useful but had great variation. These researchers did not recommend any character (factor) that they would use to predict which horses were approaching a critical condition. Blood values indicated that horses had hypertonic dehydration, so rehydration with plain water should be done rather than adding electrolytes, which will exacerbate the loss of water. Although the transported horses drank less than the penned horses, they appeared to consume enough to delay severe dehydration. All horses starting the experiment were in good, fit condition and calm, easy-to-handle individuals accustomed to extensive handling, and the researchers were concerned that less tame and more stressed horses, like those often found in slaughter transport, might be more reluctant to drink and that, under hot environmental conditions, 24 h is probably the longest period horses should be transported without access to water. Australia will soon publish Codes of Practice that will establish 36 h as the longest interval that mature non-lactating horses can be transported without unloading for 12 h of rest with access to feed and water. Friend and co-workers (1998) further stated that neither of these standards were based on research that has been published in scientific journals and indicated that the European Council of the Animal Transportation Association recently published recommendations for animals transported by road in which all horses had to be fed and offered water at least every 15 h and preferably every 6 h, with recommendations that horses not travel in groups larger than four or five and that foals and young horses be able to lie down when journeys exceed 24 h. Some commercial trailers being used to transport slaughter horses in Europe are being fitted with fold-out water troughs.

Stull (1998) reported research to help formulate US Department of Agriculture (USDA) regulations for the 1995 Senate Bill (SB 2522) entitled Humane and Safe Commercial Transportation of Horses for Slaughter Act (Anon., 1996). Nine loads (five straight-deck trailers, four potbelly/double-deck trailers) totalling 306 horses were studied on trips to packing plants with distances of 230–963 miles (370–1550 km) in 05.45–30.00 hour lengths. The differences in some pre- and post-transit factors were measured. The straight-deck trailer had greater changes in white blood count, neutrophils, lymphocytes, N : L ratio, lactate, body weight and rectal temperature. This makes the straight-deck trailer appear more stressful for transport than the potbelly trailer, but this could have been caused by poor ventilation in the particular straight-deck trailer used in the study, since the climate was described as hot and humid. Stocking density (1.24–1.54 sq. m per horse) affected changes in body weight and white blood cells (WBC), neutrophils, lymphocytes and N : L, while duration/distance of travel affected rectal temperature, neutrophils and lactate. Only 8% of the straight-deck trailer horses were injured, while 29.2% were injured in the potbelly trailer, the most prevalent location of injuries (abrasion/lacerations) being the head and the face. No horses died in transit.

A survey of trucking practices and injury to slaughter horses was conducted by Grandin *et al.* (1999), consisting of 63 trailer loads (1008 horses) arriving at two slaughter plants in Texas in July and August of 1998. Nearly half (49%) of the horses were transported on goose-neck trailers, while 42% were on double-deck (potbelly) and 9% on straight single-deck semi-trailers. The authors found that 92% of the horses arrived in good shape, while 7.7% had severe welfare problems and 1.5% were not fit for travel. Most of the severe welfare problem horses had conditions caused by owner neglect or abuse, which would have been present before transport, and only 1.8% (18 horses) had transport and marketing injuries severe enough to be rated a severe welfare problem. Examples given of origin of welfare problems were: loaded with a broken leg, emaciated, foundered, racehorses with bowed tendons and horses that were too weak to be transported. Most of the injuries acquired during transit were to the head, face, withers, back, croup and tailhead. Some moderately severe back injuries were attributed to the double-deck trailers being too low for tall horses and internal trailer ramps which are not well suited for horses. Horses in double-deck trailers appeared to unload better at night, where the authors postulated that they were attracted to the lighted barn. The authors felt that fighting (biting and kicking) due to dominance expression in loose groups was a major problem, because 13% of the carcasses had bruises caused by bites or kicks and 55% of all carcass bruises were caused by bites or kicks. Also, loads from dealers that picked up horses from more than one auction had more external injuries and carcass bruises than direct loads, probably from repeated mixing of strange horses.

These authors made additional observations during 1 day of a large horse sale (New Holland sale in Pennsylvania) in July 1998, where 168 horses were sold. At this sale, they found that all horses were individually handled with a halter or were tied up (in contrast to group slaughter shipments which are handled loose). Most of the horses arrived either on goose-neck trailer (90%) or horse trailers. Fresh abrasions were found on only five horses (4%) and all were minor, and 11.6% were classified as welfare problem animals for various reasons (skinny 3.5%, behaviour problems 7%, physically abused 1.1%). All horses were fit for travel. Prices ranged from $200 to over $1000 and the sale company did not accept horses which were severely lame or in very poor body condition. At this sale, most of the injured horses were located in the 'drop-off' pens, where dealers can unload horses bought at a previous sale for temporary rest, feeding and watering. These horses were loose in their pens and tended to fight with the dominant individuals, causing injury to the less dominant.

Overall recommendations of the authors for slaughter-horse handling and transport include the following:

1. Educate horse owners that they are responsible for horse welfare.
2. Horse associations should all have animal care guidelines.
3. Station USDA/APHIS-trained welfare inspectors in slaughter plants.
4. Fine individuals who transport horses unfit for travel.
5. Segregate aggressive mares and geldings in the same manner as stallions.
6. Improve horse identification.
7. Implement procedures to immediately euthanize horses with severe injuries, such as broken legs, when they arrive during slaughter-plant closing hours.
8. Inspect horse transport vehicles at truck weigh stations and at auctions.
9. To prevent transport of slaughter horses to Mexico or underground markets, the four currently existing horse slaughter plants should be encouraged to remain open. A lack of slaughter facilities will increase the number of horses that will die from neglect.
10. Double-deck trailers should not be used to transport tall horses.
11. Educate horse owners to improve training methods in order to prevent behaviour problems that can cause a horse to be sold for slaughter.

## Conclusions

Limited research to date indicates that, while horses do show an acute physiological stress response to transportation, with proper care the response is not long-lasting nor does it appear to interfere with either exercise or reproduction. More research is needed on the accumulated stress from extended transit, with emphasis on prevention of transit dehydration, on post-transit performance with regard to horse metabolism and muscle function, and on respiratory and immune factors. Stress in young and naïve horses needs more study. The behavioural aspects of transportation have been little studied and could be closely related to traumatic injuries acquired during loading or in transit. Experiments need to be directed at improved trailer construction for the horse, which should include being able to lower the head, as well as general engineering to improve road stability.

In the USA, USDA slaughter-horse transport standards are in the development stages, but there are still none for the regular horse industry. A greater coordination of efforts between caretakers, regulatory agents, shippers and airport management will reduce the stress of long international horse shipments. With the increased movement of horses and welfare concerns, research should proceed and transportation standards, especially concerning movement of slaughter horses and horses using road transit, should be made, either by the horse industry or by the government.

# References

Anderson, N.V., DeBowes, R.M., Nyrop, K.A. and Dayton, A.D. (1985) Mononuclear phagocytes of transport-stressed horses with viral respiratory tract infection. *American Journal of Veterinary Research* 46(11), 2272–2277.

Anon. (1996) *Safe Commercial Transportation of Horses for Slaughter Act of 1995.* S 1283, Pub L. 104–127, Title IXDC, Subtitle A, SSU901 to 905, 4 April, 1996; 110 Stat 1184.

Arnold, G.W. (1984/85) Comparison of the time budgets and circadian patterns of maintenance activities in sheep, cattle and horses grouped together. *Applied Animal Behaviour Science* 13, 19–30.

Baer, K.L., Potter, G.D., Friend, T.H. and Beaver, B.V. (1983) Observational effects on learning in horses. *Applied Animal Ethology* 11, 123–129.

Baker, A.E.M. and Crawford, B.H. (1986) Observational learning in horses. *Applied Animal Behaviour Science* 15, 7–13.

Baucus, K.L., Ralston, S.L., Nockels, C.F., McKinnon, A.O. and Squires, E.L. (1990a) Effects of transportation on early embryonic death in mares. *Journal of Animal Science* 68, 345–351.

Baucus, K.L., Squires, E.L., Ralston, S.L., McKinnon, A.O. and Nett, T.M. (1990b) Effect of transportation on the estrous cycle and concentrations of hormones in mares. *Journal of Animal Science* 68, 419–426.

Beaunoyer, D.E. and Chapman, J.D. (1987) Trailering stress on subsequent submaximal exercise performance. In: *Proceedings of 11th Equine Nutrition and Physiology Symposium.* Oklahoma State University, Stillwater, Oklahoma, pp 379–384.

Bryant, J.O. (1999) Dressage competition nurtures tomorrow's talent. *Horse Show* 62(3), 34–37.

Clark, D.K., Dellmeier, G.R. and Friend, T.H. (1988) Effect of the orientation of horses during transportation on behavior and physiology. *Journal of Animal Science* 66 (Suppl. 1), 239 (abstract).

Clark, D.K., Friend, T.H. and Dellmeier, G.R. (1993) The effect of orientation during trailer transport on heart rate, cortisol and balance in horses. *Applied Animal Behaviour Science* 38, 179–189.

Clarke, J.V., Nicol, C.J., Jones, R. and McGreevy, P.D. (1996) Effects of observational learning on food selection in horses. *Applied Animal Behaviour Science* 50, 177–184.

Covalesky, M., Russoniello, C. and Malinowski, K. (1992) Effects of show-jumping performance stress on plasma cortisol and lactate concentrations and heart rate and behavior in horses. *Journal of Equine Veterinary Science* 12(4), 244–251.

Cregier, S.E. (1982) Reducing equine hauling stress: a review. *Journal of Equine Veterinary Science* 2(6), 186–198.

Cregier, S.E. (1989) Transporting the horse: from BC to AD. *Live Animal Trade and Transport Magazine* April, 39.

Dougherty, D.M. and Lewis, P. (1993) Generalization of a tactile stimulus in horses. *Journal of the Experimental Analysis of Behavior* 59, 521–528.

Doyle, K.A. (1988) The horse in international commerce–regulatory considerations in an Australian perspective. *Journal of Equine Veterinary Science* 8(3), 227–232.

Feh, C. and de Mazières, J. (1993) Grooming at a preferred site reduces heart rate in horses. *Animal Behaviour* 46, 1191–1194.

Fernandez-Diaz, M.D.P. (1990) Effects of L-tryptophan on the stress response of Thoroughbred yearlings. Master of Science thesis of the University of Florida, pp. 82–130.

Fiske, J.C. and Potter, G.D. (1979) Discrimination reversal learning in yearling horses. *Journal of Animal Science* 49, 583–588.

Flannery, B. (1997) Relational discrimination learning in horses. *Applied Animal Behaviour Science* 54, 267–280.

Friend, T.H., Martin, M.T., Householder, D.D. and Bushong, D.M. (1998) Stress responses of horses during a long period of transport in a commercial truck. *Journal of the American Medical Association* 212(6), 838–844.

Grandin, T., McGee, K. and Lanier, J.L. (1999) Prevalence of severe welfare problems in horses that arrive at slaughter plants. *Journal of the American Veterinary Medical Asoociation* 214 (10), 1531–1533.

Haag, E. L., Rudman, R. and Houpt, K.A. (1980) Avoidance, maze learning and social dominance in ponies. *Journal of Animal Science* 50, 329–335.

Heird, J.C., Lennon, A.M. and Bell, R.W. (1981) Effects of early experience on the learning ability of yearling horses. *Journal of Animal Science* 53(5), 1204–1209.

Heird, J.C., Whitaker, D.D., Bell, R.W., Ramsey, C.B. and Lokey, C.E. (1986) The effects of handling at different ages on the subsequent learning ability of 2-year-old horses. *Applied Animal Behaviour Science* 15, 15–25.

Houpt, K.A. (1986) Stable vices and trailer problems. *Veterinary Clinics of North America: Equine Practice* 2(3), 623–633.

Houpt, K.A. (1991) *Domestic Animal Behavior for Veterinarians and Animal Scientists.* Iowa State University Press, Ames.

Houpt, K.A. and Houpt, T.R. (1988) Social and illumination preferences of mares. *Journal of Animal Science* 66, 2159–2164.

Jezierski, T., Jaworski, Z. and Górecka, A. (1999) Effects of handling on behaviour and heart rate in Konik horses: comparison of stable and forest reared youngstock. *Applied Animal Behaviour Science* 62, 1–11.

Kusunose, R. and Torikai, K. (1996) Behavior of untethered horses during vehicle transport. *Journal of Equine Science* 7(2), 21–26.

Lagerweij, E., Nelis, P.C., Weigant, V.M. and van Ree, J.M. (1984) The twitch in horses: a variant of acupuncture. *Science* 225(4667), 1172–1174.

Leadon, D., Frank, C. and Backhouse, W. (1989) A preliminary report on studies on equine transit stress. *Journal of Equine Veterinary Science* 9(4), 200–202.

Lindberg, A.C., Kelland, A. and Nicol, C.J. (1999) Effects of observational learning on acquisition of an operant response in horses. *Applied Animal Behaviour Science* 61, 187–199.

McBane, S. (1987) *Behaviour Problems in Horses*. David and Charles, North Pomfret, Vermont. 304 pp.

McCall, C.A. (1990) A review of learning behavior in horses and its application in horse training. *Journal of Animal Science* 68, 75–81.

McCann, J.S., Heird, J.C., Bell, R.W. and Lutherer, L.O. (1988a) Normal and more highly reactive horses. I. Heart rate, respiration rate and behavioral observations. *Applied Animal Behaviour Science* 19, 201–214.

McCann, J.S., Bell, R.W. and Lutherer, L.O. (1988b) Normal and more highly reactive horses. II. The effect of handling and reserpine on the cardiac response to stimuli. *Applied Animal Behaviour Science* 19, 215–226.

McClintock, S.A., Hutchins, D.R., Laing, E.A. and Brownlow, M.A. (1986) Pulmonary changes associated with flotation techniques in the treatment of skeletal injuries in the horse. *Equine Veterinary Journal* 18(6), 462–466.

Mackenzie, S.A. (1994) *Fundamentals of Free Lungeing: an Introduction to Tackless Training*. Half Halt Press, Boonsboro, Maryland, 97 pp.

Mader, D.R. and Price, E.O. (1980) Discrimination learning in horses: effects of breed, age and social dominance. *Journal of Animal Science* 50, 962–965.

Madigan, J. (1993) Evaluation of a new sling support device for horses. *Journal of Equine Veterinary Science* 13(5), 260–261.

Maier, S.F. and Seligman, M.E.P. (1976) Learned helplessness: theory and evidence. *Journal of Experimental Psychology (General)* 105, 3–46.

Mal, M.E. and McCall, C.A. (1996) The influence of handling during different ages on a halter training test in foals. *Applied Animal Behaviour Science* 50, 115–120.

Mal, M.E., Friend, T.H., Lay, D.C., Vogelsang, S.G. and Jenkins, O.C. (1991) Physiological responses of mares to short term confinement and social isolation. *Equine Veterinary Science* 11(2), 96–102.

Mal, M.E., McCall, C.A., Cummins, K.A. and Newland, M.C. (1994) Influence of preweaning handling methods on post-weaning learning ability and manageability of foals. *Applied Animal Behaviour Science* 40, 187–195.

Mars, L.A., Kiesling, H.E., Ross, T.T., Armstrong, J.B. and Murray, L. (1992) Water acceptance and intake in horses under shipping stress. *Journal of Equine Veterinary Science* 12(1), 17–20.

Miguarese, N. (1999) A smoother road to Sydney. *Horse Show* 62(4), 23.

Miller, R.M. (1991) *Imprint Training of the Newborn Foal*. Western Horsemen, Colorado Springs, Colorado, 143 pp.

Miller, R.W. (1975) *Western Horse Behavior and Training*. Dophin Books of Doubleday and Company, Garden City, New York. 305 pp.

Nestved, A. (1996) Evaluation of an immunostimulant in preventing shipping stress related respiratory disease. *Journal of Equine Veterinary Science* 16(2), 78–82.

Racklyeft, D.J. and Love, D.N. (1990) Influence of head posture on the respiratory tract of healthy horses. *Australian Veterinary Journal* 67(11), 402–405.

Robinson, N.E. (1992) *Current Therapy in Equine Medicine*, 3rd edn. W.B. Saunders, Philadelphia, 847 pp.

Russoniello, C., Racis, S.P., Ralston, S.L. and Malinowski, K. (1991) Effects of show-jumping performance stress on hematological parameters and cell-mediated immunity in horses. In: *Proceedings of 12th Equine Nutrition and Physiology Symposium*. University of Calgary, Calgary, Canada, pp. 145–147.

Sappington, B.F. and Goldman, L. (1994) Discrimination learning and concept formation in the Arabian horse. *Journal of Animal Science* 72, 3080–3087.

Sappington, B.K.F., McCall, C.A., Coleman, D.A., Kuhlers and Lishak, R.S. (1997) A preliminary study of the relationship between discrimination reversal learning and performance tasks in yearling and 2-year-old horses. *Applied Animal Behaviour Science* 53, 157–166.

Saslow, C.A. (1999) Factors affecting stimulus visibility for horses. *Applied Animal Behaviour Science* 61, 273–284.

Slade, L.M. (1987) Trailer transportation and racing performance. In: *Proceedings of 11th Equine Nutrition and Physiology Symposium*. Oklahoma State University, Stillwater, Oklahoma, pp. 511–514.

Smith, B.L., Jones, J.H., Carlson, G.P. and Pascoe, J.R. (1994) Body position and direction preferences in horses during road transport. *Equine Veterinary Journal* 26(5), 374–377.

Smith, B.L., Miles, J.A., Jones, J.H. and Willits, N.H. (1996a) Influence of suspension, tires, and shock absorbers on vibration in a two-horse trailer. *Transactions of the American Society of Agricultural Engineers* 39(3), 1083–1092.

Smith, B.L., Jones, J.H., Hornof, W.J., Miles, J.A., Longworth, K.E. and Willits, N.H. (1996b) Effects of road transport on indices of stress in horses. *Equine Veterinary Journal* 28(6), 446–454.

Stull, C.L. (1998) Health and welfare parameters of horses commercially transported to slaughter. *Journal of Animal Science* 76 (Suppl. 1), 88.

Teeter, S. (1999) Valerie Kanavy, new world endurance champion. *Horse Show* 62(2), 32–35.

Tellington-Jones, L. and Bruns, U. (1985) *An Introduction to the Tellington-Jones Equine Awareness Method*. Breakthrough Publications, Millwood, New York. 180 pp.

Traub-Dargatz, J.L., McKinnon, A.O., Bruyninckx, W.J., Thrall, M.A., Jones, R.L. and Blancquaert, A.-M.B. (1988) Effect of transportation stress on bronchoalveolar lavage fluid analysis in female horses. *American Journal of Veterinary Research* 49(7), 1026–1029.

Tyler, S.J. (1972) The behaviour and social organization of the New Forest ponies. *Animal Behaviour Monographs* 5, 85–196.

Waran, N.K. (1993) The behaviour of horses during and after transport by road. *Equine Veterinary Education* 5(3), 129–132.

Waran, N.K., Robertson, V., Cuddeford, D., Kokoszko, A. and Marlin, D.J. (1996) Effects of transporting horses facing either forwards or backwards on their behaviour and heart rate. *Veterinary Record* 139, 7–11.

Waring, G.H. (1983) *Horse Behavior: the Behavioral Traits and Adaptations of Domestic and Wild Horses, Including Ponies*. Noyes Publications, Park Ridge, New Jersey. 292 pp.

White, A., Reves, A., Godoy, A. and Martinez, R. (1991) Effects of transport and racing on ionic changes in thoroughbred race-horses. *Comparative Biochemistry and Physiology. A. Comparative Physiology* 99, 343–346.

Woodhouse, B. (1984) *Barbara's World of Horses and Ponies*. Summit Books, New York. 127 pp.

Wright, M. (1973) *The Jeffery Method of Horse Handling*. Griffin Press, Netley, South Australia, 92 pp.

# Deer Handling and Transport 17

## L.R. Matthews

*Animal Behaviour and Welfare Research Centre, Ruakura Agricultural Centre, Hamilton, New Zealand*

## Introduction

The relatively recent expansion of deer farming around the world has necessitated the development of appropriate handling facilities and practices. Deer belonging to the genus *Cervus* are the most widely distributed and numerously farmed species and include red deer (*C. elaphus*) and wapiti (elk) (*C. elaphus nelsoni*). The next most common farmed species is the fallow (*Dama dama*). Smaller numbers of other species are farmed including rusa (*C. timorensis*), sika (*C. nipon*), chital (*Axis axis*) and white-tail (*Odocoileus virginianus*).

The importance of accommodating the unique behavioural character-istics of deer into the design of safe and efficient handling facilities is well recognized (Kilgour and Dalton, 1984). In the absence of carefully controlled scientific studies, facility design has been based largely on general farmer knowledge of deer behaviour and trial and error evaluation of facilities. Nevertheless, this approach has yielded facilities that work well. This is a continually evolving process and some of the designs presented here have only just been developed.

The range of systems for handling are described in this chapter. The effectiveness of these systems and their use have been systematically assessed with reference to fundamental aspects of deer behaviour. Although there are variations in the behaviours of deer belonging to different species, the main one influencing handling relates to the degree of flightiness. The handling facilities and procedures for the less flighty species (red, wapiti and rusa) tend to be somewhat different from the more flighty ones (fallow, chital and white-tail) (English, 1992; Haigh, 1992, 1999; Woodford and Dunning, 1992). Any important distinctions are specifically mentioned.

©CAB *International* 2000. *Livestock Handling and Transport,* 2nd edn
(ed. T. Grandin)

# Basic Behaviour Patterns of Deer

Taming, sensory capacities, physical agility, social organization and learning ability are among those aspects of the basic behavioural responses of animals that need to be considered in the design of handling facilities. Apart from social organization, there are relatively few scientific studies of these responses and their influence on animal handling. Available information is derived from observational studies of animals in the wild or during farming operations and published reports (where available).

## Sensory and physical capacities

The position of the eyes on the head is similar to cattle and sheep and it can be assumed that the visual abilities of deer would be similar to these animals (Prince, 1977). Thus, deer are likely to have a wide visual field (about 300° with a blind spot to the rear and at ground level to the front of the animal when the head is raised (Hutson, 1985a) and good depth perception in a small area of binocular vision some short distance in front of the head. Inability of several deer species to detect hot flies hovering below the nose supports the notion that there is a blind spot in this area (Anderson, 1975).

Like other ruminants, deer detect moving objects readily but do not respond well to static objects (Cadman, 1966; McNally, 1977). Cattle and sheep appear to have colour vision but the extent to which deer can discriminate colours is not known. Deer are active at night and during the change of light so it must be assumed that they have excellent vision under low light conditions.

The avoidance by deer of humans approaching downwind and the use of gland secretions to mark trees and trails (McNally, 1977) indicate that deer have an acute sense of smell. Their sense of hearing is also well developed. Recorded calls of stags played to groups of hinds advances the breeding season by about 6 days (McComb, 1987). Observations by farmers that deer respond to the 'silent' dog whistle indicate that the effective hearing range extends to about 20 kHz or more.

## Physical agilities

In contrast with the traditional farm animal species, deer are very agile. Most species have little difficulty in clearing barriers of 2–2.5 m in height from a standing position and can accelerate almost instantaneously from 0 to 50 km h$^{-1}$ (Drew and Kelly, 1975). When presented with an obstacle the natural tendency for deer is to lower the head and go under or push through it rather than go over the top (Clift *et al.*, 1985). Within their natural home ranges deer prefer to move along well-defined trails that are used by successive generations (Cadman, 1966).

## Social behaviour

In the wild, deer species that are commonly farmed are group living, with highly organized social structures. Females associate with their dams and remain in matrilineal groupings throughout their lives (Clutton-Brock *et al.*, 1982). Males disperse from the females and live in bachelor herds throughout the year except during the breeding season (rut) when individual stags move into the areas occupied by the females. In favourable habitats groups comprise up to 100 or more animals (Staines, 1974).

Dominance hierarchies are a feature of both male and female groupings (Clutton-Brock *et al.*, 1982). A wide range of agnostic behaviours are used in the establishment and maintenance of these hierarchies. With males, antlers are the primary means of offence and defence. Threat gestures include lowering the head, directing the antlers towards the opponent and lateral body positioning. Animals with the largest body size and antlers tend to have the highest social ranking. During the annual velvet antler growing season and prior to antler hardening, stags use a range of other threat behaviours. These include head-high threats, which precede strikes by the forefeet or rearing on the hind-feet and boxing with the forefeet, kicking with the rear feet, biting or biting threats where the head is tilted slightly, upper lips are raised and a hissing sound is made or the tongue may be protruding and accompanied by teeth grinding (Lincoln *et al.*, 1970; Bartos, 1985). Appeasement is indicated by an outstretched neck posture, turning the head away or movement away from the aggressor. Agnostic interactions between females are similar to those seen in males during the velvet-growing season (Haigh and Hudson, 1993).

The intensity of aggressive interactions between males is strongly influenced by hormonal status. During the late winter and spring when the old antlers are cast and a new set is growing testosterone levels are low. The stags are least aggressive during this period. The shedding of velvet and hardening of the antler corresponds to rising testosterone levels which remain elevated during the rut (Darling, 1937). Aggressiveness is highest during periods of high testosterone levels, especially where animals re-join in groups after the breeding season.

Females move away from the matriarchal groups to give birth (Darling, 1937). The behaviour of the newborn in the first few weeks of life forms the basis of the group lifestyle and some of the responses to stresses in later life. After suckling, the newborn moves away from its mother and lies down in available shelter (Darling, 1937). This behaviour is characteristic of lying-out type species (Lent, 1974) and persists for the first 2–3 weeks of life. For the first few days of this period the calf freezes in response to disturbances (e.g. approaching human) (Kelly and Drew, 1976). Thereafter, a strong flight response to threatening stimuli is shown. At about 3 weeks of age the calf begins to follow its mother. It is at this time that the mother returns to the matriarchal group with the young and the foundations for leader–follower and herding behaviour typical of adult deer are established.

## Flight, domestication, taming and learning

Wild deer are well known for their large flight distance and strong flight response. Wild-caught animals and deer bred in captivity habituate rapidly to human presence and the flight distance reduces to 30–50 m or less (Clutton-Brock and Guinness, 1975; Blaxter *et al.*, 1988). Flight distances are less for animals confined in yards, when approached in familiar vehicles rather than on foot, and where handlers are associated with the feeding of supplements. The flight distance reduces to zero for animals reared by hand from birth. Hand-reared male deer should not be kept as they become extremely aggressive toward humans during the breeding season (Gilbert, 1974). Hand-reared roe deer are especially aggressive (Hemmer, 1988). Animals with a zero flight distance cannot be moved readily other than by attracting with food.

The flight responses of a herd are strongly influenced by the behaviour of the lead animals. When disturbed from a distance, deer orientate their heads toward the intruder and bunch together (Humphries *et al.*, 1990; Bullock *et al.*, 1993). Upon closer approach by the intruder, the lead animals (or most flighty ones) begin to move away from the source of the disturbance and the remainder of the group follow. If the intruder remains stationary, the leaders usually circle from the front of the group, around the outside facing back toward the intruder and then back up the centre of the mob to the front again (L. Matthews, unpublished observations). If the intruder approaches quickly the group will scatter in all directions. The flight distance for a group is reduced when extremely flighty individuals are removed.

There have been no formal studies of changes in the flightiness of deer over the generations as they have become increasingly domesticated (Kelly *et al.*, 1984) although techniques for measuring temperament and stress are being developed (Matthews *et al.*, 1994; Pollard *et al.*, 1994b; Carragher *et al.*, 1997). Nevertheless, in New Zealand it has become apparent that farmed deer are much less reactive to handling now than when deer farming began 30 years ago using wild-caught stock. Both genetic selection for less flighty animals and improved methods of training deer to handling have no doubt contributed to this effect. Regular gentle contact with the stock and hand feeding of animals (particularly after weaning) seem to be particularly important factors in the taming of deer. Hand-reared animals have been trained to come to a call over distances of 1 km. Wild animals quickly learn to avoid areas which are frequently subject to disturbance by helicopters or humans. Farmed deer learn in one trial to avoid areas where they have been handled aversively

(Pollard *et al.*, 1994a). In addition, deer quickly learn to recognize a regular handler but react with flight to unfamiliar handlers (Bull,1996).

## Behaviour Relevant to Facility Design

### Farm layout

The layout of the farm is centred around the need to confine the herd and control the feeding, breeding and handling of various sub-groups of animals. A farm typically consists of a series of fields (paddocks) linked to a handling facility via a central laneway.

The agility of deer and the need to control their seasonal activities have had a major influence on the design of farm fences and laneways. Perimeter and race fences 2 m in height will discourage most animals from attempting to leap over, as deer seemingly are unable to see the top wire. Under pressure, animals have been known to jump over 2 m fences. Outside the breeding season and when food supply is not restricted, internal fences of between 1.2 and 1.5 m in height will normally contain animals.

Full-height fences are best constructed from 13-wire high-tensile netting with vertical stay wires at 150 mm or 300 mm spacings. Ideally, the fine wires should be about 100 mm apart near the ground, increasing to about 180 mm at the top. The closer spacing at the bottom prevents escapes and entanglement of smaller animals. An even finer mesh size is required to contain newborns.

Electric fencing is being used increasingly on deer farms. Internal fences comprising about four wires to a height of 1.5 m provide a convenient movable structure that will control most animals. Electrified outriggers about 600 mm above the ground will reduce fence pacing and agnostic interactions between stags held in adjacent paddocks. Naive animals should be introduced to electric fences in the presence of trained deer. Visual barriers (e.g. trees) along fence lines may also reduce fence pacing or aggression between animals in neighbouring enclosures.

The ease of movement of deer into and out of paddocks is influenced by the position of the gateways. Deer tend to move more readily up hill than down so gateways are best located at the tops of rises. On flat areas gates work well if sited near the corners. Short wing fences leading to the gateways assist in funneling deer from larger fields. Gate widths should be a minimum of about 3.5 m wide to avoid undue constriction of the flow of the herd.

Laneways linking the paddocks and the handling complex assist the movement of stock around the farm. The design of these raceways has been influenced by several behavioural characteristics of deer. Their natural tendency to bunch (Humphries *et al.*, 1990) and move at speed is facilitated in raceways that are relatively wide, 5 m being a minimum. Wider races (up to 20 m) are more suitable for large herds and can also be used for drafting since the handler can move past deer without encroaching the flight zone. It has been suggested that deer move more readily along laneways incorporating curves (Haigh and Hudson, 1993). This has not been scientifically evaluated, and many farmers find that deer move just as readily along straight laneways.

It is important that straight sections do not lead directly into yards as this creates an apparent dead-end, causing deer to baulk. Deer flow readily into yards if they are turned through a curve on the final approach. This effect is achieved by setting the yards off to one side of the race. Deer tend to 'cut the corners' when moving into and out of yards. Impacts with the walls at corners (which causes bruising) can be reduced by avoiding the use of sharp turns or smoothing the corners.

On occasions, deer approach yards on the run. This, coupled with their poor vision of stationary objects, increases the risk of collision with fences. The visibility of wire fences can be increased by affixing vertical wooden battens some 20–40 m out from the yards and solid boarding (or hessian or plastic mesh) over the final 20 m (Fig. 17.1). It is important to have a gradual increase in visibility of the fences at the yard entrance to avoid creating the impression of a dead-end.

## Yard complex

Deer yard complexes serve four main functions – to hold, draft (sort), close handle and load out (or receive) deer. Close handling requires varying

**Fig. 17.1.**   Laneway fencing on approach to yards.

degrees of restraint depending on the procedure, the species (or breed) of deer and the individual animal. Low levels of restraint for practices such as oral administration (drenching) of anthelmintics or ear tagging can be applied manually in small working pens. Higher levels of restraint which are required for artificial insemination or removal of antlers can be applied in a cradle or purpose-built deer handler.

## Layout and construction

There are innumerable variations in facility layout and construction that permit basic farming operations to be carried out easily. In the past, yard designs were based on a completely enclosed complex containing a central drafting area with several holding pens off it. Near the drafting pen were smaller working pens (in which animals can be manually restrained), a restrainer, a weighing platform and a race leading to a loading ramp (Yerex and Spiers, 1990). Nowadays, farmed deer are generally less flighty and facilities and handling procedures a little more typical of those used with cattle are being utilized. Thus, the indoor facilities for red deer are typically very well lit and well ventilated, and groups of animals are typically handled by one person working in the pens at ground level. Other changes in modern facilities include the greater provision of outdoor holding areas and the elimination of the central drafting pen. The layout of the indoor area is now more comparable to cattle facilities, with small groups of deer brought into larger holding areas (indoors or outdoors) and then drafted off into a number of smaller pens within the facility (Fig. 17.2). Drafting is undertaken in pens by cutting animals out by hand as is common with cattle.

The design may also incorporate pens which contain one or two centrally mounted swinging gates (Fig. 17.2). In these rooms pen size can be adjusted readily to assist with restraint and drafting, and the gates are useful for pushing animals toward load out ramps or other facilities such as restrainers or weighing scales.

The behaviour of deer is critical in relation to the design and use of these yarding facilities. In contrast with other livestock, the need to reduce flightiness has been at least as influential as the need for containment in the design and construction of deer yards. Walls are built to a height of 2.25 m (2.6 m for fallow or 3 m for wapiti) to deter leaping, and are smooth-sided with no sharp corners to reduce injury and bruising in these fast-reacting animals. Secure footing is provided by concrete or other compacted, free-draining flooring material (e.g. sand).

In the past, two common design features were used to assist in reducing flightiness. Yard walls were solidly or completely close-boarded on the assumption that deer unable to see into neighbouring pens would be less reactive to handling or other disturbances in those pens. However, recent informal observations suggest that deer remain more settled when there is a

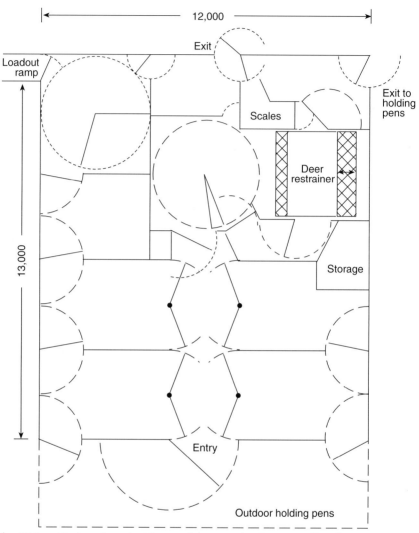

**Fig. 17.2.** Modern deer shed layout, incorporating large, small and adjustable-sized pens, working areas with one or two circular-swinging doors, and a hydraulic deer restrainer with swinging push-up doors at its entrance.

clear view of other nearby animals, approaching handlers or activity that produces unfamiliar noises (Pearse, 1992). Thus, wall designs are open boarded (75 mm spaces for fallow and red deer, 150 mm for wapiti) or mesh (50 mm × 50 mm) above a height of about 1.2 m (for red and wapiti) or 1 m (for fallow) (Fletcher, 1991; Yerex, 1991). Walls can even be open-boarded right down to ground level (provided the gaps between the boards are no more than about 50 mm up to a height of about 1.2 m). Note, it is preferable to place the boards horizontally. This allows a handler, in an emergency to climb the walls easily, and prevents flickering of light between the boards as animals move (which can occur with vertically orientated boards) and disruption to the flow of deer. In close or high pressure working areas, the walls should be solid boarded from the floor to 2.5 m (or 3 m for wapiti).

Another common procedure that has been used in the past to reduce flightiness is to reduce the level of illumination in the pens (Kelly *et al.*, 1984). Control of light levels is usually achieved by constructing a roof over the yard complex and fitting artificial lights with dimmer controls. Although there have been no scientific studies of the effect of illumination on flightiness, practical experience indicates that fractious animals and flighty species such as fallow or white-tail are easier to handle under low light conditions but more tractable deer can be handled without difficulty in well-lit facilities. There is some scientific evidence to suggest that red deer may be less disturbed if handled or penned under low-light conditions (Pollard and Littlejohn, 1994). High light levels have the advantage of allowing more easy inspection of animals for husbandry purposes. In addition, the level of arousal or aggressiveness of deer can be more readily detected should evasive actions be required. Flightiness is also influenced by the space available in a pen. Low ceilings (2 m for red deer and 2.5 m for wapiti) discourage leaping and boxing with the forelimbs and small group sizes reduce flightiness. Smaller space allowances may provide for improved welfare when animals are kept indoors on a long-term basis (Hanlon *et al.*, 1994) although on a short term basis pen size is not an issue in terms of animal wellbeing (Pollard and Littlejohn, 1996).

## Pens and races

The behaviour of deer in groups, especially when disturbed or handled, has a major influence on the design of pens and races. The ideal number and size of pens varies with the number of animals in the herd. Outdoor pens suitable for groups of up to 100 animals should provide 2–6 m² per animal depending on size and breed. Ideally, larger numbers should not be herded together at one time in a single pen. Large groups are usually broken down into smaller groups in outside holding pens soon after yarding. For ease of handling and to reduce stress, these groups should comprise no more than about 25 animals. Large stags and animals housed for longer periods or overnight require a minimum of about 1 m² per animal.

In the working area, practical experience has shown that deer are less likely to trample one another and are easiest to move amongst when group size is limited to five or six animals. To facilitate close handling and manual restraint a space allowance of up to 0.5 m² per 100 kg animal is ideal. Thus, working pens typically measure 1.5 m wide × 1.5 m or 2.0 m long. The walls of the working pen should be solid-sided from floor to ceiling to reduce the possibility of injuries. A variation of this design that allows for rapid handling and the creation of variable sized pens is based on a large pen which can be divided into segments by one or two swing gates mounted on a central pole (Fig. 17.2).

Traditionally, raceways have not been used as working areas for deer because not all animals remain settled long enough for the whole group to be processed. New systems that take advantage of several unique behavioural characteristics of deer have been developed and are eminently suitable for carrying out routine operations such as drenching, vaccinating and tagging in facilities that resemble those used for cattle.

One such system shown in Fig. 17.3 utilizes a U-shaped raceway (N. Cudby and L. Cooney, unpublished data). The wings of the U are 3–4 m long and the base is about 2 m long (Fig. 17.3a). Deer enter the 700 mm wide lane through an offset race linked with the holding pens. The operator works the deer from the inside of the U. The inside race wall (Fig. 17.3b) is solid sided to 1 m (or 1.2 m if the race is not on the same level as the operator's floor level) and above that it is curtained to 2 m. If the side wall is 1.2 m high a catwalk needs to be positioned about 200 mm above floor level. The curtain is positioned to hang to the outside of the catwalk and is closed while animals are being loaded into the race; the operator accesses the deer through slits in the curtain and works on the inside of the curtain. A ceiling consisting of pipes, open boards or mesh is placed at 2 m height to discourage jumping. The outside wall is solid to 2 m or more. Weighing and drafting of animals in the U-race can be achieved from strategically placed weigh-scales and drafting gates, respectively. About two 100 kg animals per meter of race can be handled in the U-shaped race at one time. Sliding doors located at intervals along the race can be used to facilitate animal control and drafting. Ideally, such doors should incorporate a see-through section about 1.2 m from the floor to allow animals to see each other (thereby reducing the likelihood that animals will turn around in the race). One way to position the U-race inside a deer facility is shown in Fig. 17.3c.

This type of race can also be used with wapiti, in which case the curtains can be replaced with pipe railings (running horizontally) and the ceiling may need to be raised a little.

A similar design functions well with fallow deer (Cash, 1987). In this case, the race is straight with the walls and ceiling forming a tunnel and light at one end is used to attract animals into the enclosure. The fallow tunnel-race is about 900 mm high × 310 mm wide. Operator access to the animals is via a 160 mm gap at the top of the inside wall or in the ceiling. A facility for handling white-tail is described by Haigh (1995).

Races are used in several other parts of the yarding complex. Their design varies depending on whether they are used for moving groups of animals (e.g. from holding to drafting pens, work pens or load-out ramps) or for moving single animals (e.g. on to weighing platforms or crushes). Deer move better as a group and this can be facilitated by constructing races at least 1.5 m wide so that animals can move two or three abreast (Grigor et al., 1997b). Wider races

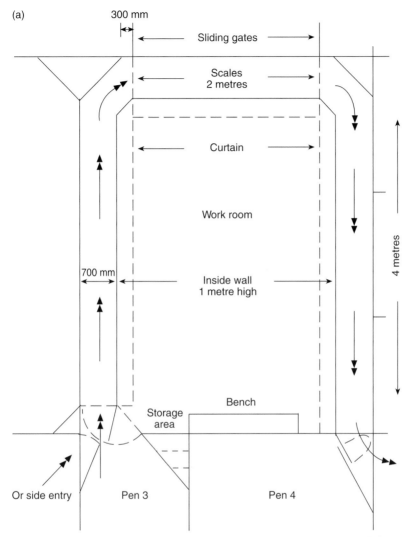

(a)

300 mm

Sliding gates

Scales
2 metres

Curtain

Work room

700 mm

Inside wall
1 metre high

4 metres

Bench

Storage
area

Or side entry          Pen 3                    Pen 4

**Fig. 17.3.** (a) A modern handling race for use with groups of up to 25 red deer. Additional sliding gates can be positioned along the race to facilitate control of the animals. Drafting can be carried out from the race by placing additional gates in the outside wall leading to neighbouring pens. (b) The cross section and dimensions of the race. Note that the curtain hangs 300 mm out from the low side wall, thereby allowing the operator easy movement and access when working the deer. If the raceway is not set below the shed floor (as shown) then the shorter side wall needs to be 1.2 m high. (c) The U-race is shown in relation to holding pens inside a deer handling facility. (*Continued overleaf.*)

should be used with larger mobs. Races for moving single animals should be 600–700 mm wide to prevent animals turning around easily. Wider races may be required for stags with a large antler spread.

(b)

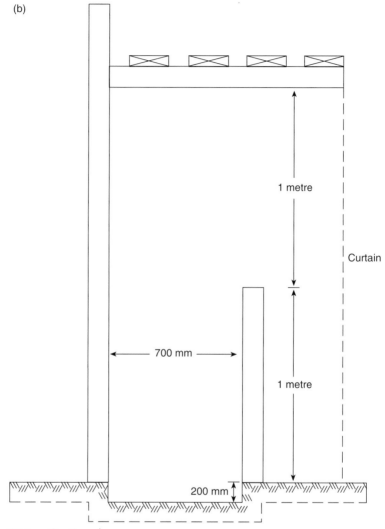

1 metre

Curtain

700 mm

1 metre

200 mm

**Fig. 17.3.** *Continued.*

As has been shown with sheep (Hutson and Hitchcock, 1978), it seems that deer move better in races with solid sides at least part way up the walls (Lee, 1992). These would limit distractions and also reduce the potential for trapping legs and causing injury. Entry to races can be encouraged by the use of forcing gates which swing around behind the animals. Whereas the need for such gates may indicate design faults in a red deer handling system, they are an essential component of safe systems for handling wapiti.

In most respects the general principles of facility design are similar for red deer and wapiti (apart from obvious differences in dimensions of enclosures

(c)

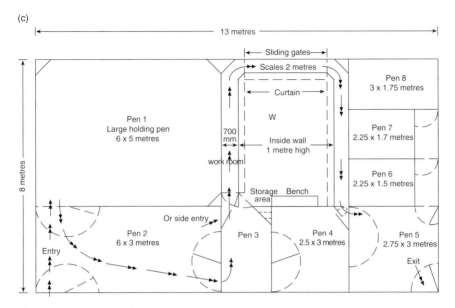

**Fig. 17.3.** *Continued.*

and robustness of construction). Notwithstanding the greater size and strength of wapiti, the tendency for these animals to move less freely than red deer when in close proximity to handlers has led to the development of wapiti-specific pen and race features that facilitate animal flow and human safety. The most advanced designs (Fig. 17.4) are comprised of a series of enclosures that function both as pens and raceways (Thorliefson *et al.*, 1997). These facilities are particularly suitable for less domesticated deer. As with red deer, the smallest enclosure units accommodate two to five animals (2.5–7 m²). These units are so designed that one of the side 'walls' is comprised of two swinging doors, and sliding doors are positioned at either end. The doors can be opened up to construct much larger pens. Reluctant animals can be moved with sliding or swing doors that can be pushed up behind animals. Within the yarding complex, gate widths of 1.2 m or more prevent animals from jamming during movement.

Walkways along the top of the pen walls provide quick access from one part of the facility to another. However, it is not desirable to handle animals from such heights as they often become too flighty and difficult to control. In raceways, animals are best worked from ground level.

Practical experience and scientific evidence (Grigor *et al.*, 1997b) shows that animals move more readily into races that incorporate curves, or where animals have a clear view of an exit. These features are particularly important in encouraging the movement of lead or single animals especially where the race terminates in a dead-end. Grigor *et al.* (1997b) showed that deer, once in

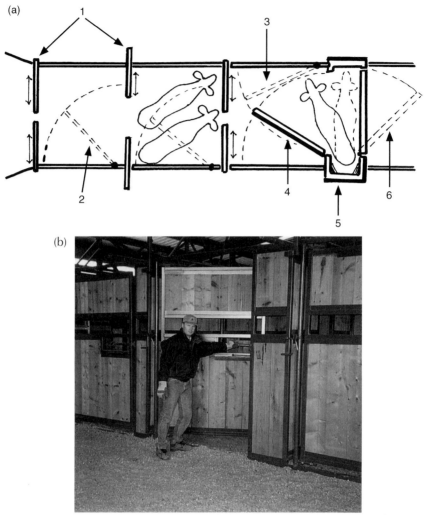

**Fig. 17.4.** (a) Elk (wapiti) handling facility. On both wapiti and red deer farms in Canada producers have developed this design which eliminates the need for a single-file race. Wapiti stay calmer because two animals stand side by side as they move through the system. All fences, gates and walls in the facility are solid. 1. Biparting sliding gates separate the compartments; 2. A pusher gate is used to move the animals to the next compartment; 3. Pusher gate to urge the animal into the padded stall; 4. Side entrance gate on the padded stall; 5. Padded stall which holds the animal for palpation, injections and artificial insemination; 6. Exit gate from stall. The padded stall shown in this diagram fits snugly but does not squeeze the animal and has numerous small doors on both the side and the rear for access to the animal. Velveting is carried out in a conventional deer restrainer which is shown in Fig. 17.5. (b) Dan Sych in Alberta, Canada demonstrates the operation of a pusher gate. All handling is carried out from the ground.

the raceway, move no more rapidly along curved races than straight laneways. However, practical experience suggests that movement through races or several pens adjacent to one another is achieved most readily if baffles, lanes or doors are arranged so as to create a zigzag pathway. Ideally, long straight races should be avoided. Where this is impractical, straight sections should be kept as short as possible and incorporate a corner or curve as close as possible to areas that terminate in a dead-end. There has been no systematic study of the effect of degree of curvature on ease of movement but anecdotal evidence suggests that corners of up to 90° and higher are effective. One report maintains that moving deer around a 135° corner at the intersection of the race and restrainer assists animal flow (Goble, 1991). Thus, curves should immediately precede entry to enclosed yards and slaughter plants, restrainers, weighing platforms, load-out ramps and transport vehicles.

There have been no scientific studies of the effect of floor slope on deer movement. Practical experience during mustering and transport indicates that deer move uphill readily but are hesitant on downward slopes of 20° or more. Given appropriate facilities, deer unload at speed from transport vehicles. Ramps at least as wide as the exit door on transporters (typically 1.2 m) and preferably wider assist unloading. Swinging doors situated along the ramp can be used to push behind animals reluctant to enter the transporter.

## Restraining devices

Restraining devices are an important component in handling complexes where a high degree of animal control is required. As with most other aspects of the deer facilities, there are a great many different restrainer designs or procedures and they are being refined continually (Haigh, 1999). Restrainer designs vary according to degree of restraint required, size of animal and behavioural characteristics of individual animals, breeds or species.

Flighty species (e.g. fallow, chital) or fractious animals invariably require high levels of restraint for most procedures requiring close handling. For less flighty species (and smaller animals) many simple operations, such as drenching, can be carried out with low levels of restraint, e.g. manually or as provided between the animals themselves when small groups (four to six deer are contained within small working pens (2–3 m²)). A second handler located near one wall can be helpful in such situations to prevent deer circling. Deer which do not stand readily when immediately next to the handler will frequently allow human contact if a second animal is manoeuvred between the handler and target animal. Animals can be discouraged from backing up by placing a hand at the rear of the deer.

Other procedures, such as shaving and injecting for disease testing, can be performed most easily under moderate levels of restraint (e.g. in the U-shaped race as described earlier) or behind a swinging door offset from a solid wall.

Reduced light levels, low ceilings and working from a slightly elevated position (e.g. from a 0.5–1 m high catwalk) can reduce fractiousness during handling. Alternatively, larger animals or breeds (wapiti) can be restrained in narrow stalls. In some designs the stall is created by closing sliding doors fore and aft of the animal standing in a race and by narrowing a movable section of one of the race walls.

In other systems the stall is located at the end of, and perpendicular to, the raceway (Fig. 17.4). The nearside 'wall' of the stall doubles as a swinging door which is opened to allow the animal to enter. In some designs the walls are padded and, when closed, are only 45 cm apart. This provides slight but firm pressure on the animal and prevents turning. In other designs, the propensity for deer to turn when isolated in narrow compartments is accommodated by leaving a much wider gap between the walls. Openings in the walls, either between pipes or in the form of removable panels, provide access ways to the restrained animal.

A high degree of restraint is required for some procedures, e.g. removal of antlers, artificial insemination, and this can be provided by drop-floor cradles and hydraulic handlers which use both mechanical and psychological aids to inhibit or prevent flight or jumping responses. Prior hand feeding of animals in the restrainer can reduce fractiousness during handling (Grandin _et al._, 1995). Secure mechanical restraint is provided by application of pressure alone or a combination of pressure together with removal of the animals' footing.

The most sophisticated, successful and versatile devices are called hydraulic workrooms or hydraulic deer handlers (Hutching, 1993) which can apply pressure, reduce footing and place a barrier over the animal's back. Typically, in its open state, the hydraulic handler consists of two well-padded walls about 1.4 m apart and 2–3 m long (see Fig. 17.5). One or more deer are walked into the pen-like space between the walls and then one of the walls is moved hydraulically toward the other to enclose the animal(s). The handler operates the device and works the deer from a raised platform. In earlier versions this platform was affixed to the moving wall, but the latest versions allow much higher acceleration of the moving wall and the operator works from the non-moving side. The walls can be adjusted up and down, which allows the opportunity to partially lift the animal off the floor and thereby reduce its flightiness.

In addition, the vertical angle of the walls can be adjusted to accommodate different body shapes, and on later models, the moving wall can be angled in the horizontal plane to form a wedge-shaped enclosure (which further restricts movement). On many models the walls incorporate removal panels which allow additional access to the animal. Several animals (with similar sized bodies) can be accommodated one behind the other in the larger work room restrainers at one time thereby increasing the efficiency of handling and reducing stress on the animals (since the animals are not isolated visually from

(a)

**Fig. 17.5.** (a) A hydraulically operated deer handler shown in its open position; this creates a small padded room about 1.4 m × 2.6 m long which is suitable for holding one or two deer. The curtains are closed during loading but can be opened after the animal(s) has (have) been restrained. The operator works from the small platform and restrains an animal by operating the moving wall, and applying steady pressure to the animal aligned between the two walls. (*Continued overleaf.*)

one another). Further efficiencies are provided by the dual-roomed type of hydraulic restrainer, which comprises two outer fixed walls and a central moving wall common to both rooms. While deer are being restrained and worked on one side, the other room is opened and loaded (Hutching, 1993).

The drop floor cradle utilizes a combination of pressure and removal of footing to restrain deer (Haigh, 1999). The cradle has a removable floor and Y-shaped walls, one of which is (usually) movable. The deer enters the cradle when the floor is upraised. When the animal is in position, the movable wall is operated to narrow the walls and contain the animal, the floor is released and the animal's weight is taken by its thorax and abdomen. The sides of the cradle can be padded to reduce the potential for bruising.

With both cradles and workrooms, additional immobilization can be achieved by placing additional pressure at the withers (Haigh, 1999) or by using a head halter.

(b)

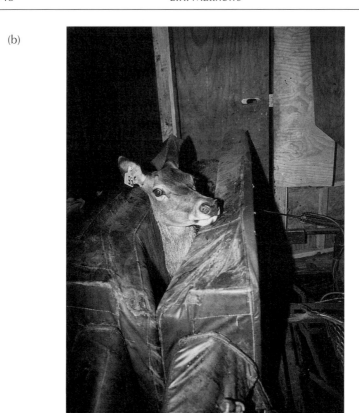

**Fig. 17.5.** *Continued.* (b) A drop-floor cradle for restraining deer.

# Transport

Practical experience gained in New Zealand and elsewhere in transporting hundreds of thousands of deer annually has led to the development of guidelines for the humane transportation of these animals (AWAC, 1994; MAFF, 1989). Recently published scientific studies, as reported below, support the view that deer can be transported safely and without undue distress provided the guidelines are followed. Transporters need to be specially constructed to take account of the physical abilities and behavioural requirements of deer.

### Transport crate design

The pressure on animals is relatively high in transport crates, therefore as with close handling facilities, they need to be smooth sided and any openings should be less than 50 mm wide if situated within 1.2 m of the floor.

The temperature inside transporters laden with deer is about 5° higher than the ambient temperature when the vehicle is moving (Harcourt, 1995). When stationary, the internal temperature rises quickly to much higher levels. Thus, ventilation slots about 100 mm wide are essential and should be located around the sides of the crate at a height of 1.4–1.6 m above the floor. Two such slots high on the crate walls are even better. There should be a 250 mm gap between the top of the internal walls of the pens and the roof to allow fresh air to circulate. Gaps for ventilation located closer to the floor should be covered with a 50 × 50 mm grid. Stoppages during transport need to be kept to a minimum to prevent excessive heat build up in the pens and hyperthermia in the deer (Harcourt, 1995).

A ceiling is required to contain the animals during transit and its height should be less than about 1.5 m to discourage aggressive behaviour and rearing. A grid (25 × 25 mm) supported 25–50 mm off the truck's deck is an appropriate flooring material for shorter journeys. On longer journeys a soft underfoot material will assist in preventing injuries to the feet.

Entranceways to the transporter and pens should be at least 1.2 m wide to assist animal flow and the doors need to be positioned so that they can be used to push up behind animals that baulk. It is useful to have the pens on the transporter arranged as modules with dimensions of about 1.2 m × 2.0 m (2.4 m$^2$) and connected to adjacent pens via swinging doors which can be opened to increase pen size. Each module is suitable for carrying six 100 kg live-weight animals. Larger animals can be accommodated in larger pens or by reducing the group size in the standard module.

Practical experience indicates that the ideal group size is six animals (100 kg weight) with a maximum of about eight. Grigor *et al.* (1998b) showed that smaller groups ($n = 5$) were less disturbed during travel than larger groups ($n = 10$). There have been several scientific studies that provide information on the effects of stocking density on the responses of deer to travel (Jago *et al.*, 1993, 1997; Waas *et al.*, 1997; Grigor *et al.*, 1998b). The densities examined ranged from about 0.4 to 1.2 m$^2$ 100 kg$^{-1}$ animal equivalent. Transport was found to be somewhat stressful in all studies, as measured by increases in plasma cortisol concentrations. Variations in space allowance had little additional effect on animal comfort or stress. At lower space allowances, animals were less able to orientate in the preferred directions (parallel or perpendicular to the direction of travel) but this did not adversely affect their ability to maintain balance or avoid falling down.

A variety of other environmental and animal factors influence the ease of handling and animal well-being during transport. In a survey of industry statistics, Jago *et al.* (1996) reported that bruising was more likely for animals transported distances greater than 200 km, for smaller or less fat deer, and for males transported during the breeding season or held overnight in lairage in late winter. Subsequent scientific studies have confirmed that there are increases in bruising rates, muscle damage (Jago *et al.*, 1997) and stress (Waas *et al.*, 1999) with increasing distance travelled but the magnitude of the effects are small, and there is little effect on levels of dehydration of journey times up to 6 h (Grigor *et al.*, 1997a, 1998d; Jago *et al.*, 1997). Biochemical signs of dehydration, as measured by increases in plasma sodium levels, become apparent 11–20 h following water withdrawal (Hargreaves and Matthews, 1995; Waas *et al.*, 1999).

Journey parameters other than distance or time travelled have more important effects on animal wellbeing. Deer are more prone to losing their footing in the first few minutes of a journey (Jago *et al.*, 1997) and on steep, winding sections of highway (Jago *et al.*, 1997; Grigor *et al.*, 1998b). Thus, careful driving is required at this time. In addition, there appears to be a greater physical challenge to animals at the rear of transporters as heart rates and plasma lactate concentrations for animals in the middle and rear pens are higher than for those at the front (Waas *et al.*, 1997). Agonistic interactions between deer on trucks are more common between recently mixed animals, and those differing greatly in body weight (Jago *et al.*, 1997). Therefore, animals allocated to the same pen on transporters should be familiar with each other and of similar sizes. Practical experience indicates that animals of different ages or gender should not be mixed in pens. Further, stags during the rut, or with velvet antlers longer than 60 mm should not be transported. Stags with hard antlers should be transported singly in pens, or preferably have the antlers removed before transport.

Behavioural measures of aversion and physiological measures of stress indicate that the process of transportation is a relatively mild stressor (in comparison with physical restraint) (Carragher *et al.*, 1997; Waas *et al.*, 1997; Grigor *et al.*, 1998a). Thus, if best practices are followed, animal well-being should not be unduly compromised by transport and associated handling.

## Lairage and Slaughter

In some countries deer are slaughtered in the field or in slaughter plants on farms and this can be a satisfactory way to avoid undue stress (Smith and Dobson, 1990), particularly in animals not familiar with close handling or selected for tractability. With field slaughter, great care needs to be taken to ensure that appropriate standards of hygiene are met (Vaarala and Korkeala, 1999). In countries where large numbers of animals are farmed, most deer are transported and slaughtered in commercial abattoirs: this process is probably more suitable for the less flighty species and those animals that have become accustomed to farming routines.

In New Zealand the slaughterhouses are purpose built for deer alone but in other countries several different species may be processed through the same facility. Ultimate muscle pH after slaughter is often used as an indicator of meat quality, with stress contributing to the development of high pH (poor quality) meat (Lawrie, 1985). Typically the pH values for deer slaughtered at commercial premises are low and within the range considered indicative of good quality meat (Smith and Dobson, 1990; Jago *et al.*, 1993, 1997; Grigor *et al.*, 1997c; Pollard *et al.*, 1999).

Practical experience indicates that the principles for good handling and facility design (as outlined above for farm facilities) apply equally well to slaughter plants. However, a number of features of lairage require special consideration, since the deer are often held for much longer periods indoors and there is a requirement to maintain a steady flow of animals to the stunning box.

Overnight lairage may increase the rate of bruising slightly at specific times of the year (e.g. during winter in males) (Jago *et al.*, 1996) but appears to have little other adverse physiological effects on deer. Grigor *et al.* (1997c) showed that lairage for up to 18 h did not lead to dehydration, depletion of muscle glycogen or high pH. There is some evidence that lairage is associated with recovery from transport as the activity of enzymes indicative of muscle damage decline with time in lairage (Jago *et al.*, 1993; Grigor *et al.*, 1997c) and activity levels return to pre-journey patterns (Grigor *et al.*, 1997c). Nevertheless, measures of other behaviours indicate that longer periods in lairage may be less than ideal as the frequency of agonistic interactions increases with time in lairage (Pollard *et al.*, 1999). Penning deer next to other species, particularly pigs, is undesirable as it leads to higher levels of aggression and heart rates in the deer (Abeyesinghe *et al.*, 1997).

The design of the race leading to the stunning box is critical for achieving a steady flow and avoiding stress in the deer. Holding pens and raceways should be designed so that each animal can remain in direct or visual contact with at least one familiar companion right up until entry to the stunning box. The use of appropriately positioned swing doors and curves, and illumination of the stunning box are the best ways to encourage movement of reluctant animals without causing undue distress. Electric goads should not be used on deer. A low ceiling constructed to permit access by the stunner will discourage rearing. In the interests of animal wellbeing, the shorter the time the animals are left in the stunning pen the better (Grigor *et al.*, 1999).

## Behaviour Relevant to Facility Use

The fundamental behaviour and stress reactions of deer have important implications not only for facility design but also for the efficiency and ease of handling the animals. These behaviours will be examined in the context of mustering, yarding and restraint. Although deer are flighty by nature, recent studies (Ingram *et al.*, 1994, 1997; Carragher *et al.*, 1997) have shown that common handling procedures (such as yarding and drafting) are not particularly stressful. Further, upon release back to pasture the animals' maintenance activities and stress hormone concentrations return rapidly to normal levels (Ingram *et al.*, 1994, 1997; Carragher, *et al.*, 1997).

## Mustering and movement

Basic behaviours relevant to efficient mustering and movement include the degree of flightiness and familiarity with the environment, and the flocking and leader–follower tendencies of deer.

Extremely flighty animals or those unfamiliar with the farm environment, are difficult to direct and are best moved by passive methods. This simply involves leaving open gates to laneways, fresh pasture or handling facilities for animals to move without encouragement from handlers. Most animals soon become familiar with the farm layout and handlers. For these tamer deer, either of two active methods of mustering are more appropriate and efficient. The simplest one utilizes the good learning abilities of deer and their tendency to follow a leader. Animals are trained to follow or approach a handler in response to visual or auditory cues. Untrained animals in the group will readily follow the leaders to new pastures or through handling facilities. Training can be carried out most readily during the hand-rearing of calves or the feeding of supplements to weaners.

The second and more common technique utilizes the natural antipredator responses of flocking and flight, together with the following tendency. As a handler approaches a herd, the animals bunch and direct their eyes and ears toward the intruder (Fig. 17.6). When the handler is near the boundary of the flight zone a proportion of the animals turn and face away (Fig. 17.6). With further advances by the handler and gradual penetration of the flight zone the leaders will begin to move and the well-developed follower response stimulates the remainder of the group to follow.

For calm and coordinated movement it is important that the handler moves steadily but slowly and remains near the boundary of the flight zone.

**Fig. 17.6.** Herding behaviour of red deer in response to a human intruder.

Occasionally, the lead animals will baulk during mustering, thereby causing the remainder of the herd to stop. To re-establish synchronized forward movement of the herd it is important for handlers to coordinate their movements with those of the lead animals. At the moment when the leader(s) has (have) returned to the front of the mob (often after having circled the group and moving back up through the centre) and is/are facing in the required direction, the handler should move forward into the flight zone. As the animals move into smaller enclosures (field to raceway to yards) their flight distance decreases and speed of travel changes, and frequent positional adjustments are required by the handler. Familiarity with the handlers and the use of one or two people helps reduce flightiness. In addition, machines (motorcycles, trucks and helicopters) and other animals (dogs and horses) that are familiar to the deer can be used successfully in the mustering process.

Sudden or deep penetration of the flight zone induces strong and less cohesive flight of the herd. In extreme cases, panic behaviour characterized by excessive and disorientated flight (Mills and Faure, 1990) may occur, thereby increasing the risk of animals running into fences and sustaining injuries. Panic responses are more likely in particularly young animals with little handling experience; flighty individuals especially if held singly, in small groups away from the normal social group; or when exposed to unfamiliar handlers, objects or noises (Bull, 1996). Unusual visual stimuli induce stronger and more sustained flight and fear responses than unfamiliar auditory stimuli (Herbold *et al.*, 1992; Hodgetts *et al.*, 1998). Through the process of social facilitation (Mills and Faure, 1990) other normally calm individuals may show similar panic behaviours. In cases where the majority of the herd is extremely disturbed, handling of the animals should be discontinued for several hours. Practical experience indicates that the flightiness of young animals during handling can be reduced by the use of older quieter animals as leaders (Pollard *et al.*, 1992), by habituating them to yards prior to handling (particularly in conjunction with feeding in the handling areas) and by mustering and moving them as a group for several hours in a large field prior to any close handling (C. Brown and L. Matthews, unpublished observations). Herrmann (1991) has shown that the provision of artificial shelters in the paddocks assists in reducing the incidence of abnormal behaviours following disturbances, although Hodgetts *et al.*'s (1998) data did not support this finding.

Previous negative handling experiences can adversely influence the ease of handling of deer. Groups kept in fields near handling facilities are more flighty and pace up and down fence lines furthest from the yards (Diverio *et al.*, 1993) particularly in the first 3–4 h after handling (Matthews *et al.*, 1990). The avoidance of the handling area and repetitive pacing suggest that some aspects of the handling process are aversive. This is supported by anecdotal observations that some animals become increasingly difficult to collect from pasture and drive toward the handling facility after repeated handling. The aversive aspects most likely result from periods of visual isolation, physical restraint or particular handling procedures (Matthews and Cook, 1991; Pollard *et al.*, 1993; Price *et al.*, 1993; Grigor *et al.*, 1998a). Minimizing stress from these factors will improve animal flow. In addition, the use of food rewards after handling and familiarization with the handling process facilitates movement in sheep (Hutson, 1985a,b) and also seems to work with deer.

## Yarding

An ideal group size for intensive handling of deer in yards is about six animals. Large mobs are usually broken down into smaller units in two steps. First, groups of about 25 animals are drafted off while the herd is contained in outdoor holding pens close to the yards. These groups are then run indoors and given access to as many holding pens as possible. The animals tend to settle out into small sized groups in the various pens which are then secured.

Drafting (sorting) techniques are similar to those used with cattle (Kilgour and Dalton, 1984). The handler enters the flight zone to induce movement and then moves forward or backward of the point of balance behind the shoulder to direct animals backward or forward, respectively. The use of short lengths of plastic pipe to extend the arms and solid wooden shields are useful in manipulating the point of balance. Apart from stags during the rut, deer should not be kept in isolation from their herd mates.

The procedures for moving animals between various sections of the yard complex are similar to those used in mustering. In well-designed facilities passive techniques work well on most occasions. Individuals or groups move readily if the doors ahead of the animals are left open – typically the handler does not need to apply pressure from behind the deer. Active techniques involving manipulation of the flight response are sometimes required, especially when moving lead animals into unfamiliar or dead-ended areas. As mentioned earlier, the use of curved entranceways assists movement. However, some animals show no flight response and may need to be pushed from behind with swing gates or shields. This situation occurs more frequently with tamer animals or wapiti breeds (Thorliefsen *et al.*, 1997). Shields are particularly effective in providing protection from strikes by the deers' front or hind feet.

Deer become reluctant to leave holding pens if they have had prior experience of aversive events in other parts of the handling facility (Grigor *et al.*, 1998a). Behavioural (Pollard *et al.*, 1994a; Grigor, 1998a) and physiological measures (Pollard *et al.*, 1993; Carragher *et al.*, 1997; Waas *et al.*, 1997) indicate that the rank ordering of the relative aversiveness of various events is: drop floor cradle; transport; social isolation; human proximity. Recently gathered data (J. Ingram and L. Matthews, unpublished data) support other studies (Pollard *et al.*, 1993; Hanlon *et al.*, 1995) which show that mixing of unfamiliar deer is a highly aversive event.

Adverse experiences have less effect on the readiness with which deer move along raceways (Matthews *et al.*, 1990; Stafford and Mesken, 1992; Grigor *et al.*, 1997b). Similarly, speed of movement along a race is not influenced by the light levels in the area ahead (Grigor *et al.*, 1997b, 1998c), although practical experience suggests it is difficult to move animals toward pens or on to transporters if there are bright shafts of light directed toward the deer. Overall then, the speed of movement of deer in races is influenced more by the strength of the flight response to the handler than the attractiveness (or aversiveness) of areas ahead of the animal.

Practical experience indicates that familiarity with the handling facility may help to reduce the aversiveness of handling. Animals fed supplements indoors or simply run through the facility a number of times are less flighty. Price *et al.* (1993) measured the heart rates of deer in yards and found that stress levels declined with increasing familiarity with the handling areas. They also recorded heart rate changes when handlers entered the holding pens. Interestingly, there were no differences in heart rate responses between familiar and unfamiliar handlers. In contrast, Bull (1996) noted in red deer at pasture that the approach of a handler wearing unfamiliar clothes was associated with a higher frequency of behaviours indicative of disturbance (more movement and less grooming) than when the handler wore familiar clothes. Interestingly, the colour of clothing influences deer vigilance and the effect varies with time of day (Bull, 1996). Deer are most alert to white clothing at night and to red in the morning.

In yarding situations, the animals are frequently worked from well within their flight zones, thereby increasing the likelihood of exaggerated flight responses, stress, threats and aggression. Maintenance of dominant behaviour by the handler over mobs (but especially stags) eases handling and lessens the risk of injury to handlers. Holding a hand high above the head, working from slightly above the deer (e.g. on catwalks) or confronting an animal with a shield are useful techniques for increasing the apparent size and dominance of the handler. Because hand-reared deer have a decreased flight response to humans (Blaxter *et al.*, 1988) it can be difficult to assert dominance over particularly aggressive deer (e.g. females with newborns, and males during the rut). It is advisable to cull hand-reared males before sexual maturity. Working the animals from outside the pens using swing gates and sliding doors reduces flightiness and aggression, especially in fallow and wapiti.

Repeated handling results in weight loss or lack of growth in deer (J. Ingram and L. Matthews, unpublished data). A number of techniques assist in reducing stress. Unfamiliar animals, or animals of different ages, sizes, sex and species should not be mixed (Hanlon *et al.*, 1995; Jago *et al.*, 1997). Individuals and small groups should remain in visual contact with other deer and be able to see and hear handlers (by talking quietly) approach. Unfamiliar and loud noises should be avoided (Hodgetts *et al.*, 1998). Ideally, stags should not be handled in the rut. If rutting stags need to be handled there is less chance of injury to themselves and handlers if they have been de-antlered, are moved in small groups only and are penned individually. Immature deer are particularly flighty and are best handled as little as possible. Any handling that is necessary should take place in the holding and working pens rather than in restrainers and races which are designed for older animals. Animals appear to be easier to handle if permitted to settle for an hour or so after mustering and drafting. We have shown that although handling is somewhat stressful, animals settle, both physiologically and behaviourally, soon after release to pasture (Carragher *et al.*, 1997). Thus, animals not required indoors or any others showing signs of excessive stress should be released. With continued handling, stressed animals may lie down. Attempts to move them are usually unsuccessful. Electric prodders should not be used. Such animals should be left in the company of others to stand before handling again. Some animals remain consistently flighty or aggressive throughout the year. In the interests of animal and handler welfare these deer should be culled.

## Restraint

With red deer, procedures requiring relatively little restraint (e.g. ear tagging, drenching) have traditionally been carried out in working pens under direct manual restraint augmented by shields and doors. There has been a move away from these methods as farmers have developed more suitable handling races for use with red, fallow and wapiti deer which ensure greater handler safety and also seem to be less stressful for deer. Procedures requiring higher levels of restraint (e.g. antler removal) are usually performed in a deer handler or under chemical restraint (Haigh and Hudson, 1993).

The basic behaviour patterns of deer can be used to advantage in several ways in animal restraint. Red deer and wapiti can be encouraged to enter handling races and restrainers by the use of curved entrances and by giving the lead animal a clear view of an exit. Single animals move readily to restrainers if they are drafted off individually from the holding pen and run through curving laneways. The tendency of fallow, white tail and chital deer to move readily from dark to light is used to advantage by dimming the light in the holding pen and lighting the entrance to the handling race or crush (Langridge, 1992; Haigh, 1999). Other sources of light shining into darkened pens need to be covered as fallow deer will persist in jumping toward these. In the handling races, animals remain settled as they are in visual and physical contact with other deer. Reducing visual stimulation either by darkening the environment or obscuring the eyes with a cloth tends to quieten animals (Jones and Price, 1992). Positioning a ceiling just above the heads of the animals prevents any jumping (Lee, 1992).

The concept of optimal pressure (Grandin, 1993) is relevant to the restraint of deer in crushes. Informal observations made while assessing the effects of analgesia (Matthews *et al.*, 1992) indicate that deer remain settled in padded crushes for up to 60 min provided the shoulders of the animal are well restrained. Animals struggle in response to pressures that are too high or low, or when the shoulders are well forward in the crush. Experience has shown that wapiti remain settled standing in squeeze races where the width of the race is narrowed but little pressure is exerted on the animal (Lee, 1992).

## Conclusion

Deer such as red and wapiti are rapidly becoming less flighty the longer they are farmed. This is most probably a result of three interacting factors: improved genetics for tractability; better methods for training animals to handling; and the development of handling facilities more appropriate to the unique behavioural characteristics of each species.

As a result of these processes, lower levels of force are required in order to move or restrain the animals, which in turn improves deer well-being during handling. No doubt handling systems will continue to evolve, with benefits for both animal and operator safety. Areas of research that will hasten this process include a greater scientific understanding of genetic factors underlying good animal temperament and of developmental processes determining rapid habituation of deer to humans and handling.

## References

Abeyesingh, S.M., Goddard, P.J. and Cockram, M.S. (1997) The behavioural and physiological responses of farmed red deer (*Cervus elaphus*) penned adjacent to other species. *Applied Animal Behaviour Science* 55 (1–2), 163–175.

Anderson, J.R. (1975) The behaviour of nose bot flies (*Cephenemyia apicata* and *C. jellisoni*) when attacking black-tailed deer (*Odocoileus hemionus columbianus*) and the resulting reactions of the deer. *Canadian Journal of Zoology* 53, 977–992.

AWAC (1994) Code of recommendations and minimum standards for the welfare of animals transported within New Zealand. *Code of Animal Welfare, No. 15*, Ministry of Agriculture, Wellington, New Zealand.

Bartos, L. (1985) Social activity and the antler cycle in red deer stags. *Royal Society of New Zealand Bulletin* 22, 269–272.

Blaxter, K.L., Kay, R.N.B., Sharman, G.A.M., Cunningham, J.M.M., Eadie, J. and Hamilton W.J. (1988) *Farming the Red Deer*. Rowett Research Institute and Hill Farming Research Organisation, Department of Agriculture and Fisheries for Scotland, Edinburgh.

Bull, R.L. (1996) The behavioural responses of farmed red deer (*Cervus elaphus*) to handling. Unpublished MSc Thesis, University of Waikato, New Zealand, 81 pp.

Bullock, D.J., Kerridge, F.J., Hanlon, A. and Arnold, R.W. (1993) Short-term responses of deer to recreational disturbances in two deer parks. *Journal of Zoology* 230(2), 327–332.

Cadman, W.A. (1966) The fallow deer. *Forestry Commission Leaflet* no. 52. HMSO, London.

Carragher, J.F., Ingram, J.R. and Matthews, L.R. (1997) Effect of yarding and handling procedures on stress responses of free-ranging red deer (*Cervus elaphus*). *Applied Animal Behaviour Science*, 51, 143–158.

Cash, R. (1987) Fallow handling systems: simplicity and reliability are vital. *The Deer Farmer* 39, 29–31.

Clift, T.R., Challacombe, J. and Dyce, P.E. (1985) Electric fencing for fallow deer. *Royal Society of New Zealand Bulletin* 22, 363–365.

Clutton-Brock, T.H. and Guinness, F.E. (1975) Behaviour of red deer (*Cervus elaphus* L.) at calving time. *Behaviour* 55, 287–300.

Clutton-Brock, T.H., Guinness, F.E. and Albon, S.D. (1982) *Red Deer: Behaviour and Ecology of Two Sexes*. Edinburgh University Press, Edinburgh.

Darling, F.F. (1937) *A Herd of Red Deer*. Oxford University Press, London.

Diverio, S., Goddard, P.J. and Gordon, I.J. (1996) Physiological responses of farmed red deer to management practices and their modulation by long-acting neuroleptics. *Journal of Agricultural Science* 126 (2), 211–220.

Diverio, S., Goddard, P.J., Gordon, I.J. and Elston, D.A. (1993) The effect of management practices on stress in farmed red deer (*Cervus elaphus*) and its modulation by long acting neuroleptics (LANS): behavioural responses. *Applied Animal Behaviour Science* 36, 363–376.

Drew, K.R. and Kelly, R.W. (1975) Handling deer run in confined areas. *Proceedings of the New Zealand Society of Animal Production* 35, 213–218.

English, A.W. (1992) Management strategies for farmed Chital deer. In: Brown, R.D. (ed.) *The Biology of Deer*. Springer-Verlag, New York, pp. 189–196.

Fletcher, T.J. (1991) Deer. In: Anderson, R.S. and Edney, A.T.B. (eds) *Practical Animal Handling*. Pergamon Press, Oxford, pp. 57–66.

Gilbert, B.K. (1974) The influence of foster rearing on adult social behaviour in fallow deer (*Dama dama*). In: Geist, V. and Walther, F. (eds) *The Behaviour of Ungulates and its Relation to Management*. IUCN Publication New Series No. 24, Morges, Switzerland, pp. 247–273.

Goble, K. (1991) Growing, yarding and harvesting of velvet. *Proceedings of the 43rd Ruakura Farmers' Conference*. Hamilton, New Zealand, pp. 146–148.

Grandin, T. (1993) Facility design in relation to behavior, stress and bruising. *Proceedings of the New Zealand Society of Animal Production* 53, 175–178.

Grandin, T., Rooney, M.B. and Phillips, M. (1995) Conditioning of Nyala (*Tragelaphus angasii*) to blood sampling in a crate with positive reinforcement. *Zoo Biology* 14, 261–273.

Grigor, P.N., Goddard, P.J., Cockram, M.S., Rennie, S.C. and MacDonald, A.J. (1997a) The effects of some factors associated with transportation on the behavioural and physiological reactions of farmed red deer. *Applied Animal Behaviour Science*, 52(1–2), 179–189.

Grigor, P.N., Goddard, P.J. and Littlewood, C.A. (1997b) The movement of farmed red deer through raceways. *Applied Animal Behaviour Science*, 52(1–2), 171–178.

Grigor, P.N., Goddard, P.J., Macdonald, A.J., Brown, S.N., Fawcett, A.R., Deakin, D.W. and Warriss, P.D. (1997c) Effects of the duration of lairage following transportation on the behaviour and physiology of farmed red deer. *Veterinary Record*, 140(1), 8–12.

Grigor, P.N., Goddard, P.J. and Littlewood, C.A. (1998a) The relative aversiveness to farmed red deer of transport, physical restraint, human proximity and social isolation. *Applied Animal Behaviour Science*, 56(2–4), 255–262.

Grigor, P.N., Goddard, P.J. and Littlewood, C.A. (1998b) The behavioural and physiological reactions of farmed red deer to transport: effects of sex, group size, space allowance and vehicular motion. *Applied Animal Behaviour Science*, 56(2–4), 281–295.

Grigor, P.N., Goddard, P.J., Littlewood, C.A. and Deakin, D.W. (1998c) Pre-transport loading of farmed red deer: effects of previous overnight housing environment, vehicle illumination and shape of loading race. *Veterinary Record*, 142(11), 265–268.

Grigor, P.N., Goddard, P.J., Littlewood, C.A. and MacDonald, A.J. (1998d) The behavioural and physiological reactions of farmed red deer to transport: effects of road type and journey time. *Applied Animal Behaviour Science*, 56(2–4), 263–279.

Grigor, P.N., Goddard, P.J., Littlewood, C.A., Warriss, P.D. and Brown, S.N. (1999) Effects of preslaughter handling on the behaviour, blood biochemistry and carcases of farmed red deer. *Veterinary Record*, 144(9), 223–227.

Haigh, J.C. (1992) Requirements for managing farmed deer. In: Brown, R.D. (ed.) *The Biology of Deer*. Springer-Verlag, New York, pp. 159–172.

Haigh, J.C. (1995) A handling system for white-tailed deer (*Odocoileus virginianus*). *Journal of Zoo Wildlife Medicine*, 26(2), 321–326.

Haigh, J.C. (1999) The use of chutes for ungulate restraint. In: Fowler, M.E. and Miller, R.E. (eds) *Zoo and Wildlife Medicine. Current Therapy 4*. W.B. Saunders, Philadelphia, pp. 657–662.

Haigh, J.C. and Hudson R.J. (1993) *Farming Wapiti and Red Deer*. Mosby, Toronto, pp. 67–98.

Hanlon, A.J., Rhind, S.M., Reid, H.W., Burrells, C. and Lawrence, A.B. (1995) Effects of repeated changes in group composition on immune response, behaviour, adrenal activity and live-weight gain in farmed red deer yearlings. *Applied Animal Behaviour Science*, 44(1), 57–64.

Hanlon, A.J., Rhind, S.M., Reid, H.W., Burrells, C., Lawrence, A.B., Milne, J.A. and McMillen, S.R. (1994) Relationship between immune response, live-weight gain, behaviour and adrenal function in red deer (*Cervus elaphus*) calves derived from wild and farmed stock, maintained at two housing densities. *Applied Animal Behaviour Science*, 41(3–4), 243–255.

Harcourt, R. (1995) Temperatures rising. *The Deer Farmer*, 125, 41–42.

Hargreaves, A.L. and Matthews, L.R. (1995) The effect of water deprivation and subsequent access to water on plasma electrolytes, haematocrit and behaviour in red deer. *Livestock Production Science* 42, 73–79.

Hemmer, H. (1988) Ethological aspects of deer farming. In: Reid, H.W. (ed.) *The Management and Health of Farmed Deer*. Kluwer, Dordrecht, pp. 129–138.

Herbold, Von H., Sachentrunk, F., Wagner, S. and Willing, R. (1992) Einfluss anthropogener stoerreize auf die herzfrequenz von Rotwild (*Cervus elaphus*) und Rehwild (*Capreolus capreolus*). *Zeitschrift Jagdwissenschaft*, 38, 145–159.

Herrmann, H.J. (1991) Aspects of the welfare of farmed red deer (*Cervus elaphus*) with results of a preliminary study of two types of environmental enrichment. Unpublished Masters thesis, University of Edinburgh, UK.

Hodgetts, B.V., Waas, J.R. and Matthews, L.R. (1998) The effects of visual and auditory disturbance on the behaviour of red deer (*Cervus elaphus*) at pasture with and without shelter. *Applied Animal Behaviour Science*, 55(3–4), 337–351.

Humphries, R.E., Smith, R.H. and Sibley, R.M. (1990) Effects of human disturbance on the welfare of park fallow deer. *Applied Animal Behaviour Science* 28, 302 (abstract).

Hutching, B. (1993) Deer handlers. *The Deer Farmer* 105, 39–47.

Hutson, G.D. (1985a) Sheep and cattle handling facilities. In: Moore, B.L. and Chenoweth, P.J. (eds) *Grazing Animal Welfare*. Australian Veterinary Association, Queensland, pp. 124–136.

Hutson, G.D. (1985b) The influence of barley food rewards on sheep movement through a handling system. *Applied Animal Behaviour Science* 14, 263–273.

Hutson, G.D. and Hitchcock, D.K. (1978) The movement of sheep around corners. *Applied Animal Ethology* 4, 349–355.

Ingram, J.R., Matthews, L.R. and McDonald, R.M. (1994) A stress free blood sampling technique for free ranging animals. *Proceedings of the New Zealand Society of Animal Production* 54, 39–42.

Ingram, J.R., Matthews, L.R., Carragher, J.F. and Schaare, P. (1997) Plasma cortisol responses to remote adrenocorticotropic hormone (ACTH) infusion in free-ranging red deer (*Cervus elaphus*). *Domestic Animal Endocrinology* 14, 63–71.

Ingram, J.R., Crockford, J.N. and Matthews, L.R. (1999) Ultradian, circadian and seasonal rhythms in cortisol secretion and adrenal responsiveness to ACTH and yarding in unrestrained red deer (*Cervus elaphus*) stags. *Journal of Endocrinology* 162, 289–300.

Jago, J.G., Matthews, L.R., Hargreaves, A.L. and van Eeken, F. (1993) Preslaughter handling of red deer: implications for welfare and carcass quality. In: Wilson, P.R. (ed.) *Proceedings of a Deer Course of Veterinarians*, No. 10. Deer Branch of the New Zealand Veterinary Association, New Zealand, pp. 27–39.

Jago, J.G., Hargreaves, A.L., Harcourt, R.G., and Matthews, L.R. (1996) Risk factors associated with bruising in red deer at a commercial slaughter plant. *Meat Science* 44 (3), 181–191.

Jago, J.G., Harcourt, R.G. and Matthews, L.R. (1997) The effect of road-type and distance transported on behaviour, physiology and carcass quality of farmed red deer (*Cervus elaphus*). *Applied Animal Behaviour Science* 51, 129–141.

Jones, A.R. and Price, S.E. (1992) Measurement of heart rate as an additional means of ascertaining the effect of disturbance in park deer. In: Bullock, D.J. and Goldspink, C.R. (eds) *Management, Welfare and Conservation of Park Deer*. Universities Federation of Animal Welfare, Potters Bar, UK, pp. 95–107.

Kelly, R.W. and Drew, K.R. (1976) Shelter seeking and sucking behaviour of the red deer calf (*Cervus elaphus*) in a farmed situation. *Applied Animal Ethology* 2, 101–111.

Kelly, R.W., Fennessy, P.F., Moore, G.H., Drew, K.R. and Bray, A.R. (1984) Management, nutrition, and reproductive performance of farmed deer in New Zealand. In: Wemmer, C.M. (ed.) *Biology and Management of the Cervidae*. Smithsonian Institution Press, Washington, DC, pp. 450–460.

Kilgour, R. and Dalton, C. (1984) *Livestock Behaviour: a Practical Guide*. Methuen, Auckland.

Langridge, M. (1992) Establishing a fallow deer farm: basic principles. In: Asher, G.W. and Langridge, M. (eds) *Progressive Fallow Deer Farming: Farm Development and Management Guide*. Ruakura Agricultural Centre, New Zealand, pp. 17–28.

Lawrie, R. A. (1985) Chemical and biochemical constitution of muscle. In: Lawrie, R.A. (ed.) *Meat Science*. Pergamon Press, Oxford, pp. 43–73.

Lee, A. (1992) Steiner's user-friendly yards make the difference. *The Deer Farmer* 95, 25–26.

Lent, P.C. (1974) Mother-infant relationships in ungulates. In: Geist, V. and Walther, F. (eds) *The Behaviour of Ungulates and its Relation to Management.* IUCN Publication New Series No. 24, Morges, Switzerland, pp. 14–55.

Lincoln, G.A., Youngson, R.W. and Short, R.V. (1970) The social and sexual behaviour of the Red Deer stag. *Journal of Reproduction and Fertility,* Suppl. 11, 71–103.

MAFF (1989) *Guidelines for the Transport of Farmed Deer.* MAFF Publications, Lion House, Alnwick, Northumberland.

Matthews, L.R., Carragher, J.F. and Ingram, J.R. (1994) Post-velveting stress in free-ranging red deer. In: Wilson, P.R. (ed.) *Proceedings of a Deer Course for Veterinarians,* No 11. Deer Branch of the New Zealand Veterinary Association, New Zealand, pp. 138–146.

Matthews, L.R. and Cook, C.J. (1991) Deer welfare research: Ruakura findings. *Proceedings of a Deer Course for Veterinarians,* New Zealand Veterinary Association, No. 8, 120–127.

Matthews, L.R., Cook, C.J. and Asher, G.W. (1990) Behavioural and physiological responses to management practices in red deer stags. *Proceedings of a Deer Course for Veterinarians,* New Zealand Veterinary Association, No. 7, 74–85.

Matthews, L.R., Ingram, J.R., Cook, C.J., Bremner, K.J. and Kirton, P.G. (1992) Induction and assessment of velvet analgesia. *Proceedings of a Deer Course for Veterinarians,* New Zealand Veterinary Association, No. 9, 69–76.

McComb, K. (1987) Roaring by red deer stags advances the date of oestrus in hinds. *Nature* 330, 648–649.

McNally, L. (1977) The senses of deer. *Deer* 4, 134–137.

Mills, A.D. and Faure, J.M. (1990) Panic and hysteria in domestic fowl: a review. In: Zyan, R. and Dantzer, R. (eds) *Social Stress in Domestic Animals.* Kluwer Academic Publishers, Dordrecht, pp. 248–272.

Pearse, A.J. (1992) Farming of wapiti and wapiti hybrids in New Zealand. In: Brown, R.D. (ed.) *The Biology of Deer.* Springer-Verlag, New York, pp. 173–179.

Pollard, J.C. and Littlejohn, R.P. (1994) Behavioural effects of light conditions on red deer in a holding pen. *Applied Animal Behaviour Science* 41(1–2), 127–134.

Pollard, J.C. and Littlejohn, R.P. (1996) The effects of pen size on the behaviour of farmed red deer stags confined in yards. *Applied Animal Behaviour Science* 47(3–4), 247–253.

Pollard, J.C., Littlejohn, R.P. and Suttie, J.M. (1992) Behaviour and weight change of red deer calves during different weaning procedures. *Applied Animal Behaviour Science* 35, 23–33.

Pollard, J.C., Littlejohn, R.P. and Suttie, J.M. (1993) Effects of isolation and mixing of social groups on heart rate and behaviour of red deer stags. *Applied Animal Behaviour Science* 38(3–4), 311–322.

Pollard, J.C., Littlejohn, R.P. and Suttie, J.M. (1994a) Responses of red deer to restraint in a y-maze preference test. *Applied Animal Behaviour Science* 39(1), 63–71.

Pollard, J.C., Littlejohn, R.P. and Webster, J.R. (1994b) Quantification of temperament in weaned deer calves of two genotypes (*Cervus elaphus* and *Cercus elaphus × Elaphurus davidianus* hybrids). *Applied Animal Behaviour Science* 41(3–4), 229–241.

Pollard, J.C., Stevenson-Barry, J.M. and Littlejohn, R.P. (1999) Factors affecting behaviour, bruising and pHμ in deer slaughter premises. *Proceedings of the New Zealand Society of Animal Production* 59, 148–151.

Price, S., Sibly, R.M. and Davies, M.H. (1993) Effects of behaviour and handling on heart rate in farmed red deer. *Applied Animal Behaviour Science* 37, 111–123.

Prince, J.H. (1977) The eye and vision. In: Swenson, M.J. (ed.) *Dukes' Physiology of Domestic Animals.* Cornell University Press, UK, pp. 696–712.

Smith, R.F. and Dobson, H. (1990) Effect of pre-slaughter experience on behaviour, plasma cortisol and muscle pH in farmed red deer. *Veterinary Record* 126 (7), 155–158.

Stafford, K.J. and Mesken, A. (1992) Electro-immobilisation in red deer. *Proceedings of a Deer Course for Veterinarians,* New Zealand Veterinary Association, No. 9, 56–68.

Staines, B.W. (1974) A review of factors affecting deer dispersion and their relevance to management. *Mammalian Review* 4, 79–91.

Thorliefson, I., Pearse, A. and Friedel, B. (1997) *Elk Farming Handbook.* North American Elk Breeders Association, Canada.

Vaarala, A.M. and Korkeala, H.J. (1999) Microbiological contamination of reindeer carcasses in different reindeer slaughterhouses. *Journal of Food Protection* 62 (2), 152–155.

Waas, J.R., Ingram, J.R. and Matthews, L.R. (1997) Physiological responses of red deer (*Cervus elaphus*) to conditions experienced during road transport. *Physiology and Behavior* 61(6), 931–938.

Waas, J.R., Ingram, J.R. and Matthews, L.R. (1999) Real-time physiological responses of red deer to translocations. *Journal of Wildlife Management* 63 (4), 1152–1162.

Woodford, K.B. and Dunning, A. (1992) Production cycles and characteristics of Rusa deer in Queensland, Australia. In: Brown, R.D. (ed.) *The Biology of Deer.* Springer-Verlag, New York, pp. 197–202.

Yerex, D. (1991) *Wapiti Behind the Wire: the Role of Wapiti in Deer Farming in New Zealand.* GP Books, Wellington, New Zealand.

Yerex, D. and Spiers, I. (1990) *Modern Deer Farm Management.* GP Books, Wellington, New Zealand.

# Poultry Handling and Transport

## Claire Weeks and Christine Nicol

*University of Bristol, Department of Clinical Veterinary
Science, Langford, Bristol BS40 5DU, UK*

## Introduction

World demand for poultry meat continues to expand. Over 40,000 million broilers were slaughtered in 1998. Turkey meat output is around 5 million tonnes (Mt) and duck meat production is expected to exceed 3 Mt before 2000. At the same time, world egg production is set to exceed 50 Mt (Watt, 1998). This has resulted in more countries producing poultry intensively and on a larger scale than ever before. The capacity of poultry houses has increased, along with larger processing plants and faster line speeds. All birds are handled and transported at least twice. There is a need for this to be carried out both efficiently and with due regard for the welfare of the individual bird. The UK and other European countries have legislation embracing the welfare of animals in transit. In many cases improving welfare has meat quality and performance benefits. Scientific research has been concentrated in two main areas: identifying and quantifying stressors in handling and transport; and developing improved systems. For this research, it is important to be able to measure the extent to which birds are stressed or frightened. In studies of transportation, the concentration of corticosterone in blood plasma is often taken as an indicator of stress, whilst the duration of an experimentally induced state of tonic immobility (TI) is thought to be a reliable measure of fear. The validity of these techniques is reviewed by Jones (1996) and Mitchell and Kettlewell (1998).

# Rearing Systems and On-farm Handling

## Chicks

Careful handling and transport of both hatching eggs and newly hatched chicks are important for subsequent performance (Meijerhof, 1997). Control of humidity as well as temperature is particularly important for eggs. Chicks are usually transferred by hand from incubator trays to lightweight, disposable containers perforated with ventilation holes. Those destined to be laying birds may first pass along a conveyor from which the males are removed for destruction. Manual sexing and sorting of meat birds may also be practised. The yolk sac reserves enable chicks to be transported for 24 h or even longer with low mortality. Chick containers are transported by either truck or aircraft, generally in controlled environments to maintain uniformly warm and yet well-ventilated conditions. Optimum temperature for chicks at normal stocking density in transport containers is 24–26°C (Meijerhof, 1997). On arrival, the chicks may be gently tipped out or removed manually.

## Broilers, turkeys and ducks

Poultry intensively reared for meat are placed in mixed- or single-sex groups of up to 60,000 birds in environmentally controlled, dimly lit (3–30 lx) houses. Birds live at stocking densities of up to 45 kg m$^{-2}$ at slaughter age on a deep litter of wood shavings, straw or similar material, with automated provision of food and water. Thus human contact and environmental stimulation are kept to a minimum.

Recently, some producers of turkeys and broilers have been adopting lighting programmes (reviewed by Buyse et al., 1996) and feeding programmes (Su et al., 1999), rather than the prevalent near-continuous lighting and ad libitum access to food. These may reduce the incidence and severity of leg problems, 'flip-overs', ascites and other consequences of genetic selection for excessively fast growth and food conversion efficiency.

Other husbandry systems, particularly in hot climates, may grow birds more slowly, in naturally ventilated and lit pole barns or with access to the outside. Catching birds in such systems is invariably harder, because the birds are more active and have more space.

Most poultry meat is grown on contract or as part of an integrated system; thus processors usually own and are responsible for the transport system. They also provide specialist catching teams to depopulate the houses. Birds are calmer and less affected by the catching process if they are handled in darkness (Duncan, 1989); thus catching in the early hours in very dim light is common. Blue lights are useful with turkeys (Siegel, personal communication to the editor). Broilers are caught by one leg, inverted and carried in bunches of three or four per hand to the waiting crates or modules (Gerrits et al., 1985).

To avoid dislocated hips and other injuries, handling by two legs is preferable, as is maintaining the bird upright (Gerrits *et al.*, 1985; Parry, 1989). Appropriate handling techniques for all species are given in Anderson and Edney (1991).

The types of container in common use for transporting poultry have been reviewed by Kettlewell and Turner (1985), Parry (1989) and Bayliss and Hinton (1990). Particularly with heavier birds, loose crate systems are laborious and thus are decreasing in popularity. Changing to modular systems, comprising a unit of compartments or sliding drawers, which can be moved by fork lift on and off the transport vehicle and right into the poultry house or lairage, may reduce birds dead on arrival (DOAs) to about a third of previous levels (Aitken, 1985; Stuart, 1985). To reduce damage to wings or legs, there should be a 2.5 cm gap between the top of the drawer and the rack to prevent birds from being caught (Grandin, 1999). Care needs to be exercised when loading the topmost drawers, particularly with heavy birds and turkeys. Manual catching is exhausting, repetitive and dirty work for the humans employed (Bayliss and Hinton, 1990). Up to 1500 broilers may be caught per person-hour in shifts of 5 h (Metheringham, 1996).

Automated catching and handling systems (reviewed by Scott, 1993) have great potential for reducing injury and distress for birds and humans alike (Lacy and Czarick, 1998). Several are now commercially available (Moran and Berry, 1992; Poultry International, 1995, 1998; Fig. 18.1), but uptake by the

**Fig. 18.1.** Automated broiler harvester. Photo courtesy of Anglia Autoflow.

industry has been minimal. The catching process, which involves gathering up the birds with rubber-fingered rotors, appears to work well in practice. An early study revealed that the heart rate of broilers caught by an automatic combine returned to base levels more quickly than the heart rate of broilers caught manually (Duncan et al., 1986). However, systems of conveying and transferring broilers to the containers need to be developed and modified to prevent fear and injury to the birds. Scott and Moran (1992) found significant increases in loss of balance, wing flapping and alarm calls by hens conveyed up or down slopes rather than horizontally. In particular, drops from one conveyor belt to another must be avoided, as they tend to cause injuries and wing flapping. Although Ekstrand (1998) did not observe the birds during catching, this is the probable cause of differences found in carcasses examined after slaughter, where twice as many wing fractures and significantly more bruising, mainly of the wings, were seen on mechanically caught birds, but insignificant differences in DOAs, compared with manual catching. In field trials, the manufacturers of the Easyload system found a halving of catching damage, from 4–6% manually to 1–3% using the machine (Poultry International, 1998). Leg bruising was significantly reduced, from 16.5% to 7.0%, by mechanical harvesting in samples of 200 birds at the plant (Lacy and Czarick, 1998), and Gracey (1986) noted DOAs averaged 0.2% for mechanically caught flocks compared with 0.3–0.6% for manually caught broilers.

The majority of broilers exhibit some degree of lameness, with estimates of up to 20–25% being substantially disabled (Julian, 1984; Kestin et al., 1992). Leg problems are also prevalent in many turkey flocks (Julian, 1998). Lameness results in significant changes to behaviour and, in particular, the number of visits to the feeders is reduced in proportion to the degree of walking disability (Weeks and Kestin, 1997). This implies a cost to the bird of lameness, which may be attributed to pain (McGeown et al., 1999; Danbury et al., 2000). Thus most catching and handling procedures are likely to cause pain, especially inverting and carrying birds by the leg(s). UK legislation, based on a European Community (EC) Directive, would also regard the more severely affected birds to be unfit to travel (MAFF, 1997). Severe clinical lameness following transport of male breeding turkeys with dyschondroplasia has also been reported (Wyers et al., 1991).

Although the potential for alterations to the rearing environment to modify the responses of birds to handling and transportation seems limited, there is an emerging consensus that fear or stress reactions can be modified by changes in commercial handling procedures per se. Jones (1992) found the TI response of both broilers and hens was reduced by gentle handling. Kannan and Mench (1996) confirmed this result in broiler chickens that were subjected to a 2-min handling treatment and then returned to their home pens, where plasma corticosterone was sampled at hourly intervals for the next 4 h. Birds that received upright handling had lower plasma corticosterone concentrations than birds that were inverted either individually or in groups of three. However, the effects of handling treatment were masked when the broilers were crated after the 2-min handling period. This suggests that crating *per se* can be stressful, as found previously by Beuving (1980), but in apparent contrast to the results of a study with end-of-lay hens (Knowles and Broom, 1993). It should be noted, however, that in Knowles and Broom's study the alternative to crating was inverted conveyance down a narrow aisle rather than return to the home pen.

## Pullets

Laying hens may be reared in cages or on litter and will usually undergo transportation at about 18 weeks of age when they reach point of lay and are taken to the egg production farm. Perforated plastic crates, generally with solid floors, are widely used, particularly when the rearer is responsible for delivery and may be using general-purpose vehicles. Narrow, modular systems that can be loaded and unloaded directly into cages and wheeled on to the transporter are favoured by professionals using dedicated vehicles. These predominate in the USA and are increasing in popularity in Europe with decreasing labour availability. The other main system is crates built as permanent fixtures on the bed of the lorry, with a central ventilation channel. Hinged openings to the outside are used to load and unload the birds, which have therefore to be carried out of and into their housing. They may also be passed in handfuls from person to person.

Pullets are relatively valuable birds with good plumage (insulation). Egg producers require them to arrive in good condition, so they tend to be handled and loaded carefully. The vehicles are usually generously ventilated, with air gaps above the floor and below the roof, air inlets in the headboard and either roof fans or a central ventilation channel the length of the trailer, including a slot in the roof.

## Hens

The majority of laying hens are still kept in battery cages, although numbers in alternative systems, such as percheries, aviaries or free range, have increased steadily over the past 10 years in northern Europe, and the EU has banned the conventional cage from 2012. After a productive year, the hens will be caught and transported, as 'spent' hens, to the slaughterhouse. Spent hens are generally purchased 'off farm'. Their low economic value reduces the care taken in handling and the investment in transport systems. Loose crates predominate, despite the labour costs of handling and cleaning them. Many types of vehicle may therefore be used, but dedicated trucks with central ventilation and side curtains are common.

Hens are removed from battery cages either individually or in groups of two or three by pulling them out by one leg despite recommendations to handle poultry by two legs (e.g. UK Codes of Recommendation). They may be struck against the cage entrance or food trough during removal. Hens may also hit cages or roof supports as they are carried down the narrow aisles of a battery house (Knowles, 1991). Depopulation is very labour-intensive, with bunches of inverted hens commonly being passed along a human chain.

Hens in many alternative systems are difficult to catch, tending to crowd and pile up at the end of aisles, creating a potential for suffocation or they flap and fly, with the risk of injury to the catchers. In some systems, the back of a tier is beyond arm's reach, resulting in birds having to be goaded or driven out, which is time-consuming and hazardous (L.J. Wilkins, 1992 personal communication). Ease of depopulation is therefore an important consideration to build into the design of any housing system.

A direct comparison of different catching and carrying methods for end-of-lay hens showed that plasma corticosterone concentrations were significantly higher when they were removed from their cages three at a time and carried in an inverted position from the house than when they were removed singly and crated before removal from the house (Knowles and Broom, 1993). However, all hens in experimental handling treatments had high concentrations of corticosterone in comparison with the control birds, which were removed individually and gently from their cages in an upright position (Knowles and Broom, 1993). Scott and Moran (1993) found that the fear levels of laying hens carried for 20 m on a flat-belt conveyor were lower than those of hens carried the same distance in an inverted position by hand or on a processing shackle. However, in the absence of a non-inverted control in this study, it is difficult to know whether the reduced fear was a consequence of upright conveyance or some other fear-reducing property of the flat-bed conveyor. As with broilers, well-designed automated handling devices would seem to have the potential to reduce trauma and fear.

## Effects of rearing experience on response to catching and transportation

The manner in which birds are raised is known to affect their subsequent fear and stress reactions in laboratory tests. Environmental enrichment provides extra stimulation in the home environment, that may affect animals' expectations about environmental complexity and enhance their ability to adapt to novelty. This may explain why birds exposed to outdoor environments and low stocking densities are generally less fearful than birds which lack such opportunities to explore and investigate their environment (Grigor et al., 1995; Sanotra et al., 1998; Scott et al., 1998). The use of enrichment stimuli, such as novel objects or even video stimuli during the rearing period can reduce the underlying fearfulness of domestic fowl, as revealed by their responses in a range of laboratory tests (Jones and Waddington, 1992; Jones, 1996).

It has frequently been suggested that appropriate environmental enrichment during the housing period might better enable birds to cope with the stressors that they will subsequently face during catching, transportation and preslaughter handling. However, the extent to which environmental enrichment can modify bird responses to the sometimes extreme stressors encountered during transportation and commercial-type handling is not clear. Nicol (1992) reported reduced levels of fear after transportation in broiler chickens that had been reared in an enriched environment. However, Scott *et al.* (1998) found no effect of enrichment when TI durations were measured after a 74-min journey, despite the fact that environmental enrichment reduced TI durations in birds caught and tested immediately. Possibly, the combined effects of commercial-type handling and transportation are so great that minor reductions in underlying fearfulness or propensity to react to stressors are simply masked.

Another approach has been to determine whether the adverse reactions that birds show to commercial handling can be reduced by giving them prior experience of regular gentle handling. Regular handling is thought to reduce birds' specific fear of humans, rather than modifying their underlying fearfulness (Jones, 1996), as the procedure fails to affect birds' responses in other contexts (Jones and Waddington, 1992; Grigor *et al.*, 1995). However, again it appears that the shock of commercial-type handling may be too great to be ameliorated by prior experience of gentle handling alone. Although Reed *et al.* (1993) found reduced levels of jumping and flapping behaviour (which resulted in reduced trauma and injury at depopulation) in end-of-lay hens that had previously experienced environmental enrichment and regular handling, Kannan and Mench (1997) detected no beneficial effects of prior handling on the plasma corticosterone concentrations of broiler chickens subjected to commercial-type handling at 7 weeks of age. Nicol (1992) found that prior gentle handling by humans either had no effect or, in combination with physical enrichment, actually resulted in an increased fear response after transportation.

## Consequences of Transportation

Transportation is an extremely stressful process for commercial poultry. Having lived in relatively uniform environments they are suddenly exposed to multiple changes, that include being handled, as discussed above, and withdrawal of food and water. They experience stimuli that may be new, such as motion, including vibration and impacts, or of greater intensity and more varied than previously, such as daylight, noise, overcrowding and temperature extremes. There are economic and welfare benefits for minimizing these stressors. The potentially adverse consequences of transportation include physical, physiological and behavioural changes.

## Death

Global figures for DOAs are unknown, but, taking a conservative estimate of 0.3%, based on European surveys (e.g. Bayliss and Hinton (1990) quote between 0.06% and 3%), some 120 million birds die between farm and factory. Whatever the exact figures, tens of millions of prematurely dead birds represent a vast economic loss to the industry and a larger cost in terms of bird welfare.

If some birds are sufficiently stressed to die, many more will be stressed close to their capacity to survive. Surveys (Warriss *et al.*, 1992a) have shown that the longer the journey time, especially over 4 h, and the longer the time between farm and slaughter, the greater the mortality rate and that time influenced mortality rates more than distance travelled (Fig. 18.2). Typical times in transit are unreported in most countries, but vary considerably. A survey of four UK broiler-processing plants by Warriss *et al.* (1990) found average time from loading to unloading was 3.6 h, with a maximum of 12.8 h. Time in transit for 90% of turkeys was under 5 h, with a maximum of 10.2 h (Warriss and Brown, 1996). Other factors which increased broiler mortality were the length of waiting time in the holding area at the processing plant, older birds (during summer months) and arrival at the processing plant in the afternoon or evening rather than the morning (Bayliss and Hinton, 1990).

## Thermal stress

Heat stress is thought to be the major contributor to both deaths (attributed to 40% of DOAs by Bayliss and Hinton (1990)) and overall transit stress. Reasons

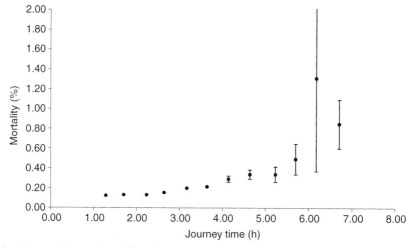

**Fig. 18.2.** The number of broilers dying increases with time in transit (Warriss *et al.*, 1992a). Figure reproduced with permission of Carfax Publishing.

for thermal stress and possible solutions will be discussed under vehicle design. It is important, however to understand the principles of heat balance. Poultry need to balance the heat produced by metabolism with heat lost to the environment.

The rate of sensible heat loss (i.e. by convection, conduction and radiation) is determined principally by the temperature gradient between the body core and the environment and by the amount of insulation between the two. Insulation is provided mainly by feathers and subcutaneous fat. Poorly feathered, wet or dirty birds lose heat faster. Wind may penetrate feathers, effectively reducing their insulation: the wind-chill effect, which may improve comfort in very hot air temperatures, but cause greater discomfort at colder air temperatures. In experimentally exposing broilers to still air temperatures between −4°C and +12°C for 3 h, Mitchell *et al.* (1997) measured falls in body temperature of up to 1.2°C, but, if the birds were wet, they became hypothermic very rapidly, down to a life-threatening 14°C below normal in the coldest air temperature.

The rate of latent heat loss by evaporation from the respiratory tract and skin depends partly on the water vapour density gradients (i.e. the degree of air saturation around the bird). It is also under physiological control. Birds in transit have little space for behavioural or postural thermoregulation. At high stocking density and without drinking-water, scope for evaporative loss is also reduced (Webster *et al.*, 1992). In practical terms, birds observed to be panting will become progressively dehydrated and increasingly heat-stressed.

## Trauma

Considerable concern has been generated in the UK by reports that 24% of laying hens acquired broken bones when they were removed from their cages at end of lay (Gregory *et al.*, 1990). The problem appeared to be due to the almost ubiquitous occurrence of disuse osteoporosis in caged hens (causing reduced bone strength), together with the rough handling many hens experience when they are removed from the cages. Recent research has therefore focused on the factors affecting bone strength, as well as less damaging methods of catching hens. Housing system influences bone strength, with birds from battery cages performing few limb movements and having weaker bones than hens from percheries and other 'alternative' systems (Knowles and Broom, 1990). Surprisingly, Gregory *et al.* (1991) found that the bone strength of battery hens which had been reared in cages was stronger than those which had been reared on deep litter. They also found that moulting was associated with a transient decrease in bone strength, with a subsequent increase during the second laying cycle. In a further study, 12.7% of hens removed from the cage by one leg sustained broken bones compared with 4.6% of hens that were removed from the cage by two legs (Gregory and Wilkins, 1992).

Bruising is a commercial problem with meat chickens and has thus attracted considerable attention (reviewed in Knowles and Broom, 1990; Nicol and Scott, 1990). It has been estimated that one in four broilers processed in the USA may have bruising of the legs, breast or wings sustained during catching and transport (Farsaie et al., 1983). In a survey of downgrading at a turkey abattoir, McEwen and Barbut (1992) found substantial levels of bruised drums; fewer leg and breast scratches were seen where birds had clipped toenails, and spur clipping reduced back scratches. They found no effect on injuries of truck design or stocking density, but increased half wing trim and bruising of drums were associated with length of time spent on the truck.

Damage to muscle cells results in an increase in the concentration of creatine kinase in blood plasma, so it may be possible to use creatine kinase concentrations as a quantitative index of injury (Mitchell et al., 1992). Apart from muscle damage, broilers also sustain a worrying number of broken bones and dislocations during the catching and transportation process. Gregory and Wilkins (1990) found that 3% of broilers had complete fractures before stunning at the processing plant and 4.5% had dislocated femurs. The dislocated femurs were probably caused by swinging the birds by one leg before placing them in the crate, as the chance of a broiler suffering a dislocation of this type increases with its body-weight. The incidence of dislocations recorded in live birds before stunning may not be a true reflection of the problem, since many dislocations result in fatal haemorrhaging. Indeed, when 1324 DOA broilers from six UK plants were examined, it was found that 27% had dislocated femurs (Gregory and Wilkins, 1992). These data indicate that the physical injury that occurs during manual catching and loading is a severe welfare problem for all poultry and should accelerate the search for more humane alternative methods.

## Fatigue

Many birds arrive at the slaughterhouse in an apparently exhausted state. It has been argued that the dehydration and depletion of body glycogen stores (Warriss et al., 1988) which occurs when broilers are subjected to food deprivation and simulated commercial transport may produce a sensation of fatigue in birds (Warriss et al., 1993). However, progressive immobility is also a correlate of increasing fearfulness (Jones, 1996) and a response to painful (nocioceptive) stimulation. This immobility may be related to learned helplessness and could be an important 'cut-off' response in transported poultry.

Sherwin et al. (1993) investigated the effects of fasting and transportation on measures of fear and fatigue in broilers. Broilers subjected to either food deprivation for 10 h or a journey of 6 h, or both, were compared with control birds which were neither fasted nor transported. When the behaviour of birds was monitored in their home pens after the treatments had ended, it was found that both fasted and transported birds were more active than controls, showing less lying behaviour for at least 12 h, suggesting that they were not particularly fatigued.

## Hunger and thirst

Newly hatched chicks are not provided with food and water until they reach the rearing unit. During transportation, which may last up to 2 days on international journeys, chicks are thus completely reliant on yolk sac metabolism. Warriss *et al.* (1992b) found that chicks deprived in this way for 48 h weighed 16.5 g less than control chicks which had access to food and water within 6 h of hatching, and showed both physiological and behavioural signs of dehydration and thirst.

The fasting regime imposed on broilers before transportation may have important implications for welfare. At the very least, it affects the amount of weight lost, which occurs after 4–6 h of fasting at a rate of 0.2–0.5% $h^{-1}$ as birds begin to metabolize body tissue (Veerkamp, 1986). Fasting is also aversive (Nicol and Scott, 1990) and may also increase stress (see below).

Broilers transported for 2, 4 or 6 h after feed withdrawal for 1–10 h had similar live and carcass weights to untransported controls (Warriss *et al.*, 1993). However, transport significantly reduced liver weight and liver glycogen concentration. Glycogen depletion of the biceps muscle increased progressively with journey time, which could have reflected muscular effort involved with maintaining balance in a moving vehicle. There was also evidence from this study that transported broilers were becoming dehydrated.

Withdrawal of food or both food and water for 24 h resulted in a 10% drop in live-weight (0.43% $h^{-1}$), of which 41% was loss in carcass weight (Knowles *et al.*, 1995). There was no evidence of significant dehydration in this study or in a survey of 800 broilers at two plants (Knowles *et al.*, 1996).

## Physiology indicative of stress

Further physiological changes are associated with crating and moving poultry after they have been caught. Increased glucagon and plasma corticosterone (Freeman *et al.*, 1984) and increased heterophil:lymphocyte ratios (Mitchell *et al.*, 1992) have been found in broilers transported for 2 or 3 h. However, the independent effects of the crating and the journey are not clear. Broom *et al.* (1990) (cited in Knowles and Broom, 1990) found that the plasma corticosterone concentrations of hens rose significantly after crating, but found no subsequent difference between hens that were left crated but stationary for 2 h and hens that were transported in a van for 2 h. Duncan (1989) found that plasma corticosterone in broilers was significantly higher if they were transported for 40 min after crating than if they were left stationary in the crate in the same vehicle for 40 min.

**Fear and aversion**

Cashman *et al.* (1989) assessed the TI duration of nearly 700 broilers on arrival at four commercial processing plants. The overall mean TI duration was 12.6 min, a level comparable to that reported after exposure to high-intensity electric shock (Gallup, 1973). Journey duration had the most significant effect on duration of TI. A strong, positive, linear relationship indicated that the birds' fear levels were primarily determined by transportation and not just the catching and loading procedures. When a similar study was conducted with 300 laying hens, there was no evidence of a positive linear relationship between journey duration and the duration of TI (Mills and Nicol, 1990). However, the TI durations of hens undergoing short journeys were higher than those found for broilers. Differences in catching procedures for broilers and battery hens are a possible explanation, with handling effects probably contributing to the TI response of hens undergoing short journeys (Mills and Nicol, 1990).

Transportation involves simultaneous exposure to many factors, including noise, motion, heat and crowding. The birds' experience of some of these separate factors has been examined by tests of preference and aversion. Mean sound levels on animal transportation lorries typically fall within the range 95–103 dbA. The response of poultry to the specific noise encountered during transportation has not been assessed. However, when given a choice between two keys which could be pecked to obtain food, hens avoided keys that activated certain sounds. A greater avoidance bias was obtained in response to the sound of chickens in a poultry shed transmitted at 100 dbA than to other sounds, including music or the sound of a train (McAdie *et al.*, 1993, 1996). It is generally thought that vibration is likely to be more aversive than noise. Nicol *et al.*, (1991) used a passive avoidance procedure and found that broiler chickens avoided exposure to short-acting circular motion and a small jolt in the horizontal plane, but did not avoid exposure to simple harmonic motion in the vertical plane. Using a related procedure, Rutter and Randall (1993) found that female broiler chickens strongly avoided sinusoidal vibration at 1.0 Hz ($1.51$ m s$^{-2}$), but showed only a mild aversive response to similar motion imposed at 0.5 Hz ($0.59$ m s$^{-2}$). Randall *et al.* (1997) examined responses to a wider range of horizontal and vertical vibrations, imposed for a 2-h period using a similar technique. They showed that both vertical and horizontal vibrations were aversive to broiler chickens, although a greater sensitivity to vertical vibration by a factor of between 1.3 and 2.5 was found at all acceleration levels. Aversion tended to increase with acceleration magnitude (0 to 5 m s$^{-2}$) and to decrease with increasing frequency (0 to 10 Hz) of the motion. As chickens find vibration below 5 Hz particularly aversive, Randall *et al.* (1997) concluded that the resonant frequencies of 1–5 Hz found on transporters are undesirable.

Although it is simpler experimentally to examine the effects of one factor in isolation, during commercial transportation many factors act simultaneously. The effects might be additive but, if they are synergistic, experimental work on single factors could underestimate bird response to combined stimuli. Abeyesinghe *et al.* (1999) designed an experiment to examine the combined effects of vibration and high air temperature. Broiler chickens chose to enter one of four colour-coded chambers for an hour in the apparatus shown in Fig. 18.3. The chambers imposed the following treatments: control, high air temperature, vertical vibration or combined high air temperature and vibration. Birds were allowed to make a total of five choices. The birds avoided the vibration, confirming previous findings that vertical vibration at 2 Hz 1 m s$^{-2}$ is aversive, but showed no apparent avoidance of the high air temperature. It may be that, despite its physiological effects on the birds, a high air temperature of 40°C is not perceived as aversive. However, when conducting choice tests, it is essential that aversion for different stimuli is measured using a common currency. In this case, the results may reflect the fact that any aversive effects of vibration are immediately apparent to the birds, whilst the effects of excessive heat may become obvious only after some time. Work currently in progress is examining this issue.

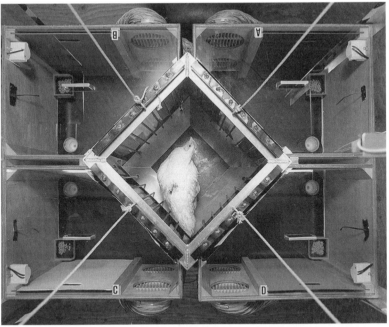

**Fig. 18.3.** Broiler chicken in the central starting chamber of a four-choice apparatus. After a short acclimatization period, the exit gates are lifted and the bird is free to enter a chamber. Once the bird enters a chamber, the gate closes behind it and the bird is confined in the chosen chamber for a period of 1 h before choosing again. Photo courtesy of Siobhan Abeyesinghe, Silsoe Research Institute.

# Transport Conditions and Vehicle Design

## Thermal

The degree of thermal stress experienced by birds in transit depends on the duration and intensity of heat and cold stressors. Side curtains are used as protection against precipitation, wind and solar radiation and, increasingly, to hide the stock from public view. However, even in winter, these often restrict ventilation too much and excessive heat and moisture levels build up around the birds (Mitchell *et al.*, 1992; Webster *et al.*, 1992; Kettlewell *et al.*, 1993).

Several studies in the temperate UK climate have measured the thermal environment in different parts of the loads of pullets, hens and broilers during winter and summer conditions and on stationary and moving vehicles of varying design (Webster *et al.*, 1992; Kettlewell *et al.*, 1993; Weeks *et al.*, 1997). These have shown that most vehicles used for transporting poultry, which are naturally ventilated, do not provide a uniform thermal environment. Studies of the aerodynamics of full-size and scale models of one design of vehicle, including a trailer, have shown that, when moving, air predominantly enters at the lower rear of the vehicle and moves forward to exit at the front of the vehicle (Baker *et al.*, 1996; Hoxey *et al.*, 1996).

Thus different parts of the load of poultry within the vehicle are over- or underventilated and birds at the main inlet point are also susceptible to becoming wet unless protected. There are large differences between conditions on moving and on stationary vehicles, again primarily due to ventilation and to speed of air movement. For example, Weeks *et al.* (1997) calculated that average air speeds immediately surrounding the birds in moving vehicles varied between 0.9 and 2.4 m s$^{-1}$ with maxima of 6.0 m s$^{-1}$. In certain positions, there was virtually no air movement to dissipate the body heat produced by the birds. Conditions in thermal 'hot' or 'cold' spots are frequently detrimental to welfare and may lead to deaths, which can be correlated with climatic conditions and excessive or inadequate ventilation (Hunter *et al.*, 1997).

This knowledge of thermal conditions experienced by live birds in road transit and the physiological consequences of these (Mitchell and Kettlewell, 1994) leads to the inescapable conclusion that controlled and uniform ventilation is essential (Weeks *et al.*, 1997; Mitchell and Kettlewell, 1998). Heat losses from model chickens among live birds in vehicles fitted with both side curtains and roof-mounted inlet fans were generally in the comfortable range, with little variation between areas of the load (Weeks *et al.*, 1997). These authors suggested that air speeds within bird crates or modules should be maintained between 0.3 and 1.0 m s$^{-1}$, except in extremely hot weather. Ventilation requirement is between 100 and 600 m$^3$ h$^{-1}$ for typical commercial loads. It is strongly recommended that all vehicles be fitted with several temperature probes placed in close proximity to the birds. Temperatures should be both recorded and linked to an in-cab monitoring and alarm system. As a guide, this should be modified according to individual loads and vehicle designs, Weeks *et al.* (1997) indicated that broilers and pullets transported at 10–15°C and poorly feathered end-of-lay birds at 22–28°C were likely to be thermally comfortable at the usual high stocking densities. In hot weather, the direct and indirect heating effects of solar radiation should be avoided by transporting at night or early in the day and parking in the shade.

The conditions at the end of the journey must also be considered. It may take 2–3 h to manually unload pullets. Broilers and spent hens may also have to wait at the processing plant (Warriss *et al.*, 1990). They may remain on the vehicle or be unloaded in modules or stacks of crates. In both instances, a well-designed lairage is preferable to remaining outside exposed to the elements. It is important that the birds themselves receive adequate ventilation; measurements in two lairages found air movements around the stacks of modules greater than 1 m s$^{-1}$ but less than 0.1 m s$^{-1}$ adjacent to the broilers (Quinn *et al.*, 1998). Temperatures and humidities among the birds consequently rose rapidly to give rise to conditions of heat stress within an hour in both winter and summer. Average body temperatures rose by 0.3°C in the first hour of lairage and by 0.1°C thereafter for the 4 h of measurement (Warriss *et al.*, 1999). The model birds used by Webster *et al.* (1992) and Weeks *et al.* (1997) indicated conditions of substantial heat and cold stress were frequently experienced by hens and broilers in lairage during loading and unloading. Thus the duration of such times needs to be kept to a minimum of preferably under 1 h. A controlled environment providing adequate ventilation while avoiding excessive wind and air movement on to the birds is highly desirable. There should also be sufficient space around each module or stack for effective air exchange and flow. Monitoring of the condition of birds and their environment is as necessary as during the journey.

**Vibration**

The fundamental frequency of most trucks used for poultry transport is 1–2 Hz, with a secondary peak at 10 Hz, which coincides with the resonance frequency of poultry viscera (Scott, 1994). More detailed measurements by Randall *et al.* (1996) found that resonant frequencies of broilers were around 15 Hz when sitting and 4 Hz when standing. Although animals can reduce the effects of vibration by moving, and by skeletal muscle tone, the scope for this in broilers with leg problems at high stocking density is very limited. Thus evidence suggests that vibration does adversely affect the birds and should be reduced, for example by using air suspension.

## Preslaughter Handling

Following arrival at the processing plant, both broiler chickens and end-of-lay hens are removed from the crates or modules and suspended by their legs from shackles for conveyance to the electric stunning bath. Many birds react to this procedure by struggling, flapping their wings and attempting to right themselves. This can lead to injury and reduces the chance that the bird will be effectively stunned prior to slaughter. Jones and Satterlee (1997) reported that covering broilers' heads with a hood immediately before shackling reduced struggling in comparison with non-hooded controls. A smaller reduction in struggling was also obtained when the birds were fitted with transparent hoods, leading Jones et al. (1998a) to suggest that the effect was due both to the tactile properties of the hoods and to their ability to impair patterned vision. Reducing ambient light intensity by itself reduced struggling in broilers shackled in groups of three (Jones et al., 1998b), but not in broilers shackled individually (Jones et al., 1998a). The authors argue that the welfare benefits of reduced injury would outweigh any possible increases in fear suggested by the increased immobility observed when hoods are fitted. Fitting birds with hoods would not be practical on most commercial processing lines, but it may be possible to design a system for a slaughterhouse based on the principles of reduced light intensity, mild tactile contact and interference with clear vision. Several plants in the USA and Canada calm chickens and turkeys on the shackle line by providing a breast rub surface made from 30 cm wide, smooth, rubber conveyor belting. The breast rub, in combination with darkness, reduces struggling and flapping.

Bird welfare would be greatly improved if the labour-intensive and stressful procedure of removing them from the containers and hanging them on shackles could be eliminated. Controlled-atmosphere (gas) stunning of chickens and turkeys is now a commercial reality with welfare and meat quality benefits, such as reduced breast muscle haemorrhaging and bone breaks (Raj et al., 1997; Hoen and Lankhaar, 1999) and may also be used for ducks which can be difficult to stun electrically (Raj et al., 1998).

Automation of shackling is being investigated, which will be easier with gas-stunned birds than with conscious ones, which may flap, struggle and experience pain when shackled (Sparrey and Kettlewell, 1994).

## Conclusions

Systems of housing, catching, handling, transport and lairage that are more humane both for poultry and for human workers need further development. In many cases, mechanization is less stressful and more efficient. We would highlight the need for improved techniques in depopulating houses. Whilst machine harvesting of broilers is being adopted commercially, conveyor systems associated with these, and which might have potential for laying-bird handling, still have drawbacks. In Europe, the political will to develop housing systems for laying hens that provide a range of facilities, including perches, dust-bathing areas and nest boxes (e.g. percheries, modified cage systems), presents a considerable challenge when birds are caught at the end of lay. Greater thought should be given to the need for depopulation and the concurrent design of appropriate handling techniques and technology when new housing systems are developed. The use of robots to collect and herd poultry is a long-term aim and a conceptual working model of a robotic duck herder has been developed in the UK (Henderson, 1999).

Research on the welfare of poultry during handling and transport has increased over the last 10 years. However, many studies are conducted on a very small scale and do not always replicate the conditions experienced commercially. Another problem is that different methods are often used to assess the response of birds to different transportation stimuli. Direct comparison of the relative aversiveness of, say, vibration and noise is therefore not possible. There is a need to develop a common research methodology.

Scientific evidence shows increasing stress and mortality in all classes of poultry as transportation time, holding time and feed- and water-deprivation time increase. Thermal stress is a major component of overall stress. Control of ventilation during transit should be provided to reduce this, as well as reducing the duration of all stages of transportation.

Hygiene regulations can conflict with animal welfare. EC hygiene regulations have forced the closure of several slaughterhouses in the UK and spent hens now have longer journeys. Improved designs of vehicles are becoming commercially available. In the future, it may be possible to inspect birds in transit by accessing on-board information records of the conditions experienced on the journey so far. The ability to access such information is a prerequisite for the sensible enforcement of welfare legislation or farm assurance standards.

# References

Abeyesinghe, S.M., Wathes, C.M., Nicol, C.J. and Randall, J.M. (1999) Avoidance of vibration and thermal stress by broiler chickens in a choice chamber. In: *Proceedings of the 33rd International Congress of the International Society for Applied Ethology*, Lillehammer, Norway.

Aitken, G. (1985) Poultry meat inspection as a commercial asset. *State Veterinary Journal* 39, 136–140.

Anderson, R.S. and Edney, A.T.B. (eds) (1991) *Practical Animal Handling*. Pergamon Press, Oxford.

Baker, C.J., Dalley, S., Yang, X., Kettlewell, P. and Hoxey, R. (1996) An investigation of the aerodynamic and ventilation characteristics of poultry transport vehicles. 2. Wind tunnel experiments. *Journal of Agricultural Engineeering Research* 65, 97–113.

Bayliss, P.A. and Hinton, M.H. (1990) Transportation of poultry with special reference to mortality rates. *Applied Animal Behaviour Science* 28, 93–118.

Beuving, G. (1980) Corticosteroids in laying hens. In: Moss, R. (ed.) *The Laying Hen and Its Environment*. Martinus Nijhoff, The Hague, pp. 65–82.

Buyse, J., Simons, P.C.M., Boshouwers, F.M.G. and Decuypere, E. (1996) Effect of intermittent lighting, light intensity and source on the performance and welfare of broilers. *World's Poultry Science Journal* 52, 121–130.

Cashman, P.J., Nicol, C.J. and Jones, R.B. (1989) Effects of transportation on the tonic immobility fear reactions of broilers. *British Poultry Science,* 30, 211–221.

Danbury, T.C., Weeks, C.A., Chambers, J.P., Waterman-Pearson, A.E. and Kestin, S.C. (2000) Preferential selection of the analgesic drug carprofen by lame broiler chickens. *Veterinary Record* 146, 307–311.

Duncan, I.J.H. (1989) The assessment of welfare during the handling and transport of broilers. In: Faure, J.M. and Mills, A.D. (eds) *The Proceedings of the Third European Symposium on Poultry Welfare*. French Branch of the World's Poultry Science Association, Tours, France, pp. 93–108.

Duncan, I.J.H., Slee, G., Kettlewell, P.J., Berry, P. and Carlisle, A.J. (1986) A comparison of the effects of harvesting broiler chickens by machine and by hand. *British Poultry Science* 27, 109–114.

Ekstrand, C. (1998) An observational cohort study of the effects of catching method on carcase rejection rates in broilers. *Animal Welfare* 7, 87–96.

Farsaie, A., Carr, L.E. and Wabeck, C.J. (1983) Mechanical harvest of broilers. *Transactions, ASAE* 26, 1650–1653.

Freeman, B.M., Kettlewell, P.J., Manning, A.C.C. and Berry, P.S. (1984) The stress of transportation for broilers. *Veterinary Record* 114, 286–287.

Gallup, G.G. (1973) Tonic immobility in chickens: is a stimulus that signals shock more aversive than the receipt of shock? *Animal Learning and Behaviour* 1, 228–232.

Gerrits, A.R., de Koning, K. and Mighels, A. (1985) Catching broilers. *Poultry* 1(5), 20–23.

Gracey, J.F. (1986) *Meat Hygiene*, 8th edn. Baillière Tindall, London, UK, pp. 455–458.

Grandin, T. (1999) *Canadian Animal Welfare Audit of Stunning and Handling in Federal and Provincial Slaughter Plants*. Grandin Livestock Handling Systems, Fort Collins, Colorado.

Gregory, N.G. and Wilkins, L.J. (1990) Broken bones in chicken, 2. Effect of stunning and processing in broilers. *British Poultry Science* 31, 53–58.

Gregory, N.G. and Wilkins, L.J. (1992) Skeletal damage and bone defects during catching and processing. In: *Bone Biology and Skeletal Disorders in Poultry*. 23rd Poultry Science Symposium, World's Poultry Science Association, Edinburgh, UK.

Gregory, N.G., Wilkins, L.J., Eleperuha, S.D., Ballantyne, A.J. and Overfield, N.D. (1990) Broken bones in domestic fowls: effect of husbandry system and stunning method in end of lay hens. *British Poultry Science* 31, 59–70.

Gregory, N.G., Wilkins, L.J., Kestin, S.L., Belyavin, C.G. and Alvey, D.M. (1991) Effect of husbandry system on broken bones and bone strength in hens. *Veterinary Record* 128, 397–399.

Grigor, P.N., Hughes, B.O. and Appleby, M.C. (1995) Effects of regular handling and exposure to an outside area on subsequent fearfulness and dispersal in domestic hens. *Applied Animal Behaviour Science* 44, 47–55.

Henderson, J. (1999) Flocking behaviour of ducks in response to predator stimuli. PhD thesis, University of Bristol.

Hoen, T. and Lankhaar, J. (1999) Controlled atmosphere stunning of poultry. *Poultry Science* 78(2), 287–289.

Hoxey, R.P., Kettlewell, P.J., Meehan, A.M., Baker, C.J. and Yang, X. (1996) An investigation of the aerodynamic and ventilation characteristics of poultry transport vehicles. I. Full-scale measurements. *Journal of Agricultural Engineeering Research* 65, 77–83.

Hunter, R.R., Mitchell, M.A. and Matheu, C. (1997) Distibution of 'dead on arrivals' within the bio-load on commercial broiler transporters: correlation with climatic conditions and ventilation regimen. *British Poultry Science* 38, S7-S9.

Jones, R.B. (1992) The nature of handling immediately prior to test affects tonic immobility fear reactions in laying hens and broilers. *Applied Animal Behaviour Science* 34, 247–254.

Jones, R.B. (1996) Fear and adaptability in poultry: insights, implications and imperatives. *World's Poultry Science Journal* 52,131–174.

Jones, R.B. and Satterlee, D.G. (1997) Restricted visual input reduces struggling in shackled broiler chickens. *Applied Animal Behaviour Science* 52, 109–117.

Jones, R.B. and Waddington, D. (1992) Modification of fear in domestic chicks *Gallus gallus domesticus* via regular handling and early environmental enrichment. *Animal Behaviour* 43, 1021–1034.

Jones, R.B., Hagedorn, T.K. and Satterlee, D.G. (1998a) Adoption of immobility by shackled broiler chickens: effects of light intensity and diverse hooding devices. *Applied Animal Behaviour Science,* 55, 327–335.

Jones, R.B., Satterlee, D.G. and Cadd, G.G. (1998b) Struggling responses of broiler chickens shackled in groups on a moving line: effects of light intensity, hoods and 'curtains'. *Applied Animal Behaviour Science* 58, 341–352.

Julian, R.J. (1984) Valgus-varus deformity of the intertarsal joint in broiler chickens. *Canadian Veterinary Journal* 25, 254–258.

Julian, R.J. (1998) Rapid growth problems: ascites and skeletal deformities in broilers. *Poultry Science* 77, 1773–1780.

Kannan, G. and Mench, J.A. (1996) Influence of different handling methods and crating periods on plasma corticosterone concentrations in broilers. *British Poultry Science* 37, 21–31.

Kannan, G. and Mench, J.A. (1997) Prior handling does not significantly reduce the stress response to pre-slaughter handling in broiler chickens. *Applied Animal Behaviour Science* 51, 87–99.

Kestin, S.C., Knowles, T.G., Tinch, A.E. and Gregory, N.G. (1992) Prevalence of leg weakness in broiler chickens and its relationship with genotype. *Veterinary Record* 131, 190–194.

Kettlewell, P.J. and Turner, M.J.B. (1985) A review of broiler chicken catching and transport systems. *Journal Agricultural Engineering Research* 31, 93–114.

Kettlewell, P., Mitchell, M. and Meehan, A. (1993) The distribution of thermal loads within poultry transport vehicles. *Agricultural Engineer* 48, 26–30.

Knowles, T.G. (1991) The welfare of hens in transit and related effects of housing system. Unpublished PhD thesis, University of Cambridge, UK.

Knowles, T.G. and Broom, D.M. (1990) The handling and transport of broilers and spent hens. *Applied Animal Behaviour Science* 28, 75–91.

Knowles, T.G. and Broom, B.M. (1993) Effect of catching method on the concentration of plasma corticosterone in end-of-lay battery hens. *Veterinary Record* 133, 527–528.

Knowles, T.G., Warriss, P.D., Brown, S.N., Edwards, J.E. and Mitchell, M.A. (1995) Response of broilers to deprivation of food and water for 24 hours. *British Veterinary Journal* 151, 197–202.

Knowles, T.G., Ball, R.C., Warriss, P.D. and Edwards, J.E. (1996) A survey to investigate potential dehydration in slaughtered broiler chickens. *British Veterinary Journal* 152, 307–314.

Lacy, M.P. and Czarick, M. (1998) Mechanical harvesting of broilers. *Poultry Science* 77, 1794–1797.

McAdie, T.M., Foster, T.M., Temple, W. and Matthews, L.R. (1993) A method for measuring the aversiveness of sounds to domestic hens. *Applied Animal Behaviour Science*, 37, 223–238.

McAdie, T.M., Foster, T.M. and Temple, W. (1996) Concurrent schedules: quantifying the aversiveness of noise. *Journal of the Experimental Analysis of Behaviour*, 65, 37–55.

McEwen, S.A. and Barbut, S. (1992) Survey of turkey downgrading at slaughter – carcass defects and associations with transport, toenail trimming, and type of bird. *Poultry Science* 71(7), 1107–1115.

McGeown, D., Danbury, T.C., Waterman-Pearson, A.E. and Kestin S.C. (1999) The effect of Carprofen on lameness in broiler chickens. *Veterinary Record* 144, 668–771.

MAFF (1997) *The Welfare of Animals (Transport) Order, 1997.* Stationery Office, London, UK.

Meijerhof, R. (1997) The importance of egg and chick transportation. *World Poultry* 13(11), 17–18.

Metheringham, J. (1996) Poultry in transit – a cause for concern? *British Veterinary Journal* 152, 247–250.

Mills, D.S. and Nicol, C.J. (1990) Tonic immobility in spent hens after catching and transport. *Veterinary Record* 126, 210–212.

Mitchell, M.A. and Kettlewell, P.J. (1994) Road transportation of broiler chickens – induction of physiological stress. *World's Poultry Science Journal* 50, 57–59.

Mitchell, M.A. and Kettlewell, P.J. (1998) Physiological stress and welfare of broiler chickens in transit: solutions, not problems! *Poultry Science* 77, 1803–1814.

Mitchell, M.A., Kettlewell, P.J. and Maxwell, M.H. (1992) Indicators of physiological stress in broiler chickens during road transportation. *Animal Welfare* 1, 92–103.

Mitchell, M.A., Carlisle, A.J., Hunter, R.R. and Kettlewell, P.J. (1997) Welfare of broilers during transportation: cold stress in winter – causes and solutions. In: Koene, P. and Blokhuis, H.J. (eds) *Proceedings of the 5th European Symposium on Poultry Welfare.* Wageningen, The Netherlands, pp. 49–52.

Moran, P., and Berry, P.S. (1992) Mechanised broiler harvesting. *Farm Buildings and Engineering* 91, 24–27.

Nicol, C.J. (1992) Effects of environmental enrichment and gentle handling on behaviour and fear responses of transported broilers. *Applied Animal Behaviour Science* 33, 367–380.

Nicol, C.J. and Scott, G.B. (1990) Transport of broiler chickens. *Applied Animal Behaviour Science* 28, 57–73.

Nicol, C.J., Blakeborough, A. and Scott, G.B. (1991) The aversiveness of motion and noise to broiler chickens. *British Poultry Science* 32, 243–254.

Parry, R.T. (1989) Technological developments in pre-slaughter handling and processing. In: Mead G.C. (ed.) *Processing of Poultry.* Elsevier, Amsterdam, the Netherlands, pp. 65–101.

Poultry International (1995) Bird-friendly automatic chicken catching. *Poultry International* August, 26–28.

Poultry International (1998) At last – fully automated livebird harvesting. *Poultry International* March 44–48.

Quinn, A.D., Kettlewell, P.J., Mitchell, M.A. and Knowles, T. (1998) Air movement and thermal microclimates observed in poultry lairages. *British Poultry Science* 39, 469–476.

Raj, A.B.M., Wilkins, L.J., Richardson, R.I., Johnson, S.P. and Wotton, S.B. (1997) Carcase and meat quality in broilers either killed with a gas mixture or stunned with an electric current under commercial processing conditions. *British Poultry Science* 38, 169–174.

Raj, A.B.M., Richardson, R.I., Wilkins, L.J. and Wotton, S.B. (1998) Carcase and meat quality in ducks killed with either gas mixtures or an electric current under commercial processing conditions. *British Poultry Science* 39, 404–407.

Randall, J.M., Cove, M.T. and White, R.P. (1996) Resonant frequencies of broiler chickens. *Animal Science* 62, 369–374.

Randall, J.M., Duggan, J.A., Alami, M.A. and White, R.P. (1997) Frequency weightings for the aversion of broiler chickens to horizontal and vertical vibration. *Journal of Agricultural Engineering Research* 68, 387–397.

Reed, H.J., Wilkins, L.J., Austin, S.D. and Gregory, N.G. (1993) The effects of environmental enrichment during rearing on fear reactions and depopulation trauma in adult caged hens. *Applied Animal Behaviour Science* 36, 39–46.

Rutter, S.M. and Randall, J.M. (1993) Aversion of domestic fowl to whole-body vibratory motion. *Applied Animal Behaviour Science* 37, 69–73.

Sainsbury, D.W.B. (1988) Broiler chickens. In: Scott: W.N. (ed.) *Management and Welfare of Farm Animals, UFAW Handbook,* 3rd edn. Baillière Tindall, London, pp. 221–232.

Sanotra, G.S., Vestergaard, K.S. and Thomsen, M.G. (1998) The effect of stocking density on walking ability, tonic immobility, and the development of tibial dyschondroplasia in broiler chicks. *Poultry Science* 77, 1844–1845.

Scott, G.B. (1993) Poultry handling: a review of mechanical devices and their effect on bird welfare. *Worlds' Poultry Science Journal* 49, 44–57.

Scott, G.B. (1994) Effects of short-term whole-body vibration on animals with particular reference to poultry. *Worlds' Poultry Science Journal* 50, 25–38.

Scott, G.B. and Moran, P (1992) Behavioural responses of laying hens to carriage on horizontal and inclined conveyors. *Animal Welfare* 1, 269–277.

Scott, G.B. and Moran, P. (1993) Fear levels in laying hens carried by hand and by mechanical conveyors. *Applied Animal Behaviour Science* 36, 337–345.

Scott, G.B., Connell, B.J. and Lambe, N.R. (1998) The fear levels after transport of hens from cages and a free-range system. *Poultry Science* 77, 62–66.

Sherwin, C.M., Kestin, S.C., Nicol, C.J., Knowles, T.G., Brown, S.N., Reed, H.J. and Warriss, P.D. (1993) Variation in behavioural indices of fearfulness and fatigue in transported broilers. *British Veterinary Journal* 149, 571–578.

Sparrey, J.M. and Kettlewell, P.J. (1994) Shackling of poultry – is it a welfare problem? *Worlds' Poultry Science Journal* 50, 167–176.

Stuart, C. (1985) Ways to reduce downgrading. *World Poultry Science* 41, 16–17.

Su, G., Sørensen, P. and Kestin, S.C. (1999) Meal feeding is more effective than early feed restriction at reducing the prevalence of leg weakness in broiler chicken. *Poultry Science*, 78, 949–955.

Veerkamp, C.H. (1986) The influence of fasting and transport on yields of broilers. *Poultry Science* 57, 619–627.

Warriss, P.D. and Brown, S.N. (1996) Time spent by turkeys in transit to processing plants. *Veterinary Record* 139, 72–73.

Warriss, P.D., Kestin, S.C., Brown, S.N. and Bevis, E.A. (1988) Depletion of glycogen reserves in fasting broiler chickens. *British Poultry Science* 29, 149–154.

Warriss, P.D., Bevis, E.A. and Brown, S.N. (1990) Time spent by broiler chickens in transit to processing plants. *Veterinary Record* 127, 617–619.

Warriss, P.D., Bevis, E.A., Brown, S.N. and Edwards, J.E. (1992a) Longer journeys to processing plants are associated with higher mortality in broiler chickens. *British Poultry Science* 33, 201–206.

Warriss, P.D., Kestin, S.C. and Edwards, J.E. (1992b) Responses of newly hatched chicks to inanition. *Veterinary Record* 130, 49–53.

Warriss, P.D., Kestin, S.C., Brown, S.N., Knowles, T.G, Wilkins, L.J., Edwards, J.E., Austin, S.D. and Nicol, C.J. (1993) The depletion of glycogen stores and indices of dehydration in transported broilers. *British Veterinary Journal* 149, 391–398.

Warriss, P.D., Knowles, T.G, Brown, S.N., Edwards, J.E., Kettlewell, P.J., Mitchell, M.A. and Baxter, C.A. (1999) Effects of lairage time on body temperature and glycogen reserves of broiler chickens held in transport modules. *Veterinary Record* 145, 218–222.

Watt Poultry Statistical Yearbook (1998) Watt, Petersfield, UK.

Webster, A.J.F, Tuddenham, A., Saville, C.A. and Scott, G.A. (1992) Thermal stress on chickens in transit. *British Poultry Science* 34, 267–277.

Weeks, C.A. and Kestin, S.C. (1997) The effect of leg weakness on the behaviour of broilers. In: Koene, P. and Blokhuis, H.J. (eds) *Proceedings of the 5th European Symposium on Poultry Welfare*, Wageningen, The Netherlands, pp. 117–118.

Weeks, C.A., Webster, A.J.F. and Wyld, H.M. (1997) Vehicle design and thermal comfort of poultry in transit. *British Poultry Science* 38, 464–474.

Wyers, M., Cherel, Y. and Plassiart, G. (1991) Late clinical expression of lameness related to associated osteomyelitis and tibial dyschondroplasia in male breeding turkeys. *Avian Diseases* 35, 408–414.

# Stress Physiology of Animals During Transport

**19**

## Toby G. Knowles and Paul D. Warriss

*School of Veterinary Science, University of Bristol, Langford, Bristol BS40 5DU, UK*

## Introduction

There is an increasing public interest in and concern for the welfare of livestock during transport. The majority of people now live in towns and cities and are no longer in day-to-day contact with farm animals. They are relatively unfamiliar with the animals and the methods of husbandry under which they are kept and, to a large extent, have an idealized picture of farming and animal production. However, there is one point in most animal production systems which is commonly open to public view – when the animals are transported. But, although necessary, transport is generally an exceptionally stressful episode in the life of the animal and one which is sometimes far removed from an idealized picture of animal welfare. So, increasing public concern for animals during transport has spurred research into their welfare, research which has attempted to quantify the severity of the stress imposed by the various stages involved in transport and to identify acceptable conditions and methods to minimize the adverse effects of transport.

Most work has concentrated on quantifying and ameliorating the effects of road transport, as this is the major mode of animal transport, and it is road transport which is the main theme of this chapter. However, there have also been research programmes targeted at other forms of transport. The export of cull sheep from Australia to the Middle East by ship can result in exceptionally high mortality rates during the sea journey. This has prompted the Australian government to fund research into the problem. An introduction to the literature covering this research can be found in Richards *et al.* (1991). A limited number of livestock, usually only those of high value, are transported by air. Recommendations for transporting live animals by air are detailed in the International Air Transport Association (IATA, 1998) *Live Animal*

*Regulations*, which are updated annually to take account of the latest research findings. These regulations are enforced by the European Union (EU), the USA and many other countries for the air transport of all live animals.

The assessment of the welfare of animals during transport in any sort of objective and scientific way requires the measurement of something, in a quantifiable and repeatable manner. Broom (1986) defined an animal's welfare as 'the state of an individual with regard to its attempts to cope with its environment'. Within this definition, an animal attempts to maintain homoeostasis through physiological and behavioural changes, and it follows that the greater the behavioural or physiological changes that are required, the more an animal is having to do to cope with the situation or environment and the poorer its welfare is likely to be. This approach provides a working basis by which welfare can be judged and is very much in line with the clinical biochemical approach to the diagnosis of disease in both human and veterinary medicine. In clinical biochemistry, one or a number of measured biochemical or haematological variables in an individual are compared with population norms in order to identify specific disorders (see, for example, Farver, 1997). This sounds fairly straightforward; however, welfare itself is not an objective, measurable thing but is an entirely human concept and, as such, cannot escape a high degree of subjective interpretation. Even when we can communicate with the animal that is being assessed, that is, within our own species, there arise differences in the assessment of the welfare of individuals, owing to differences between 'assessees' and in the opinions and backgrounds of the assessors, and these are not the only source of variability. As a whole, society's idea of what is acceptable human welfare has changed over time. How much more difficult it is, then, to try to 'second guess' the welfare of an animal with which we cannot communicate and which is unlikely to view or interpret its situation in anything approaching our own, human terms and, furthermore, for a range of people then to come to an overall agreement on the level of its welfare.

Thus, the idea of measuring the magnitude of the behavioural and/or physiological adjustments that an animal has to make to cope with its environment provides a useful structure underpinning the assessment of an animal's welfare. However well scientifically founded the measurements, their interpretation cannot escape a high degree of subjective interpretation. We might ask 'What is an unacceptable level of mortality?', when, however well animals are transported, there will always be some deaths. We can measure increasing 'hunger' and dehydration in an animal by changes in blood biochemicals, but how hungry or thirsty can that animal be allowed to become before the situation is unacceptable, when the biochemical changes that are observed increase linearly over time? During mating, play or hunting prey. many of the biochemical variables that are commonly used as measures of animal welfare reach extreme values, but most people would not consider the welfare of an animal in these situations to be impaired.

The remainder of this chapter gives an introduction to the main physiological variables that have been used to assess the stress imposed on animals by transport. As far as possible, these appear in functional groups; that is, they have been grouped as indicators of the various effects that are of interest – food deprivation, dehydration, muscular effort, etc. Following this, we summarize the best practice relating to research to date. Some further details of species-specific research on transport can be found in other chapters, but additionally there are a number of reviews in the scientific literature, which are listed in Table 19.1.

## Physiological Variables

Some commonly used physiological indicators of stress during transport are shown in Table 19.2. For a healthy, rested animal of a given species, there is a range of values for each biochemical and haematological variable within which the level of each measure for any individual would normally be expected to fall. The distribution of values found in a healthy, rested population usually forms the familiar, bell-shaped, Gaussian distribution, except for the values of enzymes, for which the distribution is positively skewed, having a greater number of higher values. Published veterinary reference ranges for variables are quoted as the range of values within which 95% of the population would be expected to fall. These limits are the 2.5 and 97.5 percentiles of any

**Table 19.1.** Recently published reviews from the scientific literature covering the road transport of livestock.

Cattle
  Tarrant, P.V. (1990)
  Warriss, P.D. (1990)
  Knowles, T.G. (1999)

Calves
  Trunkfield, H.R. and Broom, D.M. (1990)
  Knowles, T.G. (1995)

Sheep
  Knowles, T.G. (1998)

Pigs
  Warriss, P.D. (1987, 1998a,b)
  Tarrant, P.V. (1989)

Cattle, sheep and goats
  Wythes, J.R. and Morris, D.G. (1994)

Sheep and pigs
  Hall, S.J.G. and Bradshaw, R.H. (1998)

**Table 19.2.** Commonly used physiological indicators of stress during transport.

| Stressor | Physiological variable |
|---|---|
| **Measured in blood** | |
| Food deprivation | ↑ FFA, ↑ β-OHB, ↓ glucose, ↑ urea |
| Dehydration | ↑ Osmolality, ↑ total protein, ↑ albumin, ↑ PCV |
| Physical exertion | ↑ CK, ↑ lactate |
| Fear/arousal | ↑ Cortisol, ↑ PCV |
| Motion sickness | ↑ Vasopressin |
| **Other measures** | |
| Fear/arousal and physical exertion | ↑ Heart rate, ↑ respiration rate |
| Hypothermia/hyperthermia | Body temperature, skin temperature |

FFA, free fatty acids; β-OHB, β-hydroxybutyrate; PCV, packed-cell volume; CK, creatine kinase.

distribution and are approximately equivalent to ±2 standard deviations about the mean when the variable does have a Gaussian distribution. Figure 19.1 shows the frequency distributions of plasma albumin levels and of the enzyme creatine kinase (CK) from control samples obtained from cattle in a study by Knowles *et al.* (1999a). The distribution of albumin values is very close to the Gaussian curve which is superimposed on the graph, whilst the distribution of CK values is far from Gaussian. The 2.5 and 97.5 centiles of the albumin values are 34.9 and 45.4 g l$^{-1}$, respectively, and for CK 58.6 and 302.4 U l$^{-1}$. The mean and standard deviation provide a useful summary of the albumin data and the percentiles are close to the mean ±2 SD. This is not the case for the distribution of the CK values, which are strongly right-skewed.

Published reference ranges are useful in the diagnoses of a wide variety of diseases and can be useful for evaluating hypo- and hyperthermia, the degree of dehydration and, to a lesser extent, the degree of hunger arising during transport. However, it should be remembered that most of the physiological changes seen during transport are due to the action of normal homoeostatic mechanisms taking place within a healthy population of animals in response to the variety of different stressors. Thus clinical reference ranges are of limited use in evaluating welfare during transport, as the animals being transported are generally all healthy. So care should be taken in drawing any conclusions from comparisons with published normal ranges. What is really of interest is the change of a variable over time within an individual animal, as this is an indication of the scale of the response that an animal is mounting in order to cope. Because of the inherent variability between individuals, measurements are best taken at the level of the individual over time. The following is an overview of some physiological variables which can give some insight into how an animal is coping with a given situation. However, instead of moving directly to descriptions of individual biochemical and haematological markers, we start with the ultimate indicator of the inability of an animal to cope.

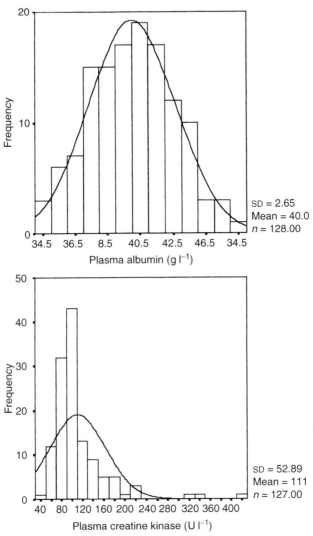

**Fig. 19.1.** The frequency histograms of plasma albumin and creatine kinase from rested cattle from a study by Knowles *et al.* (1999a). The Gaussian curves for distributions with the same means and standard deviations as the data are superimposed.

## Mortality

Mortality is a useful indicator of physiological stress. When an animal dies during transport, it is because its physiological mechanisms have failed to maintain homoeostasis. That transport is stressful is most readily quantified by the increased mortality which accompanies it. On average, in a given time period, a greater number of livestock will die if they are transported than if they are not. Of course, the animals which die initially are often those that are weaker and the overall mortality rates during road transport for most livestock are usually only fractions of 1%. This means that the rate can only be accurately estimated when large enough numbers are surveyed. However, increased mortality is interpreted by most people as an indicator of poor welfare and of the stressful nature of transport, in a way that many other types of measures are not so easily agreed upon.

If an increase in mortality is seen, even if mortality is only occurring amongst weaker animals, it is an indication that conditions are harsher and that the animals that do survive are probably facing a greater challenge. If a large enough number of animals is being considered, mortality rate is also a useful measure of the relative 'stressfulness' of methods of handling and transport. For instance, Warriss *et al.* (1992) surveyed journey times and mortality rate amongst broilers transported to slaughter and found a strong relationship between the two variables. The results showed a marked non-linear increase in bird mortality for journeys greater than 4 h, strongly suggesting that transport for longer than 4 h was undesirable. However, results like these are more easily obtained from broilers, as they have the highest mortality rates amongst the commonly transported types of livestock. The high mortality rates in broilers are in part due to the heavy commercial selection for improved growth rate and feed conversion to slaughter at 42 days, which has rather disregarded bird viability much beyond this age. The same trend can be seen occurring in pigs, which are under similar selection pressures but which have a longer reproductive cycle and are not yet at the same level of selection.

### Live-weight, β-hydroxybutyrate, free fatty acids and liver glycogen as indicators of fasting

Transport can involve extended periods without food or water and, as a consequence, there is an initial loss of live-weight, which is predominantly due to loss of gut fill; approximately 7% of body-weight in ruminants and 4% in pigs is lost during the first 18–24 h. Generally, weight loss during transport is accelerated, compared with when an animal is simply deprived of food and water and not transported. In ruminants, the main loss of gut fill takes place during the first 18–20 h of transport. Loss of gut fill in itself is unlikely to be directly deleterious to the animal. There is, however, an approximately linear loss of body-weight, measured as a decrease in carcass weight, which is due to dehydration and to the use of body reserves. This can be measured as the rate of loss of carcass weight and, within species, has been found to show quite large variation across different studies, these differences being due to the condition of the animals, the environment and the conditions of transport.

Once an animal is deprived of food and water, it has to rely on its body reserves to buffer it, until it can feed and drink again. The main energy store in the body is in the form of lipids and by far the most important of these are triacylglycerols (or triglycerides). In the form of subcutaneous fat, these can also provide thermal insulation. Triacylglycerols are mobilized by breaking them down into the constituent glycerol and fatty acids. These non-esterified fatty acids (NEFA) or free fatty acids (FFA) are transported in the blood bound to proteins. Triacylglycerols can be synthesized by many types of cell, but most synthesis takes place in the liver, adipose tissue and the small intestine. Lipolysis, the mobilization of body fat, is under hormonal control. As an animal fasts, much less glucose is available from the gut or glycogen reserves and this results in decreased levels of glucose in the blood plasma. This leads to hormone changes, increased glucagon levels and decreased insulin levels, which trigger hormone-sensitive lipase to break down adipose triacylglycerols, which are hydrolysed to FFA and glycerol. FFA can be utilized directly by most tissues and, as insoluble lipids, are bound to albumin in the plasma for transport around the body, whilst glycerol is transported dissolved in plasma water. Thus, during starvation, levels of FFA rise in the plasma, whilst actions that promote FFA synthesis suppress lipolysis and plasma FFA levels are not elevated.

The liver holds a reserve of glycogen and, during the first day of fasting, this reserve diminishes rapidly. Levels of liver glycogen can be measured by biopsy or at slaughter. Changes in liver weight can also be used as a measure of the use of these reserves. There are also reserves of glycogen within the skeletal muscle, which tend to be conserved even after several days of fasting.

During fasting, the usual metabolic pathways are modified and greater amounts of ketones are produced from FFA in the liver. Very high levels of FFA are damaging to tissues. The liver converts them to ketones. One of the main ketones is β-hydroxybutyrate (β-OHB) (or 3-hydroxybutyrate (3-OHB)). Because FFA can be utilized by most tissues, it was not clear until recently why ketones were produced. But it now appears that many tissues more easily utilize β-OHB than FFA. In fact, in some species, such as humans, ketones form the main energy source for the brain during fasting. This is not the case with the sheep or pig, where the brain still relies on glucose as the main energy source. Ketones are the main fuel of resting skeletal muscle during short-term fasting, but during long-term starvation or exercise FFA become the main energy source. There is a biological limit to the amount of FFA that can be present in the plasma, as all FFA have to be bound to albumin for transport. Levels of ketones in the plasma are not restricted in this way, which is important, as levels of plasma albumin decrease during fasting, thus reducing the amount of FFA that can be transported.

During exercise, glucose, ketones and FFA are all used as fuel. After strenuous exercise, ketone oxidation by muscles is reduced and this leads to an increase in plasma levels of FFA and β-OHB, which may be several times higher than pre-exercise levels. However, for several minutes immediately after exercise, levels may fall momentarily below pre-exercise levels as the metabolism adjusts.

## Plasma osmolality, total protein, albumin and packed-cell volume as indicators of dehydration

Water is essential to all of the processes which take place within the body, accounting for 60% of the total body-weight for most domestic animals. However, adipose tissue contains little water and 'fat' animals, such as fattened lambs and pigs, will contain a lower percentage of water. Total body water is considered, physiologically, to be made up of the extracellular fluid (ECF) volume and intracellular fluid (ICF) volume, where the ECF is all fluid outside the cells. Fluid present in the gut is sometimes considered as part of the ECF. In ruminants, the forestomach may contain a substantial amount of fluid – up to 30–60 litres in adult cattle – which, during periods of water deprivation, can act as a buffer to maintain effective circulating volume. During periods of inadequate water uptake, the water losses are balanced proportionately between the ICF and non-gut ECF; thus electrolyte balance between the two is maintained.

Packed-cell volume (PCV), total plasma protein and plasma albumin are convenient and simple measures of dehydration. PCV is the percentage of the blood volume occupied by cells (predominantly the red blood cells), the remainder of the volume being fluid. Thus, as long as there is no loss or gain of cells, PCV is a measure of the plasma volume. However, many species have a reserve of red blood cells in the spleen, which are readily released in response to excitement and stressors, so it is useful to use total plasma protein and albumin levels in addition to PCV. The assumption is made that the total amount of protein present in the plasma remains the same. Both total plasma protein and plasma albumin should show the same type of change if the effect is due to dehydration and not a dietary effect. It should be noted that the percentage changes in protein and PCV will not be the same for a given loss of plasma volume, e.g. a 10% plasma volume deficit would result in an 11% increase in protein but only, perhaps, a 2.5% increase in PCV. Osmolality can be used as a further, simple measure of plasma water content, as it is a colligative property and therefore includes all solute species. As a rough guide to the extent of dehydration, clinical signs are usually apparent when 4–6% of total body-weight of 'effective' (not including fluid in the gut) total body water has been lost, moderate dehydration is when 8–10% has been lost and severe dehydration is said to occur when losses are greater than 12% (Carlson, 1997).

## Heart rate, respiration rate, plasma cortisol and glucose as indicators of a general reaction to stress

An initial response to stress is the release of the hormones adrenaline and noradrenaline into the bloodstream from the adrenal glands. Noradrenaline is also released from sympathetic nerve endings, where it can act directly. The release of these hormones causes an acute increase in heart rate and blood pressure and stimulates hepatic glycogenolysis. This leads to an increased availability of glucose and a rise in plasma glucose levels within minutes. The effects of these hormones provide a useful measure of stress, but they have rather short half-lives in the bloodstream. In the slightly longer term, an animal's response to stress is mediated mainly through the hypophyseal–adrenal axis, a system in which neural and endocrine control systems are integrated in a such a highly complex and interdependent manner that it can only be described superficially here. Glucocorticoid hormones, produced in and released from the cortex of the adrenal glands in response to an extremely wide range of stressors, play a major role in mediating the physiological response. Cortisol is the central glucocorticoid in mammalian farm species and corticosterone in avian species. The pathway leading to the release and control of cortisol acts through the hypothalamus, pituitary and the adrenal cortex and is summarized in Fig. 19.2. The glucocorticoids play a major role in glucose metabolism, inhibit protein synthesis, initiate proteolysis and modulate immunological mediators, such as lymphokines and mediators of inflammatory reactions, causing anti-inflammatory effects. Because of the role of the brain in the release of glucocorticoids, they are widely interpreted as a measure of an animal's psychological perception of a situation, in addition to the extent of its physiological reaction.

### Creatine kinase, muscle glycogen and lactate as indicators of physical activity

The enzyme CK (also referred to as creatine phosphokinase (CPK)) is present in muscle, where it makes ATP available for contraction by the phosphorylation of ADP from creatine phosphate. It appears in the circulating plasma as a result of tissue damage and is relatively organ-specific, occurring as three isoenzymic forms, with an additional fourth variant that derives from mitochondria. Identification of the levels of isoenzymes present in the blood allows determination of the tissue which is the source and to which damage has occurred. During exercise there is increasing CK 3 (the main isoenzyme present in muscle and also known as CK-MM) activity present in the blood, as it leaks from the cells of skeletal muscle. Lactate dehydrogenase (LDH) has also been used as a measure of muscle damage, however, LDH activity is high in various tissues throughout the body and measurements are perhaps not so organ-specific.

During exercise, the main fuels for muscular contraction are glucose and fatty acids from the blood. There is also an intramuscular carbohydrate reserve in the form of glycogen, and it is when muscle glycogen stores are depleted that exhaustion has been shown to set in. The main extramuscular carbohydrate source is glycogen in the liver. Reserves of muscle and liver glycogen may be measured by biopsy but, as most transport of animals is to slaughter, it is usually assayed in muscle and liver sampled immediately after slaughter. Metabolism of glucose can take place aerobically or anaerobically; in the latter

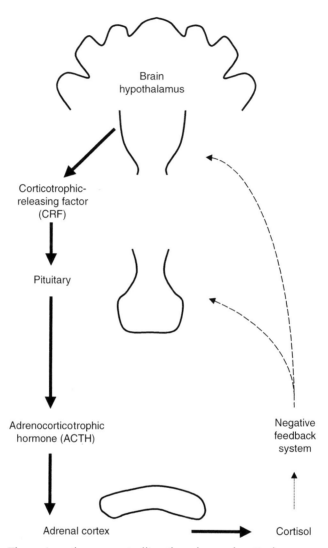

**Fig. 19.2.** The main pathways controlling the release of cortisol.

case there may be a gradual build up of lactate. Lipids can be metabolized only aerobically. In the initial few seconds of exercise, metabolism is mainly anaerobic. If the exercise is not too strenuous, aerobic metabolism of glucose and lipids takes over and lactate production decreases. The harder the exercise is, the higher is the percentage use of carbohydrate over lipid; thus lactate production is closely correlated with the intensity of exercise and may be seen as increased levels of lactate in muscle and in plasma. The degree to which the reserves of muscle glycogen are depleted at the time of slaughter has an effect on the post-mortem changes which take place in the muscle. If glycogen reserves have been depleted to any great extent, the muscle produces meat of an inferior quality, which looks dark, tends to have a less acceptable eating quality and is more prone to microbial spoilage, partly because it has a higher pH (is less acidic). Meat with this quality problem is commonly referred to as 'dark, firm and dry' (DFD) and is most prevalent amongst cattle and pigs, being less common with sheep.

## Urea

Any process which increases protein catabolism will tend to result in increased levels of plasma urea. Thus, levels of urea increase in response to stress, when levels of cortisol increase, and they will also rise as a result of food deprivation.

## Vasopressin as an indicator of travel sickness

The major role of the hormone vasopressin is to regulate body water homoeostasis, in regard to the relative osmolality of the ECF and ICF, by controlling reabsorption. Its release is mainly triggered by increased plasma osmolality and it acts by causing water retention. However, its release is relatively insensitive to changes in plasma volume; thus it normally plays little role in maintaining overall water balance (i.e. how much water is in the whole animal). Increased levels of vasopressin have been shown to be of use as an indicator of nausea and vomiting. Two types of vasopressin occur: most mammals produce arginine vasopressin, but the pig produces only lysine vasopressin.

We have been able to discuss only briefly the main biochemical and haematological indicators that have been used to evaluate animal welfare during transport. For more in-depth information, the reader should refer to one of the many textbooks which deal with clinical veterinary biochemistry. At the time of writing, Kaneko *et al.* (1997) provide a relatively up-to-date and comprehensive reference.

# The Physiological Responses of Cattle, Sheep and Pigs to Transport

## Cattle

Mortality rates amongst cattle transported by road are generally much lower than those of other forms of livestock. To a large extent, this is because the care with which animals are transported and the attention paid to their welfare are in proportion to the value of the individual animal (Hails, 1978). Over the first 18–24 h of transport, loss of body-weight can range from 3 to 11%. This is mostly due to loss of gut fill. Loss of carcass weight increases approximately linearly with transport time and has variously been reported to range from less than 1% to 8% over 48 h. Access to water can reduce both loss of body-weight and loss of carcass weight (Warriss, 1990). After 24 h of transport, there is an increase in plasma levels of β-OHB, FFA, osmolality, total protein and albumin, indicative of mobilization of food reserves and increasing dehydration (Tarrant et al., 1992; Warriss et al., 1995; Knowles et al., 1999a). After 24–31 h of transport, levels of plasma cortisol, glucose and CK are elevated. In sheep, these variables generally return to pre-transport levels after approximately 9 h of transport, but in cattle they tend to remain elevated or to increase steadily. Additionally, in cattle there is a gradual depletion of muscle glycogen (Knowles et al., 1999a) and an associated increase in the pH of the meat (Tarrant et al., 1992). These changes arise because cattle prefer to stand during transport, as they are relatively heavy animals and lying can produce considerable pressure on the parts of the body in contact with the floor of the vehicle, especially during a rough journey. The act of lying down and rising is difficult on a moving lorry at the stocking densities used for transport and there is a risk of being trampled or fallen on. The changes seen in these variables indicate that there is some physical effort involved in remaining standing and having to maintain balance against the motion of the vehicle. Despite the dangers and discomfort involved in lying, towards the end of the first 24 h of transport some cattle do lie down (Tarrant et al., 1992; Knowles et al., 1999a). This could be because of the physical effort involved with standing, although the physiological changes seen do not indicate excessive physical demand. Knowles et al. (1999a) hypothesized that the animals could possibly be in need of sleep, as those animals that did lie down displayed higher levels of plasma cortisol. Raised levels of plasma cortisol are associated with sleep deprivation in humans (Leproult et al., 1997).

Knowles et al. (1999a) offered water to cattle on board lorries for 1 h following 14 h of transport within the UK. They found that fewer than 60% of animals drank, few drank fully and activity levels rose whilst the vehicle was motionless, leading them to conclude that the stop merely prolonged transport and further exhausted the animals, rather than providing any recovery. Warriss et al. (1995) found that it took cattle 5 days to recover the live-weight lost during 15 h of transport. Knowles et al. (1999a) found little difference in the pattern of recovery following either 14, 21, 26 or 31 h of transport. Levels of plasma β-OHB, FFA, urea and glucose had recovered to pre-transport levels after 24 h in lairage with food and water freely available, as had levels of plasma cortisol. Levels of indicators of hydration took up to 72 h to return to pre-transport levels, whilst full pre-transport live-weight had not been recovered even after 72 h of lairage.

Based on the physiological indicators of fatigue and dehydration and on the behaviour of the animals, both Tarrant et al. (1992) and Knowles et al. (1999a) suggest a maximum continuous transport time of no longer than 24 h for cattle. Knowles et al. (1999a) recommend a mid-transport lairage period of, ideally, 24 h with food and water available, to allow recovery from the physical demands of transport. They considered that short mid-transport stops were unlikely to provide reasonable opportunity for rest or recovery.

However, including a lairage stop of any length provides an opportunity for cattle from different sources to exchange pathogens. Experience in the USA has shown that 200–300 kg cattle will suffer fewer post-transport health problems if they are transported for a complete 32-h journey, without any lairage stops. Whether the increased health problems are due to exposure to novel pathogens or to the inadequacy of the lairage conditions, essentially extending the stress of transport, is not known (Grandin, 1997).

Tarrant *et al.* (1992) studied the effects of three stocking densities on 600 kg cattle that were transported for 24 h. Following transport, they found that levels of plasma CK, cortisol and glucose had increased with increasing stocking density, as had the amount of bruising on the carcasses, indicative of increased physical and psychological stress and poorer welfare. They concluded that stocking densities above 550 kg m$^{-2}$ were unacceptable for this size of animal on long journeys. These results run counter to the popularly held belief within the industry that packing animals in tightly helps support them and prevents them from being jolted and bruised. Too high a stocking density was found to prevent the animals from holding a proper footing, by overly restricting their movement. The highest stocking density, and the one that was found to be unacceptable, was that which would normally be considered to represent a full load – the maximum number of animals which could be held in a pen and the gate still easily closed.

### Young calves (cattle less than 1 month of age)

Neonatal animals are generally less well adapted to cope with transport and are more vulnerable than the adult animal. The long-distance transport of very young cattle is common and usually takes place within days or weeks of birth, whilst the animal is still unweaned and is fully dependent on milk. Calf mortality during transport tends to be low; however, mortality rates following transport can be high, usually as a result of disease (Knowles, 1995). In a large-scale survey of calf mortality and husbandry within the UK, Leech *et al.* (1968) estimated the mortality of transported calves to be 160% that of calves that remained on their farm of birth. Mortality of calves transported below 1 month of age remained markedly above that of home-bred calves until 2 months after purchase. In calves under 1 month old, various authors have reported a strong negative correlation between mortality/morbidity and age when first transported (Knowles, 1995). In addition to the age at which calves are transported, the length of time that marketing takes is also important. Mormede *et al.* (1982) found less post-transport disease amongst calves whose marketing took only 13 h rather than 37 h.

The reactivity of the adrenal glands to adrenocorticotrophic hormone (ACTH) increases with age and is not fully developed in the young calf (Hartmann *et al.*, 1973). Several authors report that the increase in plasma cortisol usually seen in response to transport is not present in young calves

(Knowles, 1995). Neither do calves show the usual increase in heart rate and plasma glucose levels (Knowles *et al.*, 1997). These authors concluded that calves were unable to respond to the stress of transport because of their immaturity, and that the lack of a cortisol response was not because they were relatively 'unstressed' by the process of transportation. Using measurements of rectal temperature, Knowles *et al.* (1997, 1999b) found that, when transported during cold weather, calves found difficulty in maintaining body temperature during transport and regulating it afterwards. Loss of live-weight was greater in the cold.

During and immediately after long-distance transport, calf hauliers within Europe prefer to feed a glucose and electrolyte solution rather than milk replacer, as they report that this reduces the incidence of diarrhoea. Knowles *et al.* (1997, 1999b) found that feeding electrolyte during transport of 19–24 h provided little benefit in terms of rehydration and improvements in levels of plasma metabolites and so recommended that it was best to complete the journey without the disruption and stress of feeding. Liquid feeding of unweaned calves requires the observation, and often the handling, of each individual animal. It also requires attention to hygienic presentation of the feed, which has to be made up to the correct temperature and solution strength in order to avoid digestive problems. There was some evidence from the study of Knowles *et al.* (1999b) that feeding just cold water during transport was detrimental to the calves.

If there is sufficient room for them to do so, calves spend much of the time lying down during road transport. Knowles *et al.* (1997) found that calves spent approximately 50% of the time lying during 24 h of transport. During cold weather, the amount of the journey spent lying increased to 80–90% (Knowles *et al.*, 1999b). Following transport for 24 h Knowles *et al.* (1999b) reported that most of the commonly measured physiological variables had returned to pre-transport values after 24 h of lairage and feeding, except for live-weight and levels of plasma CK, which took up to 7 days to recover.

Overall, present evidence indicates that young calves should not be transported until they are at least over the age of 1 month, but further work is required to confirm that this age limit should not be further extended. If they are to be transported, then it is best to keep the marketing time to a minimum, to avoid feed/rest stops if transport is for no longer than 24 h, to avoid exposing the calves to cold and to avoid cross-contamination of animals from different sources. The animals should be well bedded, especially in cold weather, and transported at a stocking density which allows enough space for them all to lie down.

## Sheep

The mortality rate amongst slaughter lambs transported by road within the UK has been estimated as 0.018% (Knowles *et al.*, 1994b), as 0.10% within

South Africa (Henning, 1993) and as between 0.74 and 1.63% within Queensland, Australia (Shorthose and Wythes, 1988). In the UK those lambs which go direct from farm to slaughterhouse have an estimated mortality rate of 0.007% compared with 0.031% for those which pass through a live auction market (Knowles *et al.*, 1994a). Occasionally, mass deaths within single loads of sheep are reported. These are most often associated with a combination of high ambient temperatures and reduced ventilation on a stationary lorry. In many countries, there is a trade in cull sheep. There is anecdotal evidence that the mortality rates amongst these relatively infirm, low value animals can be high during transport.

The loss of live-weight during transport has been well documented in lambs. Wythes and Morris (1994) averaged the results from eight pieces of work and found live-weight losses of 3, 5, 7.5, 11, 12 and 14% over 6, 12, 24, 48, 72 and 96 h, respectively, with food withdrawal alone; however, losses as high as 20% after just 72 h have been reported by Horton *et al.* (1996) when also deprived of water. They also found that, following transport, food intake was depressed. Combining data from various sources, Wythes and Morris (1994) found the average rate of loss of carcass weight to be 1.7% day$^{-1}$ over 4 days when lambs were deprived of only feed and not transported, with a range from 1.3 to 2.3% day$^{-1}$. During periods of fasting and transport of up to 72 h, plasma levels of $\beta$-OHB have been found to increase linearly at a rate of approximately 0.006 mmol l$^{-1}$ h$^{-1}$ (Warriss *et al.*, 1989, Knowles *et al.*, 1995). Levels of plasma FFA tend to rise linearly with periods of fasting and transport, at a rate of approximately 20 µmol l$^{-1}$ h$^{-1}$, but peak and flatten out, with no further increase, between 18 and 24 h, whilst levels of plasma urea increase approximately linearly by 30–50% during 24 h of transport (Knowles *et al.*, 1995, 1998).

When sheep were held without food or water for 48 h at temperatures up to 35°C, Parrott *et al.* (1996) found little evidence of dehydration from measurements of plasma osmolality, but they did find evidence that the sheep were unable to maintain water balance if they consumed feed. Sheep transported for up to 24 h in the summer in the UK showed no signs of dehydration, as measured by plasma total protein, albumin and osmolality. However, sheep transported across France for 24 h, during which daytime temperatures rose above 20°C, showed signs of dehydration, with increases in plasma total protein, albumin and osmolality of approximately 10, 12 and 5% respectively (Knowles *et al.*, 1996). In accord with Parrott *et al.* (1996), Knowles *et al.* (1996) noted that feeding during and after transport tended to disrupt water balance. This has important implications for the length of mid-transport lairage stops as, after short periods of food and water deprivation, sheep are primarily interested in eating and do not drink readily or immediately (Knowles *et al.*, 1994a). A lairage stop of just 1 h, as is currently required for transport of over 14 h within Europe, is sufficient for the animals to eat but not to drink, so animals may be reloaded after having consumed a high dry-matter feed, but no water. A minimum mid-transport lairage time of 8 h has been recommended (Knowles, 1998).

Measurements of heart rate, plasma cortisol, glucose and CK have shown that it is the initial stages of transport that are most stressful to sheep (Knowles *et al.*, 1995). Heart rate peaks at loading and there is a rise in cortisol, glucose and CK levels at loading, but after 9 h of transport these variables have generally returned to approximate basal levels and the only measurable changes seen are then due to the effects of feed and water deprivation; which can be exaggerated by the conditions of transport. However, the conditions of transport are important. Sheep that are loaded at too high a stocking density to be able to lie down easily show elevated levels of plasma CK, indicative of physical fatigue caused by having to remain standing (Knowles *et al.*, 1998).

As long as they are fit, loaded at an appropriate stocking density, the ambient temperature is not extreme and the load is properly ventilated, sheep appear to cope reasonably well during transport. However, Horton *et al.* (1996) reported that, after passing through a live auction market, lambs transported for 72 h without food or water, whilst not differing in terms of performance or blood metabolites from animals simply deprived of food and water for 72 h, suffered in terms of compromised general health. This was probably a result of confinement on the lorry and exposure to unfamiliar animals and pathogens, combined with the effects of deprivation and transport *per se*. After transport, the recovery of physiological variables to pre-transport levels appears to take place in three stages (Knowles *et al.*, 1993). After 24 h of lairage, with food and water, variables usually associated with short-term stress and the variables associated with dehydration had returned to normal levels. After 96 h, there had been a well-defined recovery in live-weight and levels of most of the metabolites measured had returned to normal levels. At 144 h of lairage, a fuller recovery had taken place, levels of CK had fallen and all variables had stabilized.

Where appropriate, it is always preferable, and generally makes better economic sense, to transport carcasses rather than live animals. Transport is stressful and transport times should be kept to a minimum. After 9 h of transport, the changes in physiological variables with time tend to be linear and are of little help in determining a maximum acceptable transport time. Behavioural studies of motivation to feed have shown that sheep will begin to work for food after 10–12 h of deprivation. At present, all the evidence taken together points to an acceptable maximum journey time in the region of 24 h when transport is continuous and when food and water are not available. If a lairage stop is included in a journey, it should be for a minimum of 8 h with both food and water continuously available. However a lairage stop does increase the chance of cross-infection between animals from different sources and animals stressed by the process of transport will tend to already be immunologically compromised and vulnerable.

## Pigs

The mortality rate amongst pigs transported to slaughter within the UK has changed little over 20 years and is estimated to be 0.061% with a further 0.011% of pigs dying in the lairage pens before slaughter (Warriss and Brown, 1994). However, there are marked differences in mortality rates between different countries. The rates are particularly higher in countries where the slaughter-pig population contains a large proportion of genes from stress-susceptible breeds, such as the Pietrain and Belgian Landrace. Estimates range from 0.3 to 0.5% in Belgium and Germany (Warriss, 1998a). The other major factor influencing the mortality of pigs during transit is ambient temperature. Pigs are sensitive to high temperatures because they are poorly adapted to lose heat unless allowed to wallow, a behaviour not possible during transport. The relationship between mortality and ambient temperature is curvilinear. This is illustrated in Fig. 19.3 with data for the UK from Warriss and Brown (1994), which show that there was a marked increase in mortality when average monthly temperatures rose above 15°C. Other factors of importance are the time of last feed before loading, vehicle deck, stocking density and possibly journey time. Pigs fed too soon (< 4 h) before transport are more likely to die, as are those carried on the bottom deck, at higher densities and for longer. However, the evidence for the latter is contradictory.

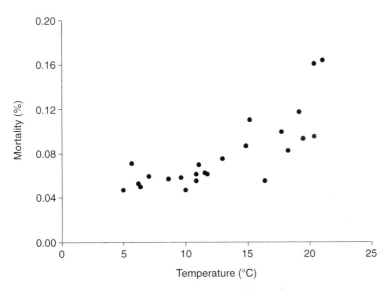

**Fig. 19.3.** The relationship between average monthly ambient temperature and the mortality of pigs transported to slaughter.

Pigs find simulated transport aversive (Ingram *et al.*, 1983), particularly the vibration associated with it (Stephens *et al.*, 1985) and if they have recently eaten a large meal. Because pigs may vomit during transport (Bradshaw and Hall, 1996; Riches *et al.*, 1996) and show increased circulating levels of vasopressin, a hormone associated with feelings of motion sickness in humans, part of their aversion may be attributable to similar feelings of sickness.

That pigs find at least some aspects of transport psychologically stressful is evidenced by increases in plasma adrenaline (Dalin *et al.*, 1993), indicating stimulation of the sympathoadrenal system, and in cortisol (see, for example, Dantzer, 1982), with corresponding depletion of adrenal ascorbic acid (Warriss *et al.*, 1983), indicating stimulation of the hypophyseal–adrenal axis. They may also find it physically stressful, based on elevations of circulating activities of the enzyme CK (Honkavaara, 1989).

The physical stress they experience will be determined by the comfort and length of the journey. It is likely to be greater if vibration levels are higher. Modern vehicles, with air suspension and driven on smooth roads, will provide more comfort than older vehicles, with traditional spring suspension systems driven on poorer-surfaced roads. Physical stress and the associated fatigue are likely to be higher if pigs stand, rather than lie down, during the journey. There is some debate about whether pigs prefer to stand or lie down. The available evidence has been reviewed by Warriss (1998b), who suggested that it pointed to the view that pigs preferred to stand on short journeys in which the conditions made it uncomfortable to lie down. These conditions could be excessive vibration or uncomfortable flooring, perhaps because of inadequate bedding. But, under comfortable conditions, many, if not all, pigs would lie down if given sufficient space, especially on longer journeys.

What is sufficient space was also discussed by Warriss (1998b). It is equivalent to a stocking density of not higher than about $235–250$ kg m$^{-2}$ for normal slaughter pigs weighing between 90 and 100 kg. For smaller pigs the space requirement would be expected to be slightly greater and for larger pigs slightly less. European Community (EC) Directive 95/29/EC requires that the loading density for pigs of around 100 kg should not exceed $235$ kg m$^{-2}$. Also, the Directive recognizes that this density may be too high under certain conditions. Breed of pig, size and physical condition of the animals or weather and journey time may mean that the space allowed has to be increased by up to 20%. Pigs carried at very high stocking densities show increased circulating levels of CK (Warriss *et al.*, 1998). The provision of appropriate amounts of space is especially important with longer journeys.

The physiological state of an animal at slaughter often affects subsequent lean-meat quality. Thus, muscle glycogen depletion can lead to elevated ultimate muscle pH in the meat. Longer transport sometimes results in muscle glycogen depletion and more meat with high ultimate pH (Malmfors, 1982; Warriss, 1987). This is seen as a higher prevalence of DFD pork.

During transport, pigs are deprived of food. They are often also deprived of water although current EU legislation (EC Directive 95/29/EC) prescribes that pigs transported for more than 8 h must have continuous access to drinking-water. There is evidence, however, that pigs drink only very small amounts of water (Lambooy, 1983; Lambooy *et al.*, 1985). Some studies (Warriss *et al.*, 1983) have indicated that pigs can become dehydrated after only short journeys, but data from other work (Becker *et al.*, 1989) have not supported this. Food deprivation leads to losses in live- and carcass weight (Warriss, 1985), liver weight and liver glycogen (Warriss and Bevis, 1987) and muscle glycogen (Warriss *et al.*, 1989). These are undesirable. However, some period of food withdrawal before transport is desirable to minimize mortality and, in the case of slaughter pigs, to facilitate hygienic carcass dressing. Four hours has been recommended (Warriss, 1994), but this may be too short a time, based on observations of vomiting during transport, although the ideal period of withdrawal is not clear (Warriss, 1998a).

Many slaughter pigs are mixed with unfamiliar animals when they are assembled for sending to slaughter. This usually leads to fighting, particularly between dominant individuals. The consequences are elevations in circulating cortisol, CK and lactate and depletion of muscle glycogen (Warriss, 1996). Mixing pigs is undesirable from the points of view of both welfare and meat quality. Nevertheless, in the UK about 40% of pigs show some evidence of fighting and between 5 and 10% evidence of severe fighting.

## Acknowledgements

Permission to reproduce copyrighted material for Fig. 19.3 was kindly granted by The Veterinary Record, 7 Mansfield Street, London. The authors thank Julie Edwards for help with the preparation of the figures.

## References

Becker, B.A., Mayes, H.F., Hahn, G.L., Nienaber, J.A., Jesse, G.W., Anderson, M.E., Heymann, H. and Hedrick, H.B. (1989) Effect of fasting and transportation on various physiological parameters and meat quality of slaughter hogs. *Journal of Animal Science* 67, 334–341.

Bradshaw, R.H. and Hall, S.J.G. (1996) Incidence of travel sickness in pigs. *Veterinary Record* 139, 503 (letter).

Broom, D.M. (1986) Indicators of poor welfare. *British Veterinary Journal* 142, 524–526.

Carlson, G.P. (1997) Fluid, electrolyte, and acid–base balance. In: Kaneko, J.J., Harvey, J.W. and Bruss, M.L. (eds) *Clinical Biochemistry of Domestic Animals*, 5th edn. Academic Press, San Diego, pp. 485–516.

Dalin, A.M., Magnusson, U., Häggendal, A. and Nyberg, L. (1993) The effects of transport stress on plasma levels of catecholamines, cortisol, corticosteroid-binding globulin, blood cell count and lymphocyte proliferation in pigs. *Acta Veterinaria Scandinavica* 34, 59–68.

Dantzer, R. (1982) Research on farm animal transport in France: a survey. In: Moss, R. (ed.) *Transport of Animals Intended for Breeding, Production and Slaughter.* Martinus Nijhoff, The Hague, the Netherlands, pp. 218–230.

Farver, T.B. (1997) Concepts of normality in clinical biochemistry. In: Kaneko, J.J., Harvey, J.W. and Bruss, M.L. (eds) *Clinical Biochemistry of Domestic Animals,* 5th edn. Academic Press, San Diego, pp. 1–19.

Grandin, T. (1997) Assessment of stress during handling and transport. *Journal of Animal Science* 75, 249–257.

Hails, M.R. (1978) Transport stress in animals: a review. *Animal Regulation Studies* 1, 289–343.

Hall, S.J.G. and Bradshaw, R.H. (1998) Welfare aspects of the transport by road of sheep and pigs. *Journal of Applied Animal Welfare Science* 1, 235–254.

Hartmann, H., Meyer, H., Steinbach, G., Deschner, F. and Kreutzer, B. (1973) Allgemeines Adaptationssyndrom (Selye) beim Kalb. 1. Normalverhalten der Blutbildwerte sowie des Glukose- und 11 -OHKS- Blutspiegels. *Archiv für Experimentelle Veterinarmedizin* 27, 811–823.

Henning, P.A. (1993) Transportation of animals by road for slaughter in South Africa. In: *Proceedings of the 4th International Symposium on Livestock Environment, 6–9 July 1993.* American Society of Agricultural Engineers, pp. 536–541.

Honkavaara, M. (1989) Influence of selection phase, fasting and transport on porcine stress and on the development of PSE pork. *Journal of Agricultural Science in Finland* 61, 415–423.

Horton, G.M.J., Baldwin, J.A., Emanuele, S.M., Wohlt, J.E. and McDowell, L.R. (1996) Performance and blood chemistry in lambs following fasting and transport. *Animal Science* 62, 49–56.

IATA (1998) *Live Animal Regulations,* 25th edn. International Air Transport Association, Montreal (http://www.iata.org/cargo).

Ingram, D.L., Sharman, D.F. and Stephens, D.B. (1983) *Journal of Physiology* 343, 2.

Kaneko, J.J., Harvey, J.W. and Bruss, M.L. (eds) (1997) *Clinical Biochemistry of Domestic Animals,* 5th edn. Academic Press, San Diego.

Knowles, T.G. (1995) A review of post transport mortality among younger calves. *Veterinary Record* 137, 406–407.

Knowles, T.G. (1998) A review of the road transport of slaughter sheep. *Veterinary Record* 143, 212–219.

Knowles, T.G. (1999) A review of the road transport of cattle. *Veterinary Record* 144, 197–201.

Knowles, T.G., Warriss, P.D., Brown, S.N., Kestin, S.C., Rhind, S.M., Edwards, J.E., Anil, M.H. and Dolan, S.K. (1993) Long distance transport of lambs and the time needed for subsequent recovery. *Veterinary Record* 133, 287–293.

Knowles, T.G., Warriss, P.D., Brown, S.N. and Kestin, S.C. (1994a) Long distance transport of export lambs. *Veterinary Record* 134, 107–110.

Knowles, T.G., Maunder, D.H.L. and Warriss, P.D. (1994b) Factors affecting the mortality of lambs in transit to or in lairage at a slaughterhouse, and reasons for carcass condemnation. *Veterinary Record* 135, 109–111.

Knowles, T.G., Brown, S.N., Warriss, P.D., Phillips, A.J., Dolan, S.K., Hunt, P., Ford, J.E., Edwards, J.E. and Watkins, P.E. (1995) The effects on sheep of transport by road for up to twenty-four hours. *Veterinary Record* 136, 431–438.

Knowles, T.G., Warriss, P.D., Brown, S.N., Kestin, S.C., Edwards, J.E., Perry, A.M., Watkins, P.E. and Phillips, A.J. (1996) The effects of feeding, watering and resting intervals on lambs exported by road and ferry to France. *Veterinary Record* 139, 335–339.

Knowles, T.G., Warriss, P.D., Brown, S.N., Edwards, J.E., Watkins, P.E. and Phillips, A.J. (1997) Effects on calves less than one month old of feeding or not feeding them during road transport of up to 24 hours. *Veterinary Record* 140, 116–124.

Knowles, T.G., Warriss, P.D., Brown, S.N. and Edwards, J.E. (1998) The effects of stocking density during the road transport of lambs. *Veterinary Record* 142, 503–509.

Knowles, T.G., Warriss, P.D., Brown, S.N. and Edwards, J.E. (1999a) Effects on cattle of transportation by road for up to 31 hours. *Veterinary Record* 145, 475–582.

Knowles, T.G., Brown, S.N., Edwards, J.E., Phillips, A.J. and Warriss, P.D. (1999b) Effect on young calves of a one hour feeding stop during a 19-hour road journey. *Veterinary Record* 144, 687–692.

Lambooy, E. (1983) Watering pigs during road transport through Europe. *Fleischwirtshaft* 63, 1456–1458.

Lambooy, E., Garssen, G.J., Walston, P., Mateman, G. and Merkus, G.S.M. (1985) Transport of pigs by car for 2 days: some aspects of watering and loading density. *Livestock Production Science* 13, 289–299.

Leech, F.B., Macrae, W.D. and Menzies, D.W. (1968) *Calf Wastage and Husbandry in Britain, 1962–63*. HMSO, London.

Leproult, R., Copinschi, G., Buxton, O. and VanCauter, E. (1997) Sleep loss results in an elevation of cortisol levels the next evening. *Sleep* 20, 865–870.

Malmfors, G. (1982) Studies on some factors affecting pig meat quality. In: *Proceedings of the 28th European Meeting of Meat Research Workers*, pp. 21–23.

Mormede, P., Soissons, J., Bluthe, R-M., Raoult, J., Legraff, G., Levieux, D. and Dantzer, R. (1982) Effect of transportation on blood serum composition, disease incidence and production traits in young calves: influence of journey duration. *Annales de Recherches Vétérinaires* 13 369–384.

Parrott, R.F., Lloyd, D.M. and Goode, J.A. (1996) Stress hormone responses of sheep to food and water deprivation at high and low ambient temperatures. *Animal Welfare* 5, 45–56.

Richards, R.B., Hyder, M.W., Fry, J., Costa, N.D., Norris, R.T. and Higgs, A.R.B. (1991) Seasonal metabolic factors may be responsible for deaths in sheep exported by sea. *Australian Journal of Agricultural Research* 42, 215–216.

Riches, H.L., Guise, H.J. and Penny, R.H.C. (1996) A national survey of transport conditions for pigs. *Pig Journal* 38, 8.

Shorthose, W.R. and Wythes, J.R. (1988) Transport of sheep and cattle. In: *Proceedings of the 34th International Congress of Meat Science and Technology*. Brisbane, pp. 122–129.

Stephens, D.B., Bailey, K.J., Sharman, D.F. and Ingram, D.L. (1985) An analysis of some behavioural effects of the vibration and noise components of transport in pigs. *Quarterly Journal of Experimental Physiology* 70, 211–217.

Tarrrant, P.V. (1989) The effects of handling, transport, slaughter and chilling on meat quality and yield in pigs – a review. *Irish Journal of Food Science and Technology.*

Tarrant, P.V. (1990) Transportation of cattle by road. *Applied Animal Behaviour Science* 28, 153–170.

Tarrant, P.V., Kenny, F.J., Harrington, D. and Murphy, M. (1992) Long distance transportation of steers to slaughter, effect of stocking density on physiology, behavior and carcass quality. *Livestock Production Science* 30, 223–238.

Trunkfield, H.R. and Broom, D.M. (1990) The welfare of calves during handling and transport. *Applied Animal Behaviour Science* 28, 135–152.

Warriss, P.D. (1985) Marketing losses caused by fasting and transport during the preslaughter handling of pigs. *Pig News and Information* 6, 155–157.

Warriss, P.D. (1987) The effect of time and conditions of transport and lairage on pig meat quality. In: Tarrant, P.V., Eikelenboom, G. and Monin, G. (eds) *Evaluation and Control of Meat Quality in Pigs.* Martinus Nijhoff Publishers, Dordrecht, the Netherlands, pp. 245–264.

Warriss, P.D. (1990) The handling of cattle pre-slaughter and its effects on carcass meat quality. *Applied Animal Behaviour Science* 28, 171–186.

Warriss, P.D. (1994) Ante-mortem handling of pigs. In: Cole, D.J.A., Wiseman, J. and Varley, M.A. (eds) *Principles of Pig Science*, pp. 425–432.

Warriss, P.D. (1996) The consequences of fighting between mixed groups of unfamiliar pigs before slaughter. *Meat Focus International* 5, 89–92.

Warriss, P.D. (1998a) The welfare of slaughter pigs during transport. *Animal Welfare* 7, 365–381.

Warriss, P.D. (1998b) Choosing appropriate space allowances for slaughter pigs transported by road: a review. *Veterinary Record* 142, 449–454.

Warriss, P.D. and Bevis, E.A. (1987) Liver glycogen in slaughtered pigs and estimated time of fasting before slaughter. *British Veterinary Journal* 143, 354–360.

Warriss, P.D. and Brown, S.N. (1994) A survey of mortality in slaughter pigs during transport and lairage. *Veterinary Record* 134, 513–515.

Warriss, P.D., Dudley, C.P. and Brown, S.N. (1983) Reduction in carcass yield in transported pigs. *Journal of the Science of Food and Agriculture* 34, 351–356.

Warriss, P.D., Bevis, E.A., Brown, S.N. and Ashby, J.G. (1989a) An examination of potential indices of fasting time in commercially slaughtered sheep. *British Veterinary Journal* 145, 242–248.

Warriss, P.D., Bevis, E.A. and Ekins, P.J. (1989b) The relationships between glycogen stores and muscle ultimate pH in commercially slaughtered pigs. *British Veterinary Journal* 145, 378–383.

Warriss, P.D., Bevis, E.A., Brown, S.N. and Edwards, J.E. (1992) Longer journeys to processing plants are associated with higher mortality in broiler chickens. *British Poultry Science* 33, 201–206.

Warriss, P.D., Brown, S.N., Knowles, T.G., Kestin, S.C., Edwards, J.E., Dolan, S.K. and Phillips, A.J. (1995) The effects on cattle of transport by road for up to fifteen hours. *Veterinary Record* 136, 319–323.

Warriss, P.D., Brown, S.N., Knowles, T.G., Edwards, J.E., Kettlewell, P.J. and Guise, H.J. (1998) The effect of stocking density in transit on the carcass quality and welfare of slaughter pigs: 2. Results from the analysis of blood and meat samples. *Meat Science* 50, 447–456.

Wythes, J.R. and Morris, D.G. (1994) *Literature Review of Welfare Aspects and Carcass Quality Effects in the Transport of Cattle, Sheep and Goats*, Parts A, B and C. Report prepared by Queensland Livestock and Meat Authority for Meat Research Corporation, Queensland Livestock and Meat Authority, Spring Hill, Queensland, Australia.

# Handling and Welfare of Livestock in Slaughter Plants

20

## Temple Grandin

*Department of Animal Science, Colorado State University, Fort Collins, CO 80523, USA*

## Introduction

Gentle handling in well-designed facilities will minimize stress levels, improve efficiency and maintain good meat quality. Cattle that became excited during restraint had tougher meat and more borderline dark cutters (Voisinet *et al.*, 1997). Rough handling or poorly designed equipment is detrimental to both animal welfare and meat quality. Tests conducted by four large US slaughter companies indicated that quiet handling and a great reduction in electric prod use in the stunning race resulted in a 10% reduction of pale, soft, exudative (PSE) pork. Careful handling prior to slaughter helps to maintain meat quality in pigs (van der Wal, 1997; Faucitano, 1998; Meisinger, 1999) Progressive slaughter-plant managers recognize the importance of good handling practices. Constant monitoring of handler performance is required to maintain high standards of animal welfare.

## How Stressful is Slaughter?

Numerous studies have been conducted to determine the relative stressfulness of different husbandry and slaughter procedures. Measurements of cortisol (stress hormone) is the most common method of evaluating handling stresses. One must remember that cortisol is a time-dependent measure. It takes approximately 15–20 min for it to reach its peak value after an animal is stressed (Lay *et al.*, 1992, 1998). Evaluations of handling and slaughter stress will be more accurate if behavioural reactions, heart rate and other blood chemistries are also measured. Adrenaline and noradrenaline have limited value in measuring slaughtering stress because both captive-bolt and electrical stunning trigger massive releases (Warrington, 1974; Pearson et al., 1977; van der Wal, 1978). If the stunning method is applied properly, the animal will be unconscious when the hormone release occurs and there will be no discomfort.

Absolute comparisons of cortisol levels between studies must be done with great caution. Cortisol levels can vary greatly between individual animals (Ray et al., 1972). Cattle that show signs of behavioural excitement usually have higher levels than calm animals. A review of many reports indicates that cortisol levels in cattle fall into three basic categories: (i) resting baseline levels; (ii) levels provoked by being held in a race or restraint in a head gate (bail) for blood testing; and (iii) excessive levels which are double or triple the farm restraint levels (Grandin, 1997a). Baseline levels vary from a low of 2 ng ml$^{-1}$ (Alan and Dobson, 1986) to 9 ng ml$^{-1}$ (Mitchell et al., 1988). Restraining extensively raised semi-wild cattle for blood testing under farm conditions elicits cortisol readings of 25–33 ng ml$^{-1}$ in steers (Zavy et al., 1992), 63 ng ml$^{-1}$ in steers and cows (Mitchell et al., 1988), 27 ng ml$^{-1}$ in steers (Ray et al., 1972) and 24–46 ng ml$^{-1}$ in weaner calves (Crookshank et al., 1979). In Brahman and Brahman-cross cattle, cortisol values ranged from 30 to 35 ng ml$^{-1}$ after 20 min of restraint in a squeeze chute (Lay et al., 1992, 1998). The levels were 12 ng ml$^{-1}$ after 5 min of restraint and rose to 23 ng ml$^{-1}$ after 10.5 min of restraint (Lay et al., 1998). In some studies, cortisol levels were expressed in nmol 1$^{-1}$ by multiplication by 0.36.

When slaughtering is done carefully, cortisol levels in cattle can be substantially lower than in farm handling conditions. Tume and Shaw (1992) reported that steers and heifers slaughtered in a small research abattoir had average cortisol levels of only 15 ng ml$^{-1}$, and cattle slaughtered in a commercial slaughter plant had levels similar to those of farm handling. β-Endorphin levels, which are another indicator of stress, were not significantly different between the two groups. Research by Gerry Means at the Lethbridge Research Center in Alberta, Canada, indicates that β-endorphin rises in response to pain, whereas cortisol is affected by psychological stress. For commercial cattle slaughter with captive-bolt stunning, the following average cortisol values have been recorded: 45 ng ml$^{-1}$ (Dunn, 1990), 25–42 ng ml$^{-1}$ (Mitchell et al., 1988), 44.28 ng ml$^{-1}$ (Tume and Shaw, 1992) and 24 ng ml$^{-1}$ (Ewbank et al., 1992). When things go wrong, the stress levels increase greatly. Cockram and Corley (1991) reported a median value of 63 ng ml$^{-1}$. One animal had a high of 162 ng ml$^{-1}$. The slaughter plant observed by Cockram and Corley (1991) had a poorly designed forcing pen and slick floors. About 38% of the cattle slipped after exiting the holding pens and 28% slipped just before entering the race. Cortisol levels also increased when delays increased waiting time in the single-file race. This was the only study where vocalizations shortly before stunning were not correlated with cortisol levels. This can probably be partly explained by earlier stress caused by the slick floors. Ewbank et al. (1992) reported the lowest average value. This may be explained by excellent handling before the stunning box.

Ewbank *et al.* (1992) found a high correlation between cortisol levels and handling problems in the stunning box. Use of a poorly designed head-restraint device, which greatly increased behavioural agitation and the time required to restrain the animal, resulted in cortisol levels jumping from 24 to 51 ng ml$^{-1}$. In the worst case, the level increased to 96 ng ml$^{-1}$. Cattle slaughtered in a badly designed restraining pen that turned them upside down had average values of 93 ng ml$^{-1}$ (Dunn, 1990). Very few sexually mature bulls have been studied, though Cockram and Corley (1991) had a few in their study. Sexually mature bulls have much lower cortisol levels than steers, cows or heifers (Tennessen *et al.*, 1984).

Less clear-cut results have been obtained in sheep. Slaughter in a quiet research abattoir resulted in much lower average levels (40 ng ml$^{-1}$) compared with a large noisy commercial plant which had dogs (61 ng ml$^{-1}$) (Pearson *et al.*, 1977). Shearing and other farm handling procedures provoke similar or slightly greater stress levels of 73 ng ml$^{-1}$ (Hargreaves and Hutson, 1990), 72 ng ml$^{-1}$ (Kilgour and de Langen, 1970) and 60 ng ml$^{-1}$ (Fulkerson and Jamieson, 1982). Two hours of restraint and isolation stress will increase cortisol levels to 100 ng ml$^{-1}$ in sheep accustomed to metabolism stalls (Apple *et al.*, 1993). Baseline levels were 22 ng ml$^{-1}$. These studies indicate that careful slaughter can be less stressful than on-farm restraint and handling. There is a need to improve practices and equipment. Preslaughter stress can greatly exceed farm stress levels when poorly designed equipment is used. However, when good facilities are combined with well-trained personnel, cattle and sheep can be induced to move through the entire system with no signs of behavioural agitation. The author has observed cattle entering a stunning restrainer like cows in a milking centre. Moving pigs through a high-speed, 1000 animals per hour, slaughter plant in a calm manner is very difficult. Some large 1000 per hour plants have installed two stunning systems to reduce stress.

## Methods to Reduce Stress

Every extra handling procedure causes increased stress and bruises. Elimination of unnecessary procedures at the slaughter plant will reduce stress. Bray *et al.* (1989) found that multiple stresses, such as washing, fasting and shearing, have a cumulative deleterious effect on meat quality. Pigs should be sorted and tattooed prior to leaving the farm. Weighing live animals at the slaughter plant can be eliminated if the marketing system is on a carcass-weight basis. Washing of cattle and sheep in Australia and New Zealand causes additional stress (Walker *et al.*, 1999). Kilgour (1988) found that washing sheep was a very stressful procedure. Washing has little effect on carcass cleanliness unless the sheep are very dirty (Glover and Davidson, 1977). Washing also increases bruising (Petersen, 1977). Reducing electric-prod use will reduce stress. Pigs moved with electric prods had significantly higher heart rates than pigs moved with a panel (Brundige *et al.*, 1998). Electric prods should be replaced with other driving aids, such as a flag made from plasticized cloth (Fig. 20.1). This works especially well for moving pigs out of pens and down alleys.

Producers must avoid breeding excitable, nervous animals that are difficult to handle. There are problems with very excitable pigs, which are almost impossible to handle quietly in a high-speed slaughter plant (Grandin, 1991a). These pigs constantly back up in the race and have excessive flocking behaviour. Shea-Moore (1998) has measured a highly significant difference in the behaviour of high-lean and fat-type pigs. High-lean pigs were more fearful and explored an open field less than the fat type. The lean pigs also baulked more and took three times longer to walk down an alley with contrasting shadows. Further research showed that high-lean-gain pigs got into significantly more fights than low-lean-gain pigs (Busse and Shea-Moore, 1999). Breeders need to select lean pigs that have a calmer temperament. Other researchers have observed that pigs from certain farms are more difficult to drive (Hunter *et al.*, 1994). Providing extra environmental stimulation in swine confinement buildings, such as rubber hoses to chew on and people walking in the pens, will produce calmer pigs that drive more easily (Grandin, 1989; Pedersen *et al.*, 1993). Playing a radio in the fattening pens at a reasonable volume level will prevent excessive startle reactions to noises such as a door slamming. It is especially important to provide environmental stimulation for pigs which will be transported a short distance to the slaughter plant. Short-transit-time animals are often more difficult to handle.

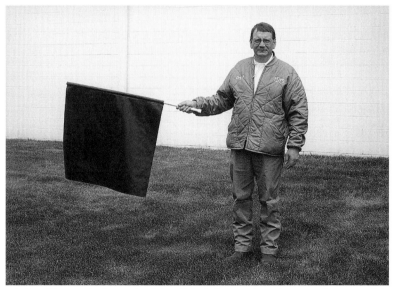

**Fig. 20.1.** Flag made from plasticized cloth for moving pigs out of pens and down alleys. They move quickly away from the rustling sound.

## Accustom Animals to Handling

The author has observed that pigs from excitable genetic lines will be easier to handle at the slaughter plant if they are trained to be accustomed to a person walking through their pens. Ten to 15 seconds per pen every day for the entire finishing period works well. In facilities where large numbers of pigs are housed in each pen, the time should be 10–15 s per 50 pigs. The person should quietly walk through each pen in a different random direction each day to teach the pigs to quietly get up and flow around them. The person should not just stand in the pen. This trains the animals to approach and chew boots instead of driving. Geverink *et al.* (1998) also reports that pigs which have been walked in the aisles during finishing on the farm are easier to handle. In another experiment, moving pigs out of their pens a month prior to slaughter improved their willingness to move (Abbott *et al.*, 1997).

Further observations in cattle slaughter plants indicates that some extensively raised cattle have never seen a person on foot. When they arrive at the slaughter plant they are difficult and dangerous to handle. This problem is most likely to occur in cattle with European Continental genetics (Grandin and Deesing, 1998). Stress would be reduced if ranchers worked to accustom their cattle to both people on foot and to people on horses. At one feedlot, excessively wild cattle had 30% dark cutters. There is more information on the benefits of getting cattle accustomed to handling in Chapter 5.

## Objective Evaluation of Animal Welfare

A study by von Wenzlawowlez *et al.* (1999) showed that many plants did not have procedures that fulfilled European animal welfare requirements. As discussed in Chapter 1, the use of an objective scoring system to continuously measure the quality of animal handling and stunning is recommended (Grandin 1997b,c, 1998a,b). The five major critical control points for monitoring welfare at slaughter plants are: (i) percentage of animals stunned correctly on the first attempt; (ii) percentage of animals that remain insensible throughout the slaughter process; (iii) percentage of animals that vocalize during handling or stunning; (iv) the percentage that slip or fall during handling or stunning; and (v) the percentage of animals prodded with an electric prod.

## Vocalization Scoring

Vocalizations in cattle and pigs are highly correlated with physiological stress measurements (Dunn, 1990; Warriss *et al.*, 1994; White *et al.*, 1995). A higher percentage of cattle vocalized when they were inverted on to their backs for ritual slaughter compared with upright restraint (Dunn, 1990). When freeze branding was compared with hot-iron branding, 23% of the hot-iron cattle vocalized and only 3% of the freeze-branded vocalized (Watts and Stookey, 1998). Isolation also increased cattle vocalization (Watts and Stookey, 1998). Grandin (1998a) observed 1125 cattle in six commercial slaughter plants during handling. The cattle were scored as being either a vocalizer or a non-vocalizer. Nine per cent vocalized (112 cattle), and all except two vocalized in direct response to an aversive event, such as electric prodding, excessive pressure from a restraint device, slipping or falling or missed stuns. Further observations have shown that leaving a bovine alone in the stunning box for too long may cause it to vocalize. In well-run slaughter plants with calm handling, 3% or less of the cattle will vocalize during handling and stunning (Grandin, 1998b). Watts and Stookey (1998) suggest that vocalization in cattle may be especially useful for measuring severe stress. The level of squealing in pigs is correlated with meat quality problems (Warriss et al., 1994). Pig squeals can also be measured with a sound meter or by determining the percentage of time pigs are quiet in the stunning pen, restrainer, race and crowd pen. As each pig is stunned, score the entire stunning area as either quiet or heard a squeal. For each stunning interval, score yes or no on squealing. Weary et al. (1998) found that, as pain increases, the frequency of high-pitched pig squeals increases.

Vocalization scoring will work well on cattle and pigs. It should not be used with sheep. Sheep moving quietly through a race will vocalize. Vocalization scoring should be done during actual handling and stunning, because animals standing in the yards sometimes vocalize to each other. Vocalization scoring should be interpreted as a statistical property of a group of animals, instead of using it to evaluate the condition of individual animals (Watts and Stookey, 1998). Vocalization scoring will not work if an electric current is used to immobilize an animal, because it prevents the animal from vocalizing. Electroimmobilization should not be used on conscious animals, because it is highly aversive (Grandin et al., 1986; Pascoe, 1986; Rushen, 1986). Electroimmobilization must not be confused with electrical stunning, which induces instantaneous unconsciousness.

Slaughter plants should use stunning methods which have been verified by scientific tests. There are a number of reviews on humane stunning methods (Grandin, 1980d, 1985/86, 1994; Eikelenboom, 1983; UFAW, 1987; Gregory, 1988, 1992; Gregory and Grandin, 1998; Devine et al., 1993). Many people that first view slaughter are concerned that stunned animals are conscious when reflexes cause the legs to move. The legs may move vigorously on a properly stunned, insensible animal. Eye reflexes, vocalizations and rhythmic breathing must be absent. When the animal is suspended upside down on the shackle, the head should hang straight down and the neck should be straight and limp. There should be no signs of an arched back. Sensible animals will attempt to right themselves by arching their backs and raising their heads.

## Reduce Noise

Observations in many slaughter plants indicate that noisy equipment increases excitement and stress. The author has observed improvements in handling and calmer cattle and pigs after a noise problem was corrected. Clanging and banging noises should be eliminated by rubber pads on gates and the shackle return should be designed to prevent sudden impact noise. Spensley *et al.* (1995) found that novel noises ranging from 80 to 89 dB increased heart rate in pigs. Intermittent noises were more disturbing to pigs than continuous noise (Talling *et al.*, 1998). Air exhausts on pneumatically powered gates should be piped outside or muffled and hydraulic systems should be engineered to minimize noise. High-pitched sounds from hydraulic systems are very disturbing to cattle. Cattle held overnight in a noisy yard close to the unloading ramp were more active and had more bruising compared with cattle in quieter pens (Eldridge, 1988). Grandin (1980a) discussed the use of music to help mask disturbing noises. The cattle became accustomed to the music in the holding pens and it provided a familiar sound when the animals approach noisy equipment The type of building construction will greatly influence noise levels. Plants with high concrete ceilings and precast concrete walls have more echoes and noise than plants built from foam-core insulation board. In one plant with precast concrete walls, the sound level was 93 dB at the restrainer when the equipment was running and the pigs were quiet. Pigs tended to become agitated because the sound increased from 88 dB in the lairage to 93 dB as the pigs approached. In a survey of 24 plants, the pork plant with the quietest handling and low pig vocalization rates had been engineered to reduce noise (Grandin, 1998b).

## Lighting and Distractions

Animals have a tendency to move from a darker place to brighter place (van Putten and Elshof, 1978; Grandin, 1982, 1996; Tanida *et al.*, 1996). Races, stunning boxes and restrainers must be illuminated with indirect, shadow-free light. Shadows, sparkling reflections, seeing people up ahead or a small chain hanging in a race will cause animals to baulk. Grandin (1996, 1998c) reviews all the little distractions that cause baulking. Quiet handling and minimal usage of electric prods is impossible if animals baulk at distractions. To find the distractions that cause baulking, a person needs to get down and look up a race from the animal's eye level. When animals are calm, they will look right at distractions that cause baulking (Fig. 20.2). One of the worst distractions is air blowing down a race into the faces of approaching animals. Distractions can also have time-of-day effects. Shadows may cause a problem when the sun is out and animals may move easily at night. Small distractions that are easy to fix can ruin the efficiency of the best system. When troubleshooting handling

**Fig. 20.2.** A calm animal will look right at a sparkling reflection or other distraction that makes it baulk.

problems, one must be very observant to determine if the problem is caused by a basic design mistake in the system or a little distraction or lighting problem that is easy to fix. Sometimes there can be more than one distraction in a system.

## Does Blood Upset Livestock?

Many people interested in the welfare of livestock are concerned about animals seeing or smelling blood. Cattle will baulk and sniff spots of blood on the floor (Grandin, 1980a,d); washing blood off facilitates movement. Baulking may be a reaction to novelty. A piece of paper thrown in the race or stunning box elicits a similar response. Cattle will baulk and sometimes refuse to enter a stunning box or restrainer if the ventilation system blows blood smells into their faces at the stunning-box entrance. They will enter more easily if an exhaust fan is used to create a localized zone of negative air pressure. This will suck smells away from cattle as they approach the stunning-box entrance.

Observations in kosher slaughter plants indicate that cattle will readily walk into a restraining box which is covered with blood. In Jewish ritual (kosher) slaughter, the throat of a fully conscious animal is cut with a razor-sharp knife. The cattle will calmly place their heads into the head restraint device and some animals will lick blood or drink it. Kosher slaughter can proceed very calmly with few signs of behavioural agitation if the restraining box is operated gently (Grandin, 1992, 1994). However, if an animal becomes very agitated and frenzied during restraint, subsequent animals often become agitated. An entire slaughter day can turn into a continuous chain reaction of excited animals. The next day, after the equipment has been washed, the animals will be calm. The excited animals may be smelling an alarm pheromone from the blood of severely stressed cattle. Blood from relatively low-stressed cattle may have little effect. However, blood from severely stressed animals which have shown signs of behavioural agitation for several minutes may elicit a fear response. Eibl-Eibesfeldt (1970) observed that, if a rat is killed instantly in a trap, the trap can be used again. The trap will be ineffective if it injures and fails to kill instantly.

Research with pigs and cattle indicates that there are stress pheromones in saliva and urine. Vieville-Thomas and Signoret (1992) and Boissy *et al.* (1998) report that pigs and cattle tend to avoid places or objects which have urine on them from a stressed animal. The stressor must be applied for 15–20 min to induce the effect. In the cattle experiment, the animals were given repeated shocks for 15 min. Rats showed a fear response to the blood of rats and mice that had been killed with carbon dioxide ($CO_2$) (Stevens and Gerzog-Thomas, 1977). Carbon dioxide causes secretion of adrenocortical steroids (Woodbury *et al.*, 1958). Observations by the author of $CO_2$ euthanasia of mice indicated that they gasp and frantically attempt to escape. Guinea-pig blood and human blood had little effect on the rats (Hornbuckle and Beall, 1974; Stevens and Gerzog-Thomas, 1977). Possibly this was due to less stress in these blood donors. $CO_2$ at high concentrations does not produce excitement in guinea-pigs (Hyde, 1962). Stevens and Gerzog-Thomas (1977) also found the alarm substance in the blood and muscle, whereas rat and mouse brain tissue did not provoke a fear response (Stevens and Gerzog-Thomas, 1977). In a choice test, rats avoided muscle and blood but there was no difference between brain tissue and water (Stevens and Saplikoski, 1973).

## Design of Lairages and Stockyards

Different countries have specific requirements; for example, truck size will determine the size of each holding pen. Space and facilities must also be designed for specialized functions, such as weighing, sorting and animal identification. Long, narrow pens are recommended in stockyards and lairages in slaughter plants (Kilgour, 1971; Grandin, 1979, 1980e, 1991b). One advantage of long, narrow pens is more efficient animal movement. Animals enter through one end and leave through the other. To eliminate 90° corners, the pens can be constructed on a 60–80° angle (Fig. 20.3). Long, narrow pens maximize lineal fence length in relation to floor areas, which may help to reduce stress (Kilgour, 1978; Grandin, 1980c,e). Cattle and pigs prefer to lie along the fence line (Stricklin *et al.*, 1979; Grandin, 1980c). Observations indicate that long, narrow pens may help reduce fighting (Kilgour, 1976). Minimum space requirements for holding fattened, feedlot steers for less than 24 h are 1.6 m$^2$ for hornless cattle and 1.85 m$^2$ for horned cattle (Grandin, 1979; Midwest Plan Service, 1987), and for slaughter-weight pigs and lambs 0.5 m$^2$. During warm weather, pigs require more space. Wild, extensively raised cattle may require additional space. However, providing too much space may increase stress, because wild cattle tend to pace in a large pen. Enough space must be provided to allow all animals to lie down at the same time.   To avoid bunching and trampling, 25 m is the maximum recommended length of each holding pen, unless block gates are installed to keep groups separated. Shorter pens are usually recommended. Pen and alley width recommendations can be found in Grandin (1991b).

## Avoid Mixing Strange Animals

To reduce stress, prevent fighting and preserve meat quality, strange animals should not be mixed shortly before slaughter (Grandin, 1983a;

**Fig. 20.3.**    Long narrow cattle pens in a 60° angle.

Tennessen *et al.*, 1984; Barton-Gade, 1985). Solid pen walls between holding pens prevent fighting through the fences. Solid fences in lairage and stockyard pens are especially important if wildlife, such as deer, elk or buffalo, are handled.

Pigs present some practical problems to keep separated. In the USA, pigs are transported in trucks with a capacity of over 200 animals. However, they are fattened in much smaller groups on many farms, and some large farms fatten pigs in groups of 200 or more. Observations at US slaughter plants indicate that mixing 200 pigs from three or four farms resulted in less fighting than mixing 6–40 pigs. One advantage of the larger group is that an attacked pig has an opportunity to escape. Price and Tennessen (1981) found a tendency towards more dark, firm, dry (DFD) carcasses and hence more stress when small groups of seven bulls were mixed, compared with larger groups of 21 bulls. Fighting between slaughter-weight pigs can be reduced by having mature boars present in the holding pens (Grandin and Bruning, 1992). Another method for reducing fighting is to feed pigs excess dietary tryptophan for 5 days before slaughter (Warner *et al.*, 1992). Further information on agonistic behaviour in pigs can be found in a review by Petherick and Blackshaw (1987).

In Denmark, the design of the pig lairage at the abattoir is very specialized. Pigs are held in long, narrow pens equipped with manual push gates. A powered push gate moves pigs up the alley to the stunner. This system was invented by T. Wichmann of the Danish Meat Research Institute. The Danes have also developed automated block gates within the long, 2 m wide pens to keep small groups of 15 pigs in separate groups (Barton-Gade, 1989; Barton-Gade *et al.*, 1993). Disadvantages of this system are the very high cost and the fact that it may not work with some of the new genetic lines of pigs which are highly excitable and difficult to handle. The Danish slaughter plant that had this system had calm, placid pigs which moved easily.

When strange bulls are mixed, physical activity during fighting increases DFD meat. The installation of either steel bars or an electric grid over the holding pens at the abattoir prevented dark cutting in bulls (Kenny and Tarrant, 1987). These devices prevent mounting. The electric grid should only be used with animals that have been fattened in pens equipped with an electric grid. In Sweden and other countries where small numbers of bulls are fattened, individual pens are recommended at the abattoir (Puolanne and Aalto, 1981). In some European slaughter plants, the holding area consists of a series of single-file races which lead to the stunner. Bulls are unloaded directly into the races and each bull is kept separated by guillotine gates. This system is recommended in situations where a farmer markets a few cattle at a time from each fattening pen. If the entire pen is marketed at one time, the animals can be penned together in a group at the plant. Another common European design is tie-up stalls with halters. Tie-up stalls should only be used with animals that are trained to lead.

## Flooring and Fence Design to Reduce Injuries

Floors must have a non-slip surface (Stevens and Lyons, 1977; Grandin, 1983a). In three out of 11 (17%) of beef plants surveyed, slippery floors caused cattle to become agitated (Grandin, 1998a). Slick floors and slipping increases stress (Cockram and Corley, 1991). For cattle, concrete floors should have deep 2.5 cm V grooves in a 20 cm square or diamond pattern. For pig and sheep abattoirs, the wet concrete should be imprinted with a stamp constructed from expanded steel mesh. The expanded steel should have a 3.8 cm long opening (Grandin, 1982). A broom finish prevents slipping when it is new (Applegate *et al.*, 1988), but practical experience has shown that it quickly wears out and the animals will fall down.

Concrete slats may be used in holding pens, but the drive alleys should have a solid concrete floor. Precast slats for cattle or swine confinement buildings will work. The slats should have a grooved surface. Slats or gratings used in pig and sheep facilities should face in the proper direction to prevent the animals from seeing light coming up through the floor.

Animals will baulk at sudden changes in floor texture or colour. Flooring surfaces should be uniform in appearance and free from puddles. In facilities that are washed, install concrete curbs between the pens to prevent water in one pen from flowing into another pen. Drains should be located outside the areas where animals walk. Livestock will baulk at drains or metal plates across an alley (Grandin, 1987). Flooring should not move or jiggle when animals walk on it. Flooring that moves causes pigs to baulk (Kilgour, 1988). Lighting should be even and diffuse to reduce shadows. Lamps can be used to encourage animals to enter races (Grandin, 1982). Additional information on the effects of vision and lighting on livestock handling is given in Chapters 5, 10, 14 and 15.

Cattle and sheep can have bruised meat even though the hide is undamaged. Bruises can occur up until the moment of bleeding. Meischke and Horder (1976) determined that stunned cattle could be bruised when they were ejected from the stunning box. Pigs are slightly less susceptible to bruising, but meat quality deteriorates when they become excited or hot. Lean pigs bruise more easily than fatter pigs. Edges with a small diameter will cause severe bruises. Steel angles or I-beams should not be used to construct pens or races as animals bumping against the edges will bruise. Round-pipe posts and fence rails are recommended. Surfaces which come into contact with animals should be smooth and rounded (Grandin, 1980e; Stevens and Lyons, 1977); exposed sharp ends of pipes should be bevelled to prevent gouging. Areas which have completely solid fences should have all posts and structural parts on the outside away from the animals. An animal rubbing against a smooth flat metal surface will not bruise. All gates should be equipped with tie-backs to prevent them from swinging into the alley. Guillotine gates should be counterweighted and padded on the bottom with conveyor belting or large-diameter hose (Grandin, 1983a). Bruising on pigs can be prevented by cutting 45 cm off the bottom of guillotine gates and replacing the bottom portion of the gate with a curtain of flexible rubber belting. The pigs think the curtain is solid and few attempt to go through it.

## Design of Races, Crowd Pens and Unloading Ramps

### Races

Single-file races work well for cattle and sheep, because they are animals that will naturally walk in single file. Cattle will walk in single file when moving from pasture to pasture. All races should have solid outer fences to prevent the animals from seeing people and other distractions outside the fence. Animal entry into the race can sometimes be facilitated if solid shields are installed to prevent approaching animals from seeing people standing by the race. A curved single-file race is especially recommended for moving cattle (Rider *et al.*, 1974; Grandin, 1980e; Figs 20.4 and 20.5). An inside radius of 5 m is ideal for cattle (additional recommendations are given in Chapter 7). Walkways for the handler should run alongside the race, and the use of overhead walkways should be avoided. In slaughter plants with restricted space, a serpentine race system can be used (Grandin, 1984). A race system at an abattoir must be long enough to ensure a continuous flow of animals to the stunner, but not be so long that animals become stressed waiting in line.

Hartung *et al.* (1997) found that pigs were less stressed in a very short 3.5 m race compared with an 11 m race. German plants run at slower speeds than US plants and a 3.5 m long race may cause more stress in a plant running 800 pigs per hour because the short race makes it more difficult to keep up with the line. In plants slaughtering 240 or fewer pigs per hour, stunning them with electric tongs in groups on the floor was less stressful than a double single-file race (Warriss *et al.*, 1994). In larger plants, floor stunning with tongs tends to get rough and careless. In a small plant, pigs stunned in small groups on the floor had better meat quality (Stuier and Olsen, 1999).

Two races are sometimes built side by side, because the pigs will enter more easily (Grandin, 1982). The outer walls of the race are solid, but the inner fence between the two races is constructed from bars. This enables the pigs to see each other and promotes following behaviour. However, this system still causes stress at the stunner, because pigs on one side have to wait. Stress could be greatly reduced by installing two stunners. This would enable two lines of pigs to move forward continuously. This concept is economically viable for large plants slaughtering over 500 pigs per hour.

A future possibility is to eliminate traditional single-file races. Multiple stunners could be installed with several parallel races (Grandin, 1991b). Funnelling the pigs into single file would be eliminated. Stunning a group of pigs with $CO_2$ gas in an elevator would eliminate handling problems associated with driving nervous excitable pigs up a single-file race. Researchers in

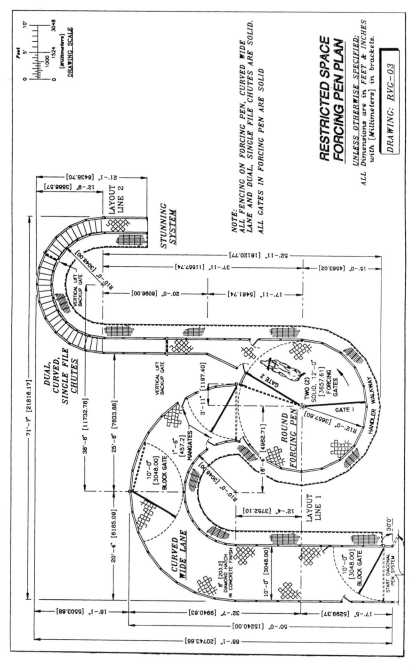

**Fig. 20.4.** Curved race system that will fit in a restricted space.

**Fig. 20.5.** Curved cattle race with solid sides at a slaughter plant.

Denmark have developed an excellent handling system for quietly moving groups of four or five pigs into a $CO_2$ chamber (Barton-Gade and Christenson, 1999; Fig. 20.6). The gate labelled D2 moves groups of 15 pigs forward and the sliding gate D5 is used to separate groups of five pigs. Gate S1 slides forward to drive pigs into the $CO_2$ chamber. Pigs remain quiet because the single-file race is eliminated (Fig. 20.6). Many handling problems would be solved, but there are concerns about the humaneness of $CO_2$ in certain breeds and genetic lines of pigs. Forslid (1987) and Ring (1988) report that $CO_2$ is humane for Yorkshire and Landrace pigs. Grandin (1988a) reports that certain breeds, such as Hampshire, have a very bad reaction and become highly agitated prior to the onset of unconsciousness. Dodman (1977) also reported variations in a pig's reaction to $CO_2$. Genetic selection is one solution to this problem. The work by Forslid (1987) measured electrical responses from the brain of purebred Yorkshires. Forslid's experiments need to be conducted in other genetic types of pigs. Even if $CO_2$ is aversive to certain pig genotypes, their overall welfare may be improved if the single-file race is eliminated. Raj *et al.* (1997) found that mixing 30% $CO_2$ with 60% argon is more humane than 90% $CO_2$.

## Crowd pens

Round crowd pens are very efficient for all species because, as the animals go around the circle, they think they are going back to where they came from (Grandin, 1998b). The recommended radius for round crowd pens is 3.5 m for

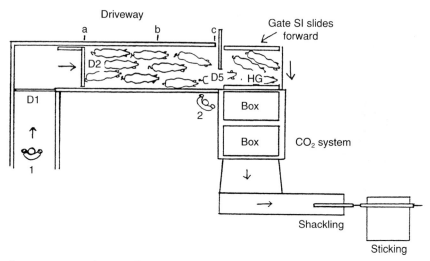

**Fig. 20.6.**    Danish group handling system for moving groups of five pigs into a $CO_2$ chamber. The single-file race is eliminated.

cattle, 1.83–2.5 m for pigs and 2.4 for sheep. For all species, solid sides are recommended on both the race and the crowd pen which leads to the race (Rider *et al.*, 1974; Brockway, 1975; Grandin, 1980c, 1982). For operator safety, gates for people must be constructed so that they can escape charging cattle. The crowd gate should also be solid to prevent animals from turning back. Wild animals tend to be calmer in facilities with solid sides. The crowd pen must be constructed on a level floor. Animals will pile up if the crowd pen is on a ramp (further information is given in Chapter 7).

Crowd pens with a funnel design work well for cattle and sheep and very poorly for pigs. A crowd pen for pigs must be designed with an abrupt entrance to the race to prevent jamming. Hoenderken (1976) devised a crowd pen constructed like a series of stair steps, which varied in width from one pig wide to two pigs wide and three pigs wide. This design works fairly well when pigs are handled rapidly in batches. It works poorly in continuous-flow systems. A round crowd pen in which two crowd gates continually revolve and with an abrupt entrance to the single-file race is being successfully used in several US pig slaughter plants. Another design is a single, offset step equal to the width of one pig. It prevents jamming at the entrance of a single-file race (Grandin, 1982, 1987). Jamming can be further prevented by installing an entrance restricter at single-file race entrances. The entrance of the single-file race should provide only 5 mm on each side of each pig. A double race should also have a single offset step to prevent jamming.

The number-one handling problem is overloading the crowd pen and the alley that leads up to it. Cattle and pigs will move more easily if the crowd pen is filled half-full and they are moved in small bunches (Fig. 20.7). Another

**Fig. 20.7.** Round crowd pen for cattle. A round crowd pen is efficient for all species because animals think they are going back to where they came from. Cattle and pigs move most easily in small groups, as shown in the photograph.

mistake is attempting to forcibly push animals with powered crowd gates. The author has observed many plants where powered crowd gates are abused. Sheep can be moved in large groups due to their intense following behaviour.

## Unloading

More than one truck unloading ramp is usually required to facilitate prompt unloading. During warm weather, prompt unloading is essential, because heat rapidly builds up in stationary vehicles. In some facilities, unloading pens will be required. These pens enable animals to be unloaded promptly prior to sorting, weighing or identification checking. After one or more procedures are performed, the animals move to a holding pen. Facilities used for unloading only should be 2.5–3 m wide to provide the animals with a clear exit into the alley (Grandin, 1980c, 1991b).

## Ramps and slopes

Ideally an abattoir stockyard should be built at truck-deck level to eliminate ramps for both unloading and movement to the stunner. This is especially important for pigs. The maximum angle for non-adjustable livestock unloading ramps is 20–25°. If possible, the ramp to the stunner should not exceed 10° for pigs, 15° for cattle and 20° for sheep. Ramp angles to the stunner should be more gradual than the maximum angles which will work for loading trucks. To reduce the possibility of falls, unloading ramps should have a flat deck at the top. This provides a level surface for animals to walk on when they first step off the truck (Stevens and Lyons, 1977; Grandin, 1979; Agriculture Canada, 1984). This same principle also applies to ramps to the stunner. A level portion facilitates animal entry into the restrainer or stunning box.

Grooved stair steps are recommended on concrete ramps (United States Department of Agriculture, 1967; Grandin, 1980c, 1991b). They are easier to walk on after the ramp becomes worn or dirty. Further recommendations are given in Chapter 7 and Grandin (1991b). However, in new, clean facilities, small pigs showed no preference between stair steps and closely spaced cleats (Phillips *et al.*, 1987, 1989). For slaughter-weight pigs, cleats should be spaced 15 cm apart (Warriss *et al.*, 1991).

## Design of Restraint Systems

Restraint devices to hold animals during stunning and slaughter have greatly improved. One of the first innovations was the V conveyor restrainer for pigs (Regensburger, 1940). It consists of two obliquely angled conveyors that form a V. Pigs ride with their legs protruding through the space at the bottom of the V. The V restrainer is a comfortable system for pigs with round, plump bodies and for sheep (Grandin, 1980c). Pressure against the sides of the pig will cause it to relax (Grandin *et al.*, 1989). However, the V restrainer is not suitable for restraining calves or extremely heavily muscled pigs with large overdeveloped hams (Lambooy, 1986). The V pinches the large hams and the slender forequarters are not supported. Some of the very lean, long pigs are also not supported properly.

Researchers at the University of Connecticut developed a laboratory prototype for a new type of restrainer system to replace V conveyor restrainers (Westervelt *et al.*, 1976; Giger *et al.*, 1977). Calves and sheep are supported under the belly and the brisket by two moving rails. This research demonstrated that animals restrained in this manner were under minimal stress. Sheep and calves rode quietly on the restrainer and seldom struggled. The space between the rails provides a space for the animal's brisket and prevents uncomfortable pressure on the sternum. The prototype was a major step forward in humane restrainer design, but many components still had to be developed to create a system which would operate under commercial conditions.

In 1986 the first double-rail restrainer was designed and installed in a large commercial calf and sheep slaughter plant by Grandin Livestock

Handling Systems and Clayton H. Landis in Souderton, Pennsylvania, USA (Grandin, 1988b, 1991c; Figs 20.8 and 20.9). The Stork Company in Holland developed a restrainer where pigs ride on a moving-centre conveyor in the early 1990s.

In 1989 the first double-rail restrainer was installed in a large cattle slaughter plant by Grandin Livestock Handling Systems and Swilley Equipment, Logan, Iowa (Grandin, 1991c, 1995). There are now 22 large cattle systems. The double-rail restrainer has many advantages compared with the V restrainer (Grandin, 1983b; see Fig. 20.8). Stunning is easier and more accurate because the operator can stand 28 cm closer to the animal. Cattle enter more easily because they can walk in with their legs in a natural

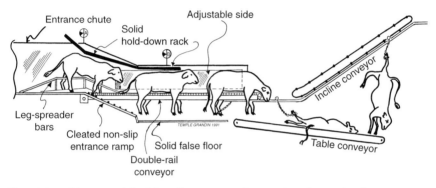

**Fig. 20.8.** Diagram of double-rail (centre-track) restrainer for cattle, sheep, pigs or calves.

**Fig. 20.9.** Large steer in the double-rail (centre-track) restrainer. Note that the conveyor is shaped to fit the animal's brisket.

position. Shackling is facilitated because the legs are spread apart and cattle ride more quietly in the double-rail restrainer.

Proper design is essential for smooth, humane operation. Incoming cattle must not be able to see light coming up from under the restrainer. It must have a false floor below the restrained animal's feet, to provide incoming cattle with the appearance of a solid floor to walk on. To keep cattle calm they must be fully restrained and settled down on the conveyor before they emerge from under the hold-down rack. Their back feet must be off the entrance ramp before they see out from beneath the hold-down rack. If the hold-down is too short, the cattle are more likely to become agitated.

### Move easily into restrainers

Animals should enter a restrainer easily. If they baulk, figure out why they are baulking instead of resorting to increased electric prodding. Lighting problems and the distractions discussed in Grandin (1996, 1998b; Chapter 7, this volume) can cause animals to refuse to enter. In one plant, adding a light at the restrainer entrance reduced baulking and electric prod use. The percentage of cattle vocalizing due to electric prodding was reduced from 8% to 0%. Designers and animal handlers should learn to use behaviour principles to control an animal. Blocking an animal's view of people with a piece of rubber belting can keep even wild cattle quiet. Details of design are very important in animal handling systems. A 0.5 m difference in the length of a shield to block an animal's vision can be the difference between calm animals and agitated, frightened animals. Cattle stay calm in the restrainer because the next animal in line can see an animal directly in front of it. Watts and Stookey (1998) found that cattle vocalized less in a race when they could see another animal in front of them less than 1 metre away.

## Conventional Stunning Boxes

A common mistake is to build stunning boxes too wide. A 76 cm wide stunning box will hold all cattle with the exception of some of the largest bulls. Stunning boxes must have non-slip floors to allow the animal to stand without slipping.

In a conventional cattle stunning box, stunning accuracy can be greatly improved by the use of a yoke to hold the head. Yokes and automatic head restraints for cattle have been developed in Australia (CSIRO, 1989; Buhot et al., 1992), New Zealand and England. Ewbank et al. (1992) found that cattle had higher stress levels when their heads were restrained. The system they observed was poorly designed and lacked a rear pusher gate. Forcing the animal's head into the restraint was difficult and took an average of 32 s. To minimize stress the yoke must be designed so that the animal will enter it willingly and it must be stunned immediately after the head is caught. Australian head-restraint equipment with a rear pusher gate works well (CSIRO, 1989). The rear pusher gate eliminates prodding. Lamps can also be used to encourage cattle to hold their heads up for stunning. New Zealand researchers have devised a humane system for electrical stunning of cattle while they are held in a head restraint (Gregory, 1993). Specifications for proper electrode placement are given in Cook *et al.* (1991).

## Ritual Slaughter

Ritual slaughter is increasing in many countries, due to increased Muslim demand for halal meat. Some Muslim religious authorities will allow head-only electrical stunning for halal slaughter, but Jewish (kosher) slaughter is always conducted on conscious animals. In some countries, such as the USA, it is legal to suspend live animals by one back leg for ritual slaughter. This cruel practice is very dangerous. Replacement of shackling and hoisting by a humane restraint device will greatly reduce accidents (Grandin, 1990). In Europe, Canada and Australia, the use of humane restraining devices is required. When ritual slaughter is being evaluated from a welfare standpoint, the animal's reactions to restraint must be separated from its reactions to throat cutting without stunning. Poorly designed or roughly operated restraint devices are probably more distressful than the throat-cut (Grandin, 1994; Grandin and Regenstein, 1994).

The first restraining device for ritual slaughter was developed in Europe over 40 years ago. The Weinberg casting pen consists of a narrow stall which slowly inverts the animal until it is lying on its back. It is less stressful than shackling and hoisting, but it is much more stressful than more modern, upright, restraint devices (Dunn, 1990). Animals restrained in the Weinberg casting pen had much higher levels of vocalizations and cortisol (stress hormone) compared with cattle restrained in the upright restraining pen. An improved rotating pen called the Facomia pen is now available. It holds the animal's head and body more securely than the old-fashioned Weinberg. However, it is probably more stressful than the best upright restraint. Cattle resist inversion. Inverted cattle twist their necks in an attempt to right their heads, and they may aspirate greater amounts of blood.

A major innovation in ritual restraint equipment was the ASPCA (American Society for the Prevention of Cruelty to Animals) pen (Marshall *et al.*, 1963). It consists of a narrow stall with solid walls with an opening in the front for the animal's head. A lift under the belly prevents the animal from collapsing after the throat-cut (Fig. 20.10). Proper design and operation are essential (Grandin, 1992, 1994; Grandin and Regenstein, 1994). The belly lift should not lift the animal off the floor. Some older models of this pen cause excessive stress because the animal is lifted off the floor by the belly lift. A stop should be installed to restrict belly lift travel to 28 in (71 cm). The air or

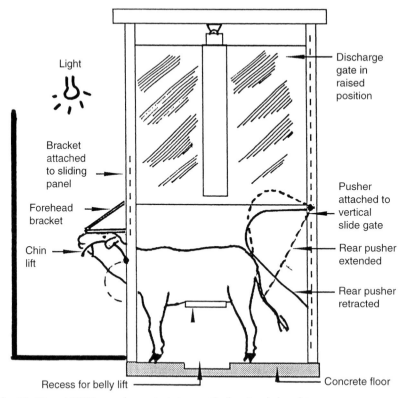

**Fig. 20.10.**    ASPCA pen for restraining cattle for ritual slaughter.

hydraulic pressure which operates the rear pusher gate should be reduced to prevent excessive pressure on the animal's rear. The head holder must have a stop or a bracket to prevent excessive bending of the neck and a pressure-limiting device to prevent the application of excessive pressure. A 25 cm (10 in) wide forehead fold-down bracket covered with rubber belting will help prevent discomfort to the animal. Flow controls or speed reducers should be installed in the hydraulic or pneumatic system to prevent sudden jerky movement.

Further developments in ritual slaughter equipment are the use of a mechanical head holder on a V restrainer and ritual slaughter of calves on the double rail (Grandin, 1988b). The author has developed a head-holding device for the large-cattle double-rail restrainer (Fig. 20.11). It is being used in a large commercial slaughter plant for both ritual slaughter and captive-bolt stunning. A slot in the forehead bracket is provided for captive-bolt stunning and it could be modified for electric stunning. The research team at the University of Connecticut has also developed a small, inexpensive restrainer to hold calves and sheep during ritual slaughter (Giger *et al.*, 1977). For larger calves, a miniature ASPCA pen could be built.

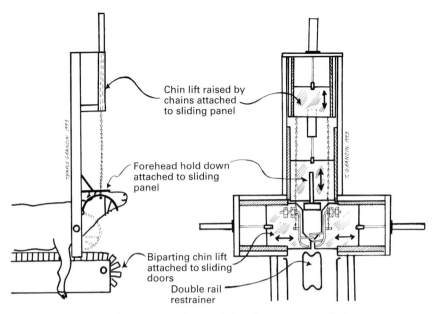

**Fig. 20.11.** Head-holding device for ritual slaughter or captive bolt stunning.

## Behaviour Principles of Restraint

During work on restraint systems at four different kosher slaughter plants, the author developed four behavioural principles of restraint.

**1.** Block vision. The animal must see a lighted place to move into, but solid panels or curtains should be used to prevent it from seeing people.
**2.** Slow steady movement. Parts of an apparatus that press against an animal must move with slow, steady movement. Sudden jerky motion scares.
**3.** Optimum pressure. A device must hold an animal tightly enough for it to feel held but not so tightly that it causes discomfort.
**4.** Does not trigger righting reflex. The device should hold the animal in a comfortable upright position. If it slips or feels unbalanced, it may struggle.

## Conclusions

Carefully conducted slaughter is less stressful than on-farm handling and restraint. When different systems are being evaluated, the variables of basic equipment design must be separated from rough handling and distractions which make animals baulk. Correcting these problems will usually improve animal movement in all types of equipment.

Engineers who design equipment must pay close attention to design details. Layout mistakes, such as bending a race too sharply at the junction between the race and the forcing pen, will cause baulking. Pigs handled in a race system with layout mistakes had increased stress (Weeding *et al.*, 1993). Management must become increasingly sensitive to animal welfare. The single most important factor which determines how animals are treated is the attitude of management. New developments in equipment design will make quiet humane handling easier, but all systems must have good management to go with them. Management must closely supervise both employee behaviour and equipment maintenance.

# References

Abbott, T.A., Hunter, E.J., Guise, J.H. and Penny, R.H.C. (1997) The effect of experience of handling on pigs willingness to move. *Applied Animal Behaviour Science* 54, 371–375.

Agriculture Canada (1984) *Recommended Code of Practice for Care and Handling of Pigs.* Publication 1771/E, Agriculture Canada, Ottawa.

Alan, M.G.S. and Dobson, H. (1986) Effect of various veterinary procedures on plasma concentrations of cortisol, luteinizing hormone and prostaglandin $E_2$ metabolite in the cow. *Veterinary Record* 118, 7–10.

Apple, J.K., Minton, J.E., Parsons, K.M. and Unruh, J.A. (1993) Influence of repeated restraint and isolation stress and electrolyte administration on pituitary–adrenal secretions, electrolytes and other blood constituents of sheep. *Journal of Animal Science* 71, 71–77.

Applegate, A.L., Curtis, S.E., Groppel, J.L., McFarlane, J.M. and Widowski, T.M. (1988) Footing and gait of pigs on different concrete surfaces. *Journal of Animal Science* 66, 334–341.

Barton-Gade, P. (1985) Developments in the pre-slaughter handling of slaughter animals. In: *Proceedings, European Meeting of Meat Research Workers.* Albena, Bulgaria, pp. 1–6.

Barton-Gade, P. (1989) Pre-slaughter treatment and transportation in Denmark. In: *Proceedings, International Congress of Meat Science and Technology.* Copenhagen, Denmark.

Barton-Gade, P. and Christenson, L. (1999) *Transportation and Pre-stun Handling $CO_2$ Systems.* Report No. 53.310, Danish Meat Institute, Roskilde, Denmark.

Barton-Gade, P., Blaabjerg, L. and Christensen, L. (1993) A new lairage system for slaughter pigs. *Meat Focus International* 2, 115–118.

Boissy, A., Terlow, C. and LeNeindre, P. (1998) Presence of pheromones from stressed conspecifics increases reactivity to aversive events in cattle, evidence for the existence of alarm substances in urine. *Physiology and Behavior* 4; 489–495.

Bray, A.R., Graafhuis, A.E. and Chrystall, B.B. (1989) The cumulative effect of nutritional shearing and preslaughter washing stresses on the quality of lamb meat. *Meat Science* 25, 59–67.

Brockway, B. (1975) *Planning a Sheep Handling Unit.* Farm Buildings Centre, Kenilworth, UK.

Brundige L., Okeas, T., Doumit, M. and Zanella, A.J. (1998) Loading techniques and their effect on behavior and physiological responses of market pigs. *Journal of Animal Science* 76 (Suppl. 1), 99 (abstract).

Buhot, J.W., Egan, A.F., Wharton, R.A. and Bowtell, K.D. (1992) An automated system for dressing bovines. In: *38th Congress of Meat Science and Technology*. INRA, Thiex, St Gene Champanelle, France.

Busse, C.S. and Shea-Moore, M.M. (1999) Behavioral and physiological responses to transportation stress. *Journal of Animal Science* 77 (Suppl. 1), 147 (abstract).

Cockram, M.S. and Corley, K.T.T. (1991) Effect of pre-slaughter handling on the behaviour and blood composition of beef cattle. *British Veterinary Journal* 147, 444–454.

Cook, C.J., Devine, C.E., Gilbert, K.V., Tavener, A. and Day, A.M. (1991) Electro-encephalograms in young bulls following upper cervical vertebrae-to-brisket stunning. *New Zealand Veterinary Journal* 39, 121–125.

Crookshank, H.R., Elissalde, M.H., White, R.G., Clanton, D.C. and Smalley, H.E. (1979) Effect of transportation and handling of calves on blood serum composition. *Journal of Animal Science* 48, 430–435.

CSIRO (1989) *Cattle Head Capture Unit, Meat Research Newsletter*, CSIRO Meat Research Laboratory, Cannon Hill, Brisbane, Queensland, Australia.

Devine, C.E., Cook, C.J., Maasland, S.A. and Gilbert, K.V. (1993) The humane slaughter of animals. In: *39th International Congress of Meat Science and Technology*. Agriculture-Canada, Lacombe Research Station, Alberta, Canada, pp. 223–228.

Dodman, N.H. (1977) Observations on the use of the Wernberg dip-lift carbon dioxide apparatus for pre-slaughter anaesthesia of pigs. *British Veterinary Journal* 133, 71–80.

Dunn, C.S. (1990) Stress reactions of cattle undergoing ritual slaughter using two methods of restraint. *Veterinary Record* 126, 522–525.

Eible-Eibesfeldt, I. (1970) *Ethology: the Biology of Behavior*. Holt Rhinehart and Winston, New York, p. 236.

Eikelenboom, G. (ed.) (1983) *Stunning of Animals for Slaughter*. Martinus Nijhoff, Boston, Massachusetts.

Eldridge, G.A. (1988) The influence of abattoir lairage conditions on the behavior and bruising of cattle. In: *Proceedings, 34th International Congress of Meat Science and Technology*. Brisbane, Australia.

Ewbank, R., Parker, M.J. and Mason, C.W. (1992) Reactions of cattle to head restraint at stunning: a practical dilemma. *Animal Welfare* 1, 55–63.

Faucitano, L. (1998) Preslaughter stressors effect on pork: a review. *Muscle Foods* 9, 293–303.

Forslid, A. (1987) Transient neocortical, hippocampal and amygdaloid EEG silence induced by one minute infiltration of high concentration $CO_2$ in swine. *Acta Physiologica Scandinavica* 130, 1–10.

Fulkerson, W.J. and Jamieson, P.A. (1982) Patterns of cortisol release in sheep following administration of synthetic ACTH and imposition of various stressor agents. *Australian Journal of Biological Science* 35, 215–222.

Geverink, N.A., Kappers, A., Van de Burgwal, E., Lambooij, E., Blokhuis, J.H. and Wiegant, V.M. (1998) Effects of regular moving and handling on the behavioral and physiological responses of pigs to pre-slaughter treatment and consequences for meat quality. *Journal of Animal Science* 76, 2080–2085.

Giger, W., Prince, R.P., Westervelt, R.G. and Kinsman, D.M. (1977) Equipment for low stress animal slaughter. *Transactions of the American Society of Agricultural Engineers* 20, 571–578.

Glover, A.F. and Davidson, R.M. (1977) *Report on Cutter Lamb Survey*. New Zealand Meat Producers Board, Wellington, New Zealand.

Grandin, T. (1979) Designing meat packing plant handling facilities for cattle and hogs. *Transactions of the American Society of Agricultural Engineers* 22, 912–917.

Grandin, T. (1980a) Livestock behavior as related to handling facility design. *International Journal of the Study of Animal Problems* 1, 33–52.

Grandin, T. (1980b) Bruises and carcass damage. *International Journal of the Study of Animal Problems* 1, 121–137.

Grandin, T. (1980c) Designs and specifications for livestock handling equipment in slaughter plants. *International Journal of the Study of Animal Problems* 1, 178–299.

Grandin, T. (1980d) Mechanical, electrical and anaesthetic stunning methods for livestock. *International Journal for the Study of Animal Problems* 1, 242–263.

Grandin, T. (1980e) Observations of cattle behavior applied to the design of cattle handling facilities. *Applied Animal Ethology* 6, 10–31.

Grandin, T. (1982) Pig behavior studies applied to slaughter plant design. *Applied Animal Ethology* 9, 141–151.

Grandin, T. (1983a) Welfare requirements of handling facilities. In: Baxter, S.H., Baxter, M.R. and MacCormack, J.A.D. (eds) *Farm Animal Housing and Welfare*. Martinus Nijhoff, Boston, Massachusetts, pp. 137–149.

Grandin, T. (1983b) *System for Handling Cattle in Large Slaughter Plants*. Paper No. 83–4506, American Society of Agricultural Engineers, St Joseph, Michigan, USA.

Grandin, T. (1984) Race system for slaughter plants with 1.5 m radius curves. *Applied Animal Behaviour Science* 13, 295–299.

Grandin, T. (1985/86) Cardiac arrest stunning of livestock and poultry. In: Fox, M.W. and Mickley, L.D. (eds) *Advances in Animal Welfare Science*. Martinus Nijhoff, Boston, Massachusetts, pp. 1–30.

Grandin, T. (1987) Animal handling. *Veterinary Clinics of North America Food Animal Practice* 3, 323–338.

Grandin, T. (1988a) Possible genetic effect on pig's reaction to $CO_2$ stunning. In: *Proceedings 34th International Congress of Meat Science and Technology*. CSIRO Meat Research Laboratory, Cannon Hill, Queensland, Australia.

Grandin, T. (1988b) Double rail restrainer conveyor for livestock handling. *Journal of Agricultural Engineering Research* 41, 327–338.

Grandin, T. (1989) Effect of rearing environment and environmental enrichment on behavior and neural development of young pigs. Doctoral dissertation, University of Illinois, Urbana-Champaign, Illinois.

Grandin, T. (1990) Humanitarian aspects of shehitah in the United States. *Judaism* 39, 436–446.

Grandin, T. (1991a) Handling problems caused by excitable pigs. In: *Proceedings 37th International Congress of Meat Science and Technology*. Federal Center for Meat Research, Kulmbach, Germany.

Grandin, T. (1991b) Principles of abattoir design to improve animal welfare. In: Matthew, J. (ed.) *Progress in Agricultural Physics and Engineering*. CAB International, Wallingford, UK.

Grandin, T. (1991c) *Double Rail Restrainer for Handling Beef Cattle*. Paper No. 91–5004, American Society Agricultural Engineers, St Joseph, Michigan.

Grandin, T. (1992) Observations of cattle restraint devices for stunning and slaughtering. *Animal Welfare* 1, 85–91.

Grandin, T. (1994) Euthanasia and slaughter of livestock. *Journal of the American Veterinary Medical Association,* 204, 1354–1360.

Grandin, T. (1995) Restraint of livestock. In: *Proceedings of the Animal Behavior and the Design of Livestock and Poultry Systems International Conference.* Northeast Regional Agricultural Engineering Service, Cooperative Extension, Cornell University, Ithaca, New York, pp. 208–223.

Grandin, T. (1996) Factors that impede animal movement at slaughter plants. *Journal of the American Veterinary Medical Association* 209, 757–759.

Grandin, T. (1997a) Assessment of stress during handling and transport. *Journal of Animal Science* 75, 249–257.

Grandin, T. (1997b) *Good Management Practices for Animal Handling and Stunning.* American Meat Institute, Washington, DC.

Grandin, T. (1997c) *Survey of Handling and Stunning in Federally Inspected Beef, Pork, Veal and Sheep Slaughter Plants.* ARS Research Project No. 3602–32000–002–08G, United States Department of Agriculture, Washington, DC.

Grandin, T. (1998a) Objective scoring of animal handling and stunning practices in slaughter plants. *Journal of the American Veterinary Medical Association* 212, 36–93.

Grandin, T. (1998b) Solving livestock handling problems in slaughter plants. In: Gregory, N.G. (ed.) *Animal Welfare and Meat Science.* CAB International, Wallingford, UK.

Grandin, T. (1998c) The feasibility of using vocalization scoring as an indicator of poor welfare during slaughter. *Applied Animal Behaviour Science* 56, 121–128.

Grandin, T. and Bruning, J. (1992) Boar presence reduces fighting in slaughter weight pigs. *Applied Animal Behaviour Science* 33, 273–276.

Grandin, T. and Deesing, M. (1998) Genetics and behavior during restraint and herding. In: Grandin T. (ed.) *Genetics and the Behavior of Domestic Animals,* Academic Press, San Diego, California, pp. 113–144.

Grandin, T. and Regenstein, J.M. (1994) Religious slaughter and animal welfare: a discussion for meat scientists. In: *Meat Focus International.* CAB International, Wallingford, UK, pp. 115–123.

Grandin, T., Curtis, S.E. and Widowski, T.M. (1986) Electro-immobilization versus mechanical restraint in an avoid–avoid choice test. *Journal of Animal Science* 62, 1469–1480.

Grandin, T., Dodman, N. and Shuster, L. (1989) Effect of naltrexone on relaxation induced by flank pressure in pigs. *Pharmacology, Biochemistry and Behavior* 33, 839–842.

Gregory, N.G. (1988) Humane slaughter. In: *34th International Congress of Meat Science and Technology.* CSIRO, Meat Research Laboratory, Brisbane, Australia.

Gregory, N.G. (1992) Pre-slaughter handling, stunning and slaughter. In: *38th International Congress of Meat Science and Technology.* INRA, Thiex, Genes Champanelle, France.

Gregory, N.G. (1993) Slaughter technology, electrical stunning in large cattle. *Meat Focus International* 2, 32–36.

Gregory, N.G. and Grandin, T. (1998) *Animal Welfare and Meat Science.* CAB International, Wallingford, UK.

Hargreaves, A.L. and Hutson, G.D. (1990) The stress response in sheep during routine handling procedures. *Applid Animal Behaviour Science* 26, 83–90.

Hartung, V.J., Floss, M., Marahrens, M., Nowak, B. and Fedlhusen, F. (1997) Stress response of slaughter pigs in two different access systems to electrical stunning. *DTW Disch Tierarztl Wochenschr* 104(2), 66–68.

Hlavacek, R.J. (1963) Method for restraining animals. US Patent 3,115,670.

Hornbuckle, P.A. and Beall, T. (1974) Escape reactions to the blood of selected mammals by rats. *Behavioral Biology* 12, 573–576.

Hoenderken, R. (1976) Improved system for guiding pigs for slaughter to the restrainer. *Die Fleischwirtschaft* 56(6), 838–839.

Hornbuckle, P.A. and Beall, T. (1974) Escape reactions to the blood of selected mammals by rats. *Behavioral Biology* 12, 573–576.

Hunter, E.J., Weeding, C.M., Guise, H.J., Abbott, T.A. and Penny, R.H. (1994) Pig welfare and carcass quality: a comparison of the influence of slaughter handling systems in two abattoirs. *Veterinary Record* 135, 423–425.

Hyde, J.L. (1962) The use of solid carbon dioxide for producing short periods of anesthesia in guinea pigs. *American Journal of Veterinary Research* 23, 684–685.

Kenny, F.J. and Tarrant, P.V. (1987) The behavior of young Friesian bulls during social regrouping at an abattoir: influence of an overhead electrified wire grid. *Applied Animal Behaviour Science* 18, 233–246.

Kilgour, R. (1971) Animal handling in works: pertinent behavior studies. In: *Proceedings of the 13th Meat Industry Research Conference*. Hamilton, New Zealand, pp 9–12.

Kilgour, R. (1976) The behaviour of farmed beef bulls. *New Zealand Journal of Agriculture* 132(6), 31–33.

Kilgour, R. (1978) The application of animal behavior and the humane care of farm animals. *Journal of Animal Science* 46, 1479–1486.

Kilgour, R. (1988) Behavior in the pre-slaughter and slaughter environments. In: *Proceedings, International Congress of Meat Science and Technology*, Part A. Brisbane, Australia, pp. 130–138.

Kilgour, R. and de Langen, H. (1970) Stress in sheep from management practices. *Proceedings, New Zealand Society of Animal Production* 30, 65–76.

Lambooy, E. (1986) Automatic electrical stunning of veal calves in a V-type restrainer. In: *Proceedings, 32nd European Meeting of Meat Research Workers, Ghent, Belgium*. Paper 2:2, pp. 77–80.

Lay, D.C., Friend, T.H., Randel, R.D., Bowers, C.L. Grissom, K.K. and Jenkins, O.C. (1992) Behavioral and physiological effects of freeze and hot iron branding on crossbred cattle. *Journal of Animal Science* 70, 330–336.

Lay, D.C., Friend, T.H., Randel, R.D., Bowers, C.L., Grissom, K.K., Nevendorff and Jenkins, O.C. (1998) Effects of restricted nursing on physiological and behavioral reactions of Brahman calves to subsequent restraint and nursing. *Applied Animal Behaviour Science* 56, 109–119.

Marshall, M., Milburg, E.E. and Shultz, E.W. (1963) Apparatus for holding cattle in position for humane slaughtering. US Patent 3,092,871.

Meischke, H.R.C. and Horder, J.C. (1976) A knocking box effect on bruising in cattle. *Food Technology in Australia* 28, 369–371.

Meisinger, D. (1999) *A System for Assuring Pork Quality*, National Pork Producers Council, Des Moines, Iowa.

Midwest Plan Service (1987) *Structures and Environment Handbook*, 10th edn. Iowa State University, Ames, Iowa, p. 319.

Mitchell, G., Hattingh, J. and Ganhao, M. (1988) Stress in cattle assessed after handling, transport and slaughter. *Veterinary Record* 123, 201–205.

Pascoe, P.J. (1986) Humaneness of electro-immobilization unit for cattle. *American Journal of Veterinary Research* 10, 2252–2256.

Pearson, A.J., Kilgour, R., de Langen, H. and Payne, E. (1977) Hormonal responses of lambs to trucking, handling and electric stunning. *New Zealand Society of Animal Production* 37, 243–249.

Pedersen, B.K., Curtis, S.E., Kelley, K.W. and Gonyou, H.W. (1993) Well being of growing finishing pigs: environmental enrichment and pen space allowance. In: Collins, E. and Boon, C. (eds) *Livestock Environment IV*. American Society of Agricultural Engineers, St Joseph, Michigan, pp. 43–150.

Petersen, G.V. (1977) Factors associated with wounds and bruises in lambs. *New Zealand Veterinary Journal* 26, 6–9.

Petherick, J.C. and Blackshaw, J.K. (1987) A review of the factors influencing aggressive and agonistic behavior in the domestic pig. *Australian Journal of Experimental Agriculture* 27, 605–611.

Phillips, P.A., Thompson, B.K. and Fraser, D. (1987) *Ramp Designs for Young Pigs*. Technical Paper No. 87–4511, American Society of Agricultural Engineers, St Joseph, Michigan.

Phillips, P.A., Thompson, B.K. and Fraser, D. (1989) The importance of cleat spacing in ramp design for young pigs. *Canadian Journal of Animal Science* 69, 483–486.

Price, M.A. and Tennessen, T. (1981) Preslaughter management and dark cutting in carcasses of young bulls. *Canadian Journal of Animal Science* 61, 205–208.

Puolanne, E. and Aalto, H. (1981) The incidence of dark cutting beef in young bulls in Finland. In: Hood, D.E. and Tarrant, P.V. (eds) *The Problem of Dark Cutting Beef*. Martinus Nijhoff, The Hague, pp. 462–475.

Raj, A.B., Johnson, S.P., Wotton, S.B. and McInstry, J.L. (1997) Welfare implications of gas stunning of pigs: the time to loss of somatosensory evolved potentials and spontaneous electrocorticogram of pigs during exposure to gasses. *Veterinary Journal* 153, 329–339.

Ray, D.E., Hansen, W.J., Theures, B. and Stott, G.H. (1972) Physical stress and corticoid levels in steers. *Proceedings Western Section American Society of Animal Science* 23, 255–259.

Regensburger, R.W. (1940) Hog stunning pen. US Patent 2,185,949.

Rider, A., Butchbaker, A.F. and Harp, S. (1974) *Beef Working, Sorting and Loading Facilities*. Technical Paper No. 74–4523, American Society of Agricultural Engineers, St Joseph, Michigan.

Ring, C. (1988) Two aspects of $CO_2$ stunning method for pigs: animal protection and meat quality. In: *34th International Congress of Meat Science and Technology*. CSIRO Meat Research Laboratory, Cannon Hill, Queensland, Australia.

Rushen, J. (1986) Aversion of sheep to electro-immobilization and physical restraint. *Applied Animal Behaviour Science* 15, 315.

Shea-Moore, M. (1998) The effect of genotype on behavior in segregated early weaned pigs tested in an open field. *Journal of Animal Science* 76 (Suppl. 1), 100 (abstract).

Spensley, J.C., Wathes, C.M., Waron, N.K. and Lines, J.A. (1995) Behavioral and physiological responses of piglets to naturally occurring sounds. *Applied Animal Behaviour Science* 44, 277.

Stevens, D.A. and Gerzog-Thomas, D.A. (1977) Fright reactions in rats to conspecific tissue. *Physiology and Behavior* 18, 47–51.

Stevens, D.A. and Saplikoski, N.J. (1973) Rats reactions to conspecific muscle and blood evidence for alarm substances. *Behavioral Biology* 8, 75–82.

Stevens, R.A. and Lyons, D.J. (1977) *Livestock Bruising Project: Stockyard and Crate Design.* National Materials Handling Bureau, Department of Productivity, Canberra, Australia.

Stricklin, W.R., Graves, H.B. and Wilson, L.L. (1979) Some theoretical and observed relationships of fixed and portable spacing behavior in animals. *Applied Animal Ethology* 5, 201–214.

Stuier, S. and Olsen, E.V. (1999) Drip loss dependent on stress during lairage and stunning. In: *International Congress of Meat Science and Technology, Yokohama, Japan,* Paper No. 4-P29, pp. 302–303.

Talling, J.C., Waran, N.K., Wathes, C.M. and Lines, J.A. (1998) Sound avoidance by domestic pigs depends on characteristics of the signal. *Applied Animal Behaviour Science* 58, 255–266.

Tanida, H., Miura, A., Tanaka, T. and Yoshimoto, T. (1996) Behavioral responses of piglets to darkness and shadows. *Applied Animal Behaviour Science* 49:173–183.

Tennessen, T., Price, M.A. and Berg, R.T. (1984) Comparative responses of bulls and steers to transportation. *Canadian Journal of Animal Science* 64, 333–338.

Tume, R.K. and Shaw, F.D. (1992) Beta-endorphin and cortisol concentrations in plasma of blood samples collected during exsanguination of cattle. *Meat Science* 31, 211–217.

UFAW (1987) *Proceedings Symposium on Humane Slaughter of Animals.* Universities Federation Animal Welfare, Potters Bar, UK.

United States Department of Agriculture (1967) *Improving Services and Facilities at Public Stockyards.* Agriculture Handbook 337, Packers and Stockyards Administration, United States Department of Agriculture, Washington, DC.

van der Wal, P.G. (1978) Chemical and physiological aspects of pig stunning in relation to meat quality: a review. *Meat Science* 2, 19–30.

van der Wal, P.G. (1997) Causes of variation in pork quality. *Meat Science* 46, 319–327.

van Putten, G. and Elshof, W.J. (1978) Observations of the effects of transport on the well being and lean quality of slaughter pigs. *Animal Regulation Studies* 1, 247–271.

Vieville-Thomas, R.K. and Signoret, J.P. (1992) Pheromonal transmission of aversive experiences in domestic pigs. *Journal of Chemical Endocrinology* 18, 1551.

Voisinet, B.D., Grandin, T., O'Connor, S.F., Tatum, J.D. and Deesing, M.J. (1997) *Bos indicus* cross feedlot cattle with excitable temperaments have tougher meat and a higher incidence of borderline dark cutters. *Meat Science* 46, 367–377.

von Wenzlawowlez, M., von Hollenben, K., Kofer, J. and Bostelmann, N. (1999) Welfare monitoring at slaughter plants in Styria (Austria). In: *International Congress of Meat Science and Technology, Yokomaha, Japan.* Paper No. 2-P3, pp. 64–65.

Walker, P., Warner, R., Weston, P. and Kerr, M. (1999) Effect of washing cattle pre-slaughter on glycogen levels of m. semitendinosus and m. semimembranosus. In: International Congress of Meat Science and Technology, Yokohama, Japan, Paper No. 4-P27, pp. 298–299.

Warner, R.D., Eldridge, G.A., Barnett, J.L., King, R.H. and Winfield, C.G. (1992) Preslaughter behavior and meat quality of pigs fed excess dietary tryptophan. In: *38th International Congress of Meat Science and Technology.* Department of Technology de la Viande, INRA, Thiex 63122, St Genes Chapanelle, France.

Warrington, P.D. (1974) Electrical stunning: a review of the literature. *Veterinary Bulletin* 44, 617–633.

Warriss, P.D., Bevis, E.A., Edwards, J.E., Brown, S.N. and Knowles, T.G. (1991) Effects of angle of slope on the care with which pigs negotiate loading ramps. *Veterinary Record* 128, 419–421.

Warriss, P.D., Brown, S.N. and Adams S.J.M. (1994) Relationship between subjective and objective assessment of stress at slaughter and meat quality in pigs. *Meat Science* 38, 329–340.

Watts, J.M. and Stookey, J.M. (1998) Effects of restraint and branding on rates and acoustic parameters of vocalization in beef cattle. *Applied Animal Behaviour Science* 62, 125–135.

Weary, D.M., Braithwaite, L.A. and Fraser, D. (1998) Vocal response to pain in piglets. *Applied Animal Behaviour Science* 56, 161–172.

Weeding, C.M., Hunter, D.J., Guise, H.J. and Penny, H.C. (1993) Effects of abattoir and slaughter handling systems on stress indicators in pig blood. *Veterinary Record* 133, 10–13.

Westervelt, R.G., Kinsman, D., Prince, R.P. and Giger, W. (1976) Physiological stress measurement during slaughter of calves and lambs. *Journal of Animal Science* 42, 831–834.

Weyman, G. (1987) *Unloading and Loading Facilities at Livestock Markets*. Council of National and Academic Awards, Environmental Studies, Hatfield Polytechnic, Canberra, Australia.

White, R.G., DeShazer, J.A., Tressler, C.J., Borcher, G.M, Davey, S., Waninge, A., Parkhurst, A.M., Milanuk, M.J. and Clems, E.T. (1995) Vocalizations and physiological response of pigs during castration with and without anesthetic. *Journal of Animal Science* 73, 381–386.

Woodbury, D.M., Rollins, L.T., Gardner, M.D., Hirschi, W.L., Hogan, J.R., Rallison, M.L., Tanner, G.S. and Brodie, D.A. (1958) Effects of carbon dioxide on brain excitability and electrolytes. *American Journal of Physiology* 190, 79–94.

Zavy, M.T., Juniewicz, P.E., Phillips, W.A. and Von Tungeln, D.L. (1992) Effects of initial restraint, weaning and transport stress on baseline and ACTH stimulated cortisol responses in beef calves of different genotypes. *American Journal of Veterinary Research* 53, 551–557.

# Index*

---

* Compiled by Mark Deesing.